Mechanics of Solids

Mechanics of Solids

Dr. S. Christian Johnson
Dean and Professor
Excel Engineering College
Pallakkapalayam, Komarapalayam - 637 303
Namakkal Dt.

Publishing for future

SCITECH PUBLICATIONS (INDIA) PVT. LTD.

www.scitechpublications.com

SCITECH PUBLICATIONS (INDIA) PVT. LTD.
Plot No: 52, D.No: 2/619,
First Main Road, VGN Nagar,
Iyyappanthangal, Chennai - 600 056.
E-mail: scitech@airtelmail.in
Visit us at www.scitechpublications.com

Regd. Office : 7/3c, Madley Road
T.Nagar, Chennai - 600 017.

Branches:

HYDERABAD: scitech_hyd@yahoo.co.in
KOLKATA: scitech_kol@yahoo.co.in
NEW DELHI: scitech_del@yahoo.co.in
PUNE: scitech.pune@yahoo.com

978 93 85983 08 5

Published by M. R. Purushothaman for Scitech Publications (India) Pvt. Ltd., Plot No: 52, D.No: 2/619, First Main Road, VGN Nagar, Iyyappanthangal, Chennai - 600 056. Printed at Padmavathi Offset, Chennai - 600 002.

Dedicated

to

My Parents

Preface

This book on "***Mechanics of Solids***" is prepared keeping the curriculum for affiliated colleges of Anna University. Fundamental concepts and basic principles are presented in an easy to understand language so as to help the students to grasp the subject even by self-study. Salient features of the book are as below.

- Stress and strain at a point, various forms of stresses, elastic constants and their relationships, Stress strain behavior of ductile and brittle materials, Thin wall cylinders including thermal stresses.
- Internal forces like shear forces and bending moments including stresses associated with them including theory of simple bending.
- Concept of slope and deflection in beams including computations of such values using different methods in statically determinant beams.
- Theory of torsion, stresses and strains associated with twisting including springs.
- Two dimensional state of stresses, Principal planes and stresses associated with such planes, Locating the principal plane and principal stresses by analytical and geometrical methods including analysis of plane trusses. .

For examination point of view several short and long answer questions with exercise problems have been included. I am thankful to Prof. Dr. A. K. Natesan, Chairman, Dr. N. Mathan Karthick, Vice Chairman, Excel Group Institutions and Prof. Dr. V. K. Shunmughanaathan for their support and encouragement towards bringing out this book. I would like to express my sincere thanks to the faculty members of Civil Department, Prof. Paul Nayanar, Prof. S. Arun Kumar and Prof. D. Mythili who have proof read and helped me. I sincerely acknowledge the help and support rendered by my parents, my wife Mrs. Leethiyal, my daughters, especially my eldest daughter Er. C.J. Blessy Grant, B.E., for her continuous inspiration in preparing this book. My appreciation and gratefulness to my publisher are expressed here for their cooperation in bringing out this book in a short period. Special thanks to Mr. M.R. Purushothaman and Mr. V. Ramesh and their colleagues at Scitech Publications (India) Pvt. Ltd for their cooperation in bringing out this book in an admirable manner. Suggestions for the improvement of this book will be thankfully acknowledged and incorporated in the next edition.

July, 2016 **Dr. S. Christian Johnson**

Contents

Chapter 1

STRESS AND STRAIN

Stress and Strain at a point-Tension, Compression and Shear stress- Hooke's Law Relationship among elastic constants - Stress Strain Diagram for Mild Steel. TOR steel, Concrete - Ultimate stress - Yield Stress and factor of safety - Thermal Stresses Thin Cylinders and Shells - Strain Energy due to Axial force - Resilience - Stresses due to impact and suddenly applied load - Compound Bars.

1.1 Introduction

Mechanics of Solids or Mechanics of Materials or Strength of Materials is a subject which deals with the behaviour of objects withstanding stresses and strains.

1.2 Stress at a Point

When a body is subjected to an external force, it will undergo deformation against which the body will resist by an internal force. This internal resisting force per unit area is generally called as stress.

Stress is denoted by Greek letter called sigma "σ". Its unit is normally referred by Mega Pascal (MPa.) 1Mpa = $1N/mm^2$, 1 Pascal = $1N/m^2$.

$$\text{Stress} = \frac{\text{Force}}{\text{Area}} \quad \text{i.e.} = \sigma = \frac{P}{A}$$

where $\sigma \rightarrow$ Stress in N/m^2

$P \rightarrow$ Load/Force in Newtons(N)

$A \rightarrow$ Area in square metres(m^2)

In Short,

Stress is defined as force divided by cross sectional area of any specimen subjected to that force.

1.2.1 Various forms of stresses

Basic type of stresses that exists in structure can be grouped under the following heads. They are

- **Normal stress :** If we consider a rod subjected to an axial force "F" as in Figure g1.1, then the normal stress is computed using the following formula. The normal stress is usually denoted by the Greek letter "σ" which is pronounced as sigma. Figure 1.2 shows normal stress, which is compressive in nature.

$$\sigma = \frac{F}{A}$$

- **Shear stress :** If we consider the same rod and the applied forces are parallel to the rod's cross sectional area as in Figure 1.3, then the induced stress is called as shear stress. The shear stress is normally denoted by the a Greek letter τ which is pronounced as tau.
- **Simple stress :** If a body is subjected to an external force in only one direction, then the stress developed in the body is called simple stress.
- **Compound stress :** If a body is subjected to an external force in more than one direction, then the stress developed in the body is called compound stress.

Normal stresses

The Stresses developed across the cross section of a structural member, if the forces are acting perpendicular to the cross sectional area is called normal stress. This may be either tensile or compressive in nature.

This is also known as uniaxial state of stress, because the stresses acts only in one direction as in Figure 1.1 or 1.2. However, such a state of uniaxial stress rarely exists, but biaxial and tri-axial state of stresses where either the two mutually perpendicular normal stresses acts or three mutually perpendicular normal stresses acts as shown in the Figure 1.4 are commonly existing.

Tensile, Compressive and Shear Stresses

When the force vectors are directed away from each other, the bar is stretched. The normal stress associated with this type of force is referred to as tensile stress because the forces puts the body in tension. Conversely, if the force vectors are directed towards each other, the bar is compressed. The normal stress associated with this force configuration is referred to as compressive stress because the forces place the body in compression. These two force configurations are illustrated in Figure 1.1 and 1.2.

- **Tensile stress**: If a structural member is subjected to an equal and opposite axial pull, the member is said to undergo a tensile stress at any cross section.

$$\text{Tensile stress}(\sigma) = \frac{\text{Tensile load}}{\text{Area}} = \frac{P}{A}$$

Fig .1.1 Tensile Force

- **Compressive stress**: If a structural member is subjected to an equal and opposite axial push, then the member is said to undergo a compressive stress at any cross section.

$$\text{Compressive Stress}(\sigma) = \frac{\text{Tensile load}}{\text{Area}} = \frac{P}{A}$$

Fig .1.2 Compressive Force

- **Shear stresses**: If the force are applied parallel to the cross sectional area (tangential to the cross section) then the stress induced in the cross section of the structural member is called Shear stress.

Let us consider now the situation, where the cross - sectional area of a block of material is subject to a distribution of forces which are parallel, rather than normal,

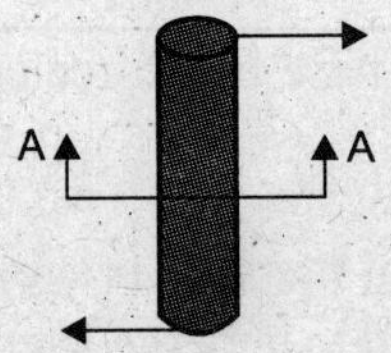

Fig .1.3 Shear Force

to the area concerned as in Figure 1.3. The resulting internal force intensities are known as shear stress, which is normally calculated using the following formula. The greek symbol τ(tau)(suggesting tangential) is used to denote shear stress.

$$\text{Shear stress}(\tau) = \frac{P}{A}$$

Where P is the shear force and A the area over which it acts. As we know that the particular stress generally holds good only at a point therefore we can define shear stress at a point as

$$\tau = \lim_{\delta A \to O} \frac{\delta F}{\delta A}$$

However, it must be borne in mind that the stress (resultant stress) at any point in a body is basically resolved into two components one acts perpendicular (Normal Stress) and other acting parallel (shear stress (or) tangential stress).

Bearing Stress

When one object presses against another as in Figure 1.5, bearing stress is said to occur in the interface area. They are in fact compressive stresses in nature.

Other Stresses

They are combination of stresses e.g. bending stress is a combination of tensile, compressive and shear stresses. Torsional stress, as encountered in twisting of a shaft is a shearing stress.

Stress Units

In SI unit system with F expressed in Newtons (N) and area (A) in square meters (m^2), the stress σ will be expressed in N/m^2. This unit is called Pascal (Pa). However, one finds that the pascal is an exceedingly small quantity and that, in practice, multiples of this unit must be used, namely, the kilopascal (KPa), the

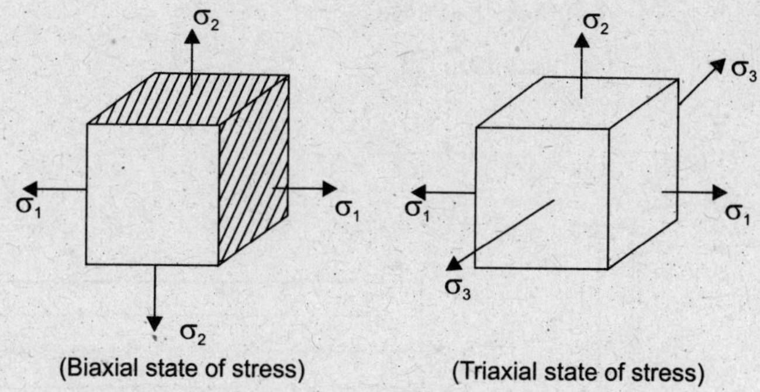

Fig .1.4 Biaxial and Triaxial state of stress

megapascal (MPa)(or) the gigapascal (GPa).

$$1KPa = 10^3 Pa$$
$$1MPa = 10^6 Pa$$
$$1GPa = 10^9 Pa$$

In U.S. unit system, the force F is usually express in pounds (lb) or kilopounds (kip), and the cross sectional area A is square inches (in^2). The stress σ will then be expressed in pounds per square inch (psi) or kilopounds per square inch (ksi).

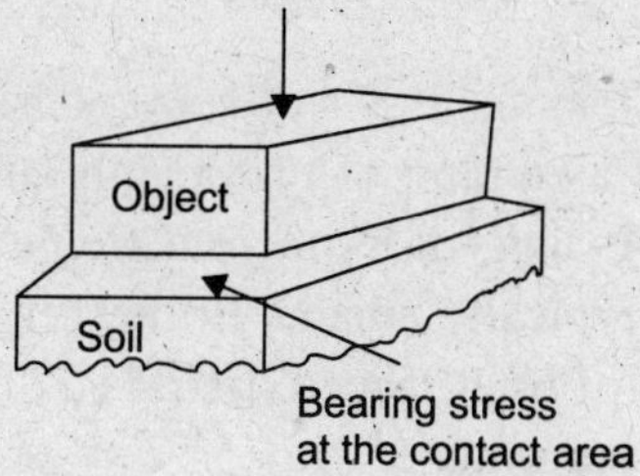

Fig .1.5 Bearing Stress

1.3 Strain at a Point

When a body is said to be stressed by an external force, its dimensions will change. If original length of bar is stated as "L" and change in length by an amount "δL". then the strain produced is the ratio between change in length and original length. Strain is normally denoted by a Greek letter called sigma "ε" (Refer Figure 1.6)

$$\varepsilon = \frac{\delta_L}{L}$$

Where, δ_L - Change in length

L - Original Length

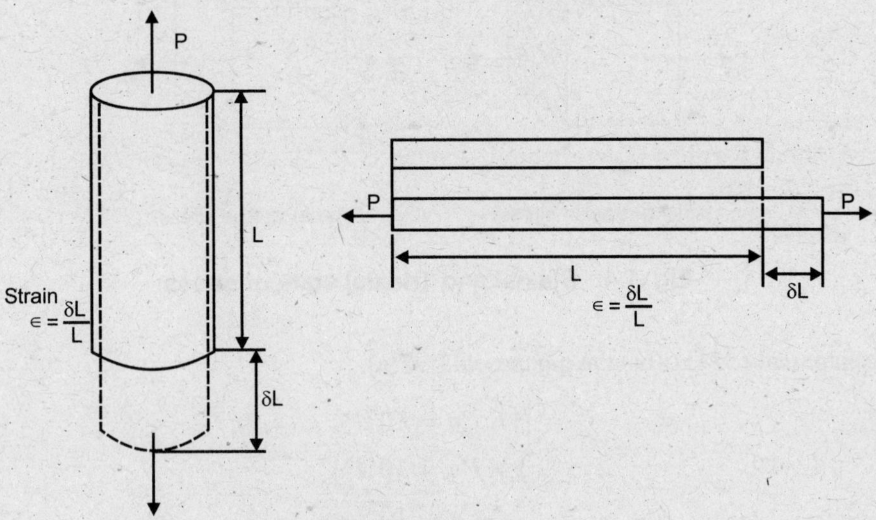

Fig .1.6 Concept of linear strain

Strain Units

As strain is a ratio of two lengths, it is a dimensionless quantity. It is customary, however, to express strain as a ratio of two length units. In the SI unit system, strain is usually expressed in units of μ (micrometre/metre) because, as mentioned before, deformations are typically small. In the English unit system, strain is usually expressed in units of in/in. Since strain is a dimensionless quantity; it is sometimes expressed as a percentage.

As the extensions of materials under load are very small, it is normal to express the strain in the form of strain $\times 10^6$ i.e., Micro strain. Its symbol then is $m\ \hat{i}$.

1.3.1 Various Types of Strains

- **Tensile Strain :** If a structural member is subjected to tensile force, the length of the member is increased. The ratio of increase in length to the original Length is known as tensile strain.
- **Compressive Strain :** If a structural member is subjected to an compressive force, the length of the member is shortened. The ratio of decrease in length

to the original length is known as compressive strain.

$$\in = \frac{\text{Decrease in length}}{\text{Original length}}$$

$$\in = \frac{\delta L}{L}$$

- **Shear Strain :** If a structural member is subjected to a shear stress, it undergo deformation accompanied by as change in subtended angle of the sides as shown in Figure 1.7.

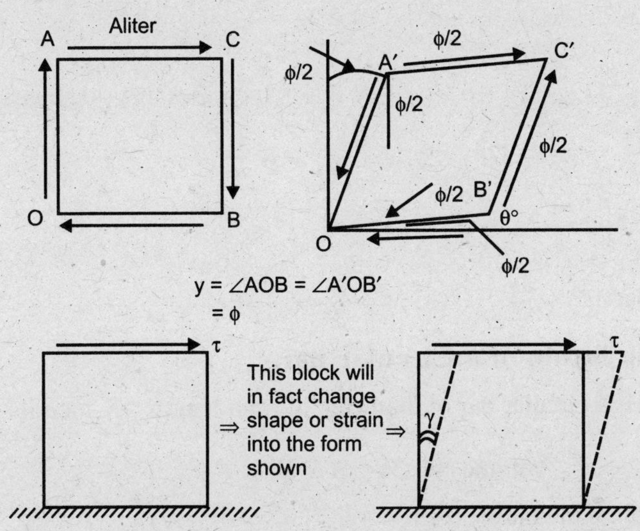

Fig .1.7 Concept of Shear strain

The tangent of the angle through which two adjacent sides rotate relative to their initial position is termed shear strain. Shear strain is measured in radians and hence is non - dimensional i.e. it has no unit and it is referred by a symbol "γ" called a gamma.

Volumetric Strain

The ratio of change in volume to the original volume due to the application of load is called volumetric strain.

$$\text{Volumetric Strain}(\varepsilon v) = \frac{\text{Change is Volume}}{\text{Original Volume}} = \frac{\delta v}{v}$$

Volumetric Strain of a Rectangular Bar

If we consider a rectangular bar with dimensions length(ℓ), breadth (b) and thickness (t), its volume (v) = ℓbt.

Then, change in volume $(\delta v) = \ell b \cdot \delta t + \ell t \cdot \delta b + bt \cdot \delta \ell$

$$\text{Volumetric strain} = \frac{\text{Change in volume}}{\text{Original volume}} \left(\frac{\delta v}{v}\right)$$

Now substituting for 'δv' and 'v', we get

$$\begin{aligned}
\frac{\delta v}{v} &= \frac{\delta t}{t} + \frac{\delta b}{b} + \frac{\delta l}{l} \\
&= \varepsilon + \varepsilon_1 + \varepsilon_1 \\
&= \varepsilon + \varepsilon(-\mu) + \varepsilon(-\mu) \text{ [as } \varepsilon \text{ increases as } \varepsilon_1 \text{ decreases]} \\
\frac{\delta v}{v} &= \varepsilon[1 - 2\mu] \\
\varepsilon v = \frac{\delta v}{v} &= \frac{P}{AE}[1 - 2\mu]
\end{aligned}$$

[we know that $\delta\ell = \frac{P\ell}{AE}$ $\quad \therefore \epsilon = \frac{\delta\ell}{\ell} = \frac{P}{AE}$]

Volumetric Strain of a Circular Bar

If we consider a circular bar of diameter "d" and length "ℓ", then its volume is

$$\text{Volume } (V) = \pi/4 \times d^2 \times \ell$$

$$\text{Change in volume } (\delta v) = \pi/4 \times d^2 \times \delta l + \pi/4 \times 2 \times dl \times \delta d$$

Substituting for 'δV' and 'V' we get,

$$\begin{aligned}
\text{Volumetric strain} = \frac{\delta v}{v} &= \frac{\delta l}{l} + 2 \times \frac{\delta d}{d} \\
&= \varepsilon + 2\varepsilon \\
&= \varepsilon - 2\mu \in \\
&= \varepsilon(1 - 2\mu) \\
\frac{\delta v}{V} &= \varepsilon(1 - 2\mu)
\end{aligned}$$

1.4 Hookes Law

Hookes law states that, when a material is loaded within its elastic limit the stress is directly proportional to the strain.

i.e, stress α strain (within its elastic limit)

i.e., $\sigma \alpha \in$; $\therefore \sigma = E \in$

where, E = Young's modulus (or) modulus of elasticity (N/mm^2)

σ = Stress in N/mm^2

ε = Strain

1.5 Change in Length in Terms of Material Parameters

When we consider a body subjected to a tensile stress, We know that,

$$\text{Stress, } (\sigma) = \frac{\text{load}}{\text{Area}} = \frac{P}{A} \tag{1}$$

$$\text{Strain, } (\in) = \frac{\delta\ell}{\ell} \tag{2}$$

From Hooke's law

$$\text{Youngs modulus}(E) = \frac{\text{Stress}}{\text{Strain}} = \frac{\sigma}{\in} = \frac{P/A}{\varepsilon}$$

$$E = \frac{P}{A \in}$$

$$\in = \frac{P}{AE} \tag{3}$$

Equating (2) and (3) equation

$$\frac{\delta\ell}{\ell} = \frac{P}{AE}$$

$$\text{We get} \quad \delta\ell = \frac{P\ell}{AE}$$

$$\boxed{\text{Change in length } \delta\ell = \frac{P\ell}{AE}}$$

1.6 Change in Length for Uniformly Varying Circular(Uniformly Tapering)Cross Section

If we consider a uniformly varying circular rod with diameter tapering from d_1 at one end to d_2 at other end, then the change in length in such a bar when a force 'P' is applied in the direction as shown in Figure 1.8 can be calculated from the following formula.

$$\delta L = \frac{4PL}{\pi E d_1 d_2}$$

Where

$\delta L \Rightarrow$ Change in length of uniformly varying section.

$P \Rightarrow$ Force

$L \Rightarrow$ Length of uniformly varying bar.

$d_1 \Rightarrow$ Bigger Dia of the bar.

$d_2 \Rightarrow$ Smaller dia of bar.

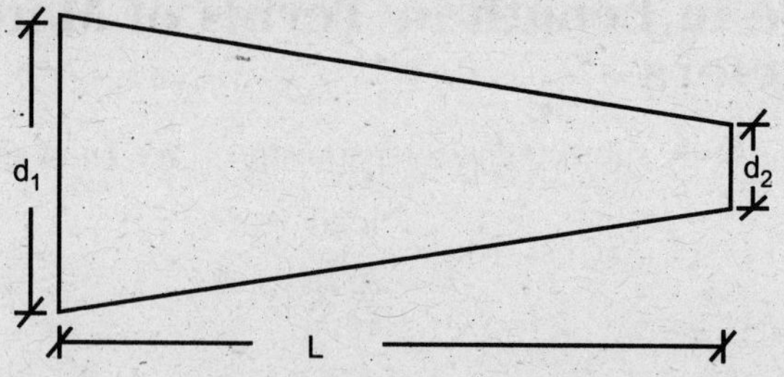

Fig .1.8 Uniformly tapering circular section

1.7 Stresses in Bars of Varying Section

Let us consider a non-uniform cross section of member designated by AB, BC and CD having cross sectional areas of A_1, A_2 and A_3 with lengths L_1, L_2 and L_3 as shown in Figure 1.9.

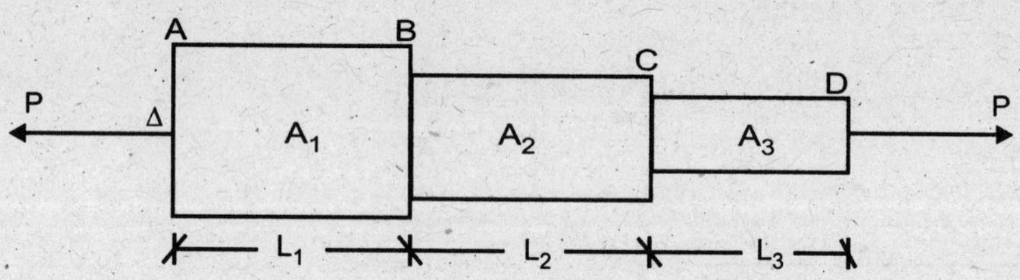

Fig .1.9 Bars of varying cross - sections

$$\text{Stress in portion } AB = \frac{P}{A1},$$

$$\text{Stress in portion } BC = \frac{P}{A2},$$

$$\text{Stress in portion } CD = \frac{P}{A3}$$

$$\text{Elongation of } AB\,(\delta L_1) = \frac{PL_1}{A_1E_1}$$

$$\text{Elongation of } BC\,(\delta L_2) = \frac{PL_2}{A_2E_2}$$

$$\text{Elongation of } CD\,(\delta L3) = \frac{PL_3}{A_3E_3}$$

$$\therefore \text{ Total elongation} = \delta L_1 + \delta L_2 + \delta L_3 = \frac{PL_1}{A_1E_1} + \frac{PL_2}{A_2E_2} + \frac{PL_3}{A_3E_3}$$

As force 'P' and young's modulus 'E'' is constant, the above equation can be written as.

$$(\delta L) = \frac{P}{E}\left[\frac{L_1}{A_1} + \frac{L_2}{A_2} + \frac{L_3}{A_3}\right]$$

1.8 Principle of Superposition

In certain occasions, a structural member may be subjected to external axial forces not only at its ends, but also at some of its inter junction points along the length of the body. Under such cases, the forces are split up and their effects are considered on individual sections. Then the total deformation is calculated by adding the deformation of the individual sections. Such a principle of finding out the resultant deformation is known as principle of superposition. As per the principle of superposition, the total deformation is calculated using the following formula.

$$(\delta L) = \frac{P_1L_1 + P_2L_2 + P_3L_3 + \cdots}{AE}$$

Applying for uniformly varying section, We get,

$$(\delta L) = \frac{4PL}{\pi E d_1 d_2}$$

Where,	P	=	Force
	L	=	Length of the rod
	d1	=	Diameter of the rod in the head end
	d2	=	Diameter of the rod in the tail end

1.9 Change in Length for Uniformly Varying Rectangular Cross Section

In certain occasions, uniformly tapering section varies from a square section at one end to another square section at other ends. In some other occasions instead of square section, rectangular section may exist as shown in Figure 1.10.

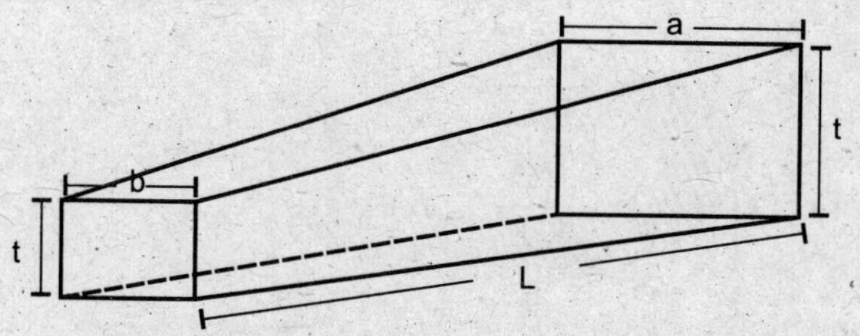

Fig .1.10 Uniformly tapering rectangular section

$$\delta_L = \frac{PL}{Et(a-b)} \log_e (a/b) \qquad \text{(Note : a > b)}$$

Where P = Force
L = length of the rod
a = Width of the bar in the head end
b = Width of the bar in the tail end
t = Thickness of the bar
E = Young's modulus in N/mm^2

1.10 Stresses In Composite Sections

In structural constructions, composite members are used at large these days. A composite member is one which is composed of two or more different materials which are joined together as in Figure 1.11. The whole system is elongated or compressed as a single unit. In such case, the following are the two governing principles.

Principle 1

Elongation or contraction in each bar is equal. As a result the strains induced in those bars are also equal.

i.e., Change in length of bar (1) = Change in Length of bar (2)

$$\text{i.e.} \quad \frac{P.L1}{A_1E_1} = \frac{P.L_2}{A_1E_2}$$

Principle 2

The sum of loads carried by individual materials of a composite member is equal to the total load applied on the member.

Total load, (P) = Load carried by bar (1) + Load carried by bar (2)

$P = P_1 + P_2$

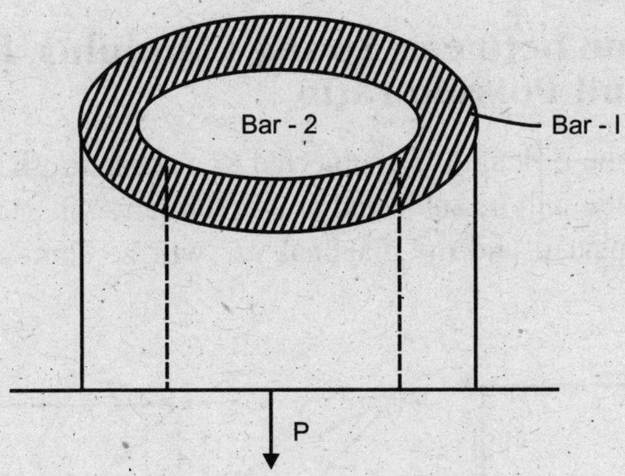

Fig .1.11 Composite Bar

1.11 Elastic Constants

When we consider the elastic behavior of an isotropic materials under, normal, shear and hydrostatic loading, it will be interesting to know that a total of four elastic constants namely young's Modulus (E), Rigidity Modulus(G), Bulk Modulus (K), and Poisson's ratio (μ) or (1/m) are important. These elastic constants are not independent to the others rather if any two of them are known, the other two can be found.

Let us define these elastic constants

1. **Young's Modulus or Modulus of Elasticity** $\Rightarrow$ It is ratio between Tensile stress and tensile strain (i.e.) youngs modulus

   E = Tensile Stress / Tensile strain

2. **Shear Modulus or Modulus of rigidity** $\Rightarrow$ It is the ratio between shear stress and shear strain i.e. shear modulus.

   G = Shear stress / Shear strain

3. **Poisson's ratio** : It is the ratio between lateral and longitudinal strain.

   (1/m) or μ = lateral strain / longitudinal strain

4. **Bulk Modulus of elasticity** : It is the ratio between direct stress and volumetric strain.

   K = Direct stress / Volumetric strain

Where

Volumetric strain = sum of linear strain in x, y and z direction.

Direct stress = stress which cause the change in volume.

1.11.1 Relation between Young's Modulus, Rigidity Modulus and Poisson ratio

Let us consider a cube of length 'L' subjected to a shear stress (τ). Due to such shear stresses the cube will be subjected to a deformation in such a way that the diagonal BD is elongated and the diagonal AC will be shortened as shown in Figure. 1.12.

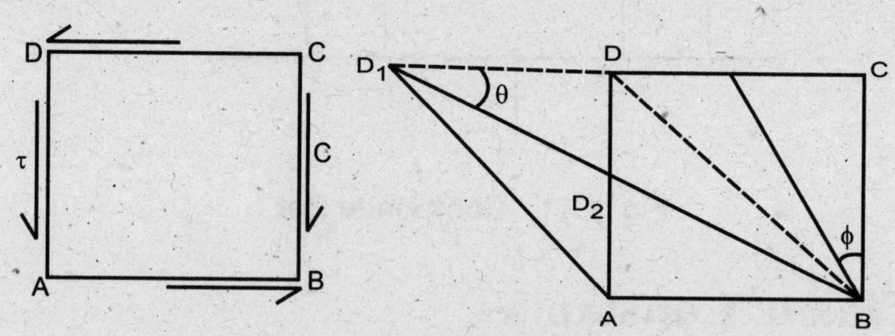

Fig .1.12 Shape change of cube element

Therefore strain on the diagonal 'BD'

Longitudinal strain = Change in length / original length

$$= \frac{BD_1 - BD}{BD} = \frac{D_1D_2}{BD}$$

$$\Longrightarrow \frac{DD_1 \cos 45^\circ}{BD} \qquad \left[\because \cos 45^\circ = \frac{D_1D_2}{D_1D}\right].$$

$$\Rightarrow \frac{DD_1 \cos 45^\circ}{AD/\cos 45^\circ} \qquad \left(\because \frac{AD}{\cos 45^\circ} = BD\right).$$

$$i.e, \frac{AD}{BD} = \cos 45^\circ$$

$$= \frac{DD_1 \cos 45^\circ}{AD/\frac{1}{\sqrt{2}}}$$

$$= \frac{DD_1}{AD \times \sqrt{2}} \times \frac{1}{\sqrt{2}}$$

$$\boxed{\epsilon_1 = \frac{DD_1}{2AD}}$$

We know that,

$$\tan\phi = \frac{DD1}{AD} \Rightarrow \phi = \frac{DD1}{AD} \qquad [\because \phi \text{ is very small}]$$

Substituting ϕ value in Eq $\in_1 = \frac{DD_1}{2AD}$ $\left(\because \frac{DD_1}{AD} = \phi\right)$

$$\therefore \quad \in_1 = \frac{1}{2}\phi = \in = \frac{\phi}{2} \tag{1}$$

Longitudinal strain, $\in_1 = \dfrac{\phi}{2}$

Hence it is to be understood that Longitudinal strain is half of the shear strain in the case.

Linear Strain or longitudinal strain $\in_1 = \dfrac{\phi}{2}$.

we know that,

Modulus of Rigidity, $(G) = \dfrac{\text{Shearing Stress}}{\text{Shearing Strain}}$

$$\therefore \quad \phi = \frac{\tau}{G}$$

Substituting of value (τ/G)in equation (1)

Linear Strain

$$G_1 = \frac{\phi}{2} = \frac{\tau/G}{2} = \frac{\tau}{2G}$$

$$\therefore \boxed{\text{Linear Strain}, \in_1 = \frac{\tau}{2G}} \tag{2}$$

Longitudinal strain on the diagonal BD is due to tensile stress and in given by Young's Modulus.

Young's Modulus, $E = \dfrac{\text{Tensile Stress}}{\text{Tensile Strain (Longitudinal)}}$

$$E = \frac{\tau}{\in \ell}$$

$$\boxed{\text{Tensile Strain (Longitudinal)}, (\in \ell) = \frac{\tau}{E}}$$

Tensile strain on the diagonal BD is due to compressive stress on the diagonal AC which is given by Poisson's ratio.

$$\text{Poisson's Ratio}, \frac{1}{m} = \frac{\text{Lateral Strain}}{\text{Longitudinal strain (or) Tensile Strain}}$$

$$\Rightarrow \text{Lateral Strain} = \text{Poisson ratio} \times \text{Tensile Strain}(\in \ell)$$

$$\therefore \text{Lateral Strain}, (e_t) = \frac{1}{m} \times \frac{\tau}{E}$$

$$\boxed{\text{Lateral Strain, } e_t = \frac{1}{m} \times \frac{\tau}{E}}$$

$$\text{Total strain} = \left\{\begin{matrix}\text{longitudinal strain (or)}\\ \text{Tensile strain}\end{matrix}\right\} + \{\text{Lateral Strain}\}$$

$$= \in \ell + \in t$$

$$= \frac{\tau}{E} + \frac{1}{m}\left(\frac{\tau}{E}\right)$$

$$\text{Total strain} = \frac{\tau}{E}\left(1 + \frac{1}{m}\right) \qquad (3)$$

Now equating the total strains(2) & (3)

$$= \frac{\tau}{2G} = \frac{\tau}{E}\left(1 + \frac{1}{m}\right)$$

$$= \frac{1}{2G} = \frac{1}{E}\left(1 + \frac{1}{m}\right)$$

$$\therefore \boxed{E = 2G\left(1 + \frac{1}{m}\right)}.$$

1.11.2 Relation between Young's modulus, Bulk Modulus and Poisson ratio

Let us consider a cube ABCDEFGH is which is subjected to three mutually perpendicular tensile stresses of equally equal intensity as shown in the Figure 1.13

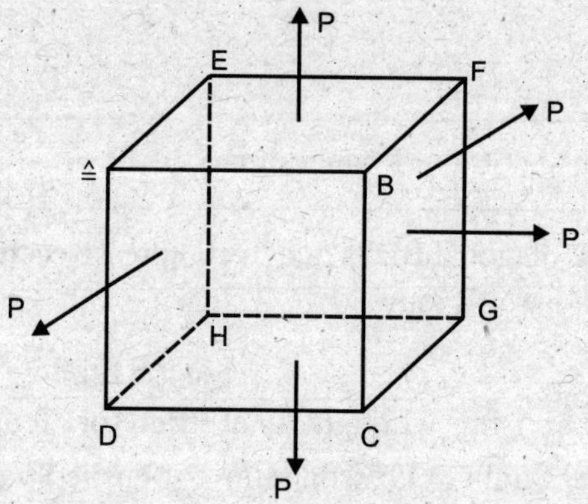

Fig .1.13 Young's modulus, Bulk modulus, and Poisson ratio

Let σ = Stress on each face

E = Young's Modulus

K = Bulk Modulus

$\frac{1}{m}$ (or) μ = Poisson's ratio

L = Length

Now consider the deformation of one side (AB) of cube under the action of three mutually perpendicular stresses. This side will suffer the following three strains.

1. Tensile Strain of AB is equal to $\frac{\sigma}{E}$ due to stresses on the faces AEHD and BFGC.

   $$\text{Young's modulus (E)} = \frac{\text{Tensile stress}}{\text{Tensile strain}} = \frac{\sigma}{\in}$$

   $$\text{Tensile Strain } \epsilon_1 = \frac{\sigma}{E}$$

2. Compressive lateral strain of AB is equal to $\frac{1}{m} \times \frac{\sigma}{E}$ due to stresses on the faces AEFB and DHGC Poisson's ratio $\left(\frac{1}{m}\right)$.

3. Compressive lateral strain of AB is equal to $\frac{1}{m} \times \frac{\sigma}{E}$ due to stresses on the faces ABCD and EFGH.

   $\therefore$ Total strain of AB is given by,

$$\frac{dL}{L} = \frac{\sigma}{E} - \frac{1}{m}\left(\frac{\sigma}{E}\right) - \frac{1}{m}\left(\frac{\sigma}{E}\right) \qquad (\because \text{ Tensile strain } +ve, \text{ Lateral strain } -ve)$$

$$\frac{dL}{L} = \frac{\sigma}{E}\left(1 - \frac{1}{m} - \frac{1}{m}\right)$$

$$= \frac{\sigma}{E}\left(1 - \frac{2}{m}\right)$$

$$\boxed{\frac{dL}{L} = \frac{\sigma}{E}\left(1 - \frac{2}{m}\right)}$$

We know that, original volume of cube $v = L^3$ Differentiating with respect to 'L'.

$$dv = 3L^2dL$$
$$\frac{dv}{v} = \frac{3L^2dL}{v}$$
$$\frac{dv}{v} = \frac{3L^2dL}{L^3} \qquad [\because v = L^3]$$
$$\frac{dv}{v} = \frac{3dL}{L}$$

Substituting $\frac{dL}{L}$ value,

$$\frac{dv}{v} = 3\left(\frac{\sigma}{E}\left(1 - \frac{2}{m}\right)\right)$$
$$\frac{dv}{v} = \frac{3\sigma}{E}\left(1 - \frac{2}{m}\right)$$

We know that

$$\left(\frac{\because \text{ Direct Stress}}{\text{Volumetric Strain}}\right) = \frac{\sigma}{\frac{dv}{v}} = k$$
$$= \frac{\sigma}{\frac{3\sigma}{E}\left(1 - \frac{2}{m}\right)}$$
$$K = \frac{E}{3\left(1 - \frac{2}{m}\right)}$$
$$E = 3k\left(1 - \frac{2}{m}\right)$$

1.11.3 Relation between E, G and K

The relationship between E, G and K can be easily determined by eliminating from the already derived relations

$$E = 2G(1 + \mu) \qquad (1)$$
$$E = 3K(1 - \mu) \qquad (2)$$

on equating (1) & (2)

Following relationship may be obtained

$$\boxed{E = \frac{9GK}{(3K + G)}}$$

1.12 Stress - Strain Diagrams for Mild Steel (Ductile Material)

The capacity of materials to allow large deformations or large extensions without failure is termed as ductility. The materials with high ductility are termed as ductile materials.

When a material like mild steel is subjected to a uniaxial load, some contraction or expansion always takes place. Thus, dividing the applied force by the corresponding actual area of the specimen at that instant will give a stress called true stress and the corresponding strain is called true strain.

A typical tensile test stress- strain curve for the mild steel is shown in Figure 1.14, when it is tested in a universal testing machine (UTM).

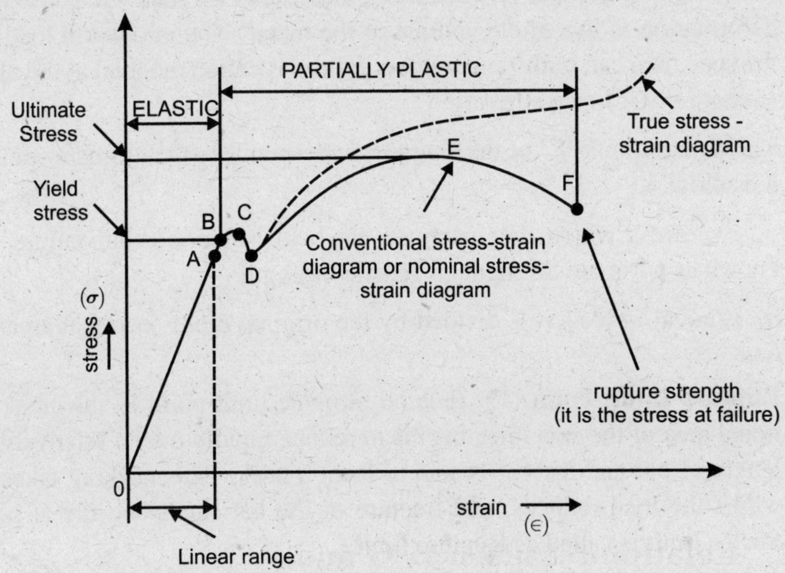

Fig .1.14 Stress- Strain curve of a mild steel

1.12.1 Salient Points of the Stress -Strain curve of a mild steel

- **Proportionality Limit Point (A):** In the stress - strain curve, it is seen that the stress is proportional to strain or elongation is proportional to the load giving a straight line relationship upto a point A. Up to that limit called proportionality limit, stress is proportional to strain.
- **Elastic Limit Point (B):** For a short period beyond the point A, the material

may still be elastic in the sense that the deformations are completely recovered when the load is removed. The limiting point B is termed as Elastic Limit.

- **Yield Limits Points (C) and (D):** Beyond the elastic limit plastic deformation occurs and strains are not totally recoverable. There will be thus permanent deformation or permanent set when load is removed. These two points are termed as upper and lower yield points respectively. The stress at the yield point is called the yield strength.

  It is seen in the a stress - Strain curve, that the yield point is so near that proportional limit that for most purpose the two may be taken as one. However, it is much easier to located the former.

- **Ultimate Limit Point (E):** A further increase in the load will cause marked deformation in the whole volume of the metal. The maximum load which the specimen can withstand without failure is called the load at the ultimate strength or Ultimate stress.

  The highest point 'E' of the diagram corresponds to the ultimate strength of a material.

  σ_{ult} = Stress which the specimen can with stand without failure and is known as Ultimate Strength or Tensile Strength.

  σ_u is equal to load at E divided by the original cross-sectional area of the bar.

- **Rupture Limit Point (F):** Beyond ultimate limit point E, the cross - sectional area of the specimen begins to reduce rapidly over a relatively small length of bar and the bar is said to form a neck. This necking takes place whilst the load reduces, and fracture of the bar finally occurs at point F. Such a limit is called as Rupture limit.

Percentage Elongation

The ductility of a material in tension can be characterized by its elongation and by the reduction in area at the cross section where fracture occurs.

It is the ratio of the extension in length of the specimen after fracture to its initial gauge length, expressed in percent.

$$\delta = \frac{L_f - L_i}{L_i} \times 100$$

L_f = Final length of the specimen = gauge length of specimen after fracture (or the distance between the gauge marks at fracture)

L_i = Initial length of the specimen = gauge length before fracture (i.e. initial gauge length)

For 50 mm gauge length, mild steel will have percentage elongation in the order of 10% to 40%.

Percentage reduction in area

$(A - A_f/A) \times 100$

$A-$ Initial cross sectional area of the specimen

A_f- Final area of specimen after fracture

1.13 Stress - Strain Diagram for a TOR Steel /High Yield Strength Deformed Bars

Material like TOR steel (Low Carbon steel) or HYSD bars do not possess a well define yield points and the stress strain diagram will be as shown in Figure 1.15.

For such materials in order to find out the yield strength, a line is drawn parallel to the straight line portion of initial stress diagram by off setting this by an amount equal to 0.2% of the strain as shown as shown in Figure 1.15.

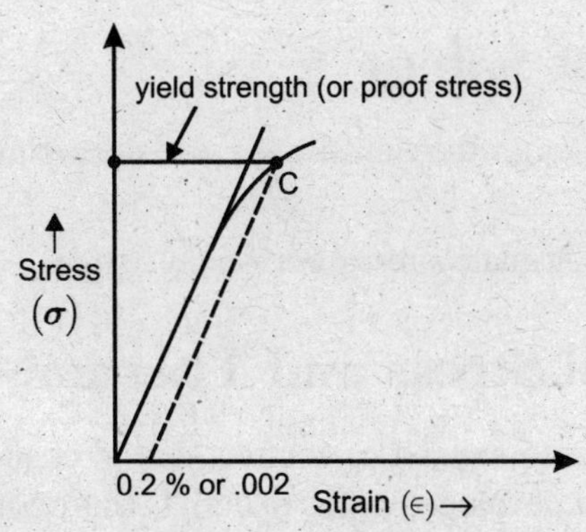

Fig .1.15 Stress- strain diagram of Tor steel or HYSD bars

1.14 Stress Strain Diagram for Concrete / Brittle Materials

A brittle material is one which exhibits a relatively small extensions or deformations to fracture, so that the partially plastic region of the tensile test graph is much reduced. This type of graph is shown by the cast iron or steels with high carbon contents or concrete.

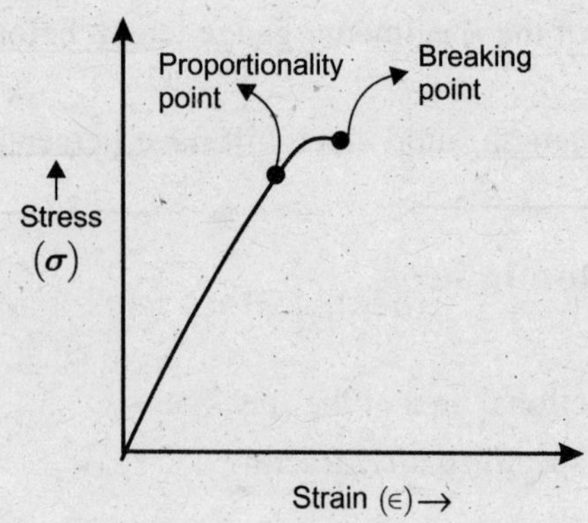

Fig .1.16 Stress - strain diagram of a brittle material (Concrete)

In case of brittle materials like concrete, the elongation and reduction in area of the specimen is very less. The stress - strain curve for concrete have only two points, one is limit of proportionality and another point is breaking point or rupture point. The stress - Strain curve of concrete is parabolic as shown in Figure 1.19.

1.15 Factor of Safety

It is defined as the ratio of ultimate tensile stress to the permissible stress (working stress).

Factor of safety = Ultimate stress / Permissible stress.

1.16 Thermal Stress and Thermal Strain

When a material is free to expand or contract due to change in temperature, no stress and strain will be developed in the material. But when the same material is rigidly fixed at both the ends, the change in length is prevented. As a result of this a stress will be developed in that material. Such a stress is known as Temperature Stress or Thermal Stress and the corresponding strain is known as Thermal Strain.

Determination of Thermal Stress and Strain

Let us consider of rod AB of length 'L' fixed at both ends A and B as in Figure 1.17. When the temperature of the rod is raised it tends to expand by α TL. Where α is the coefficient of linear expansion.

If the fixity at the end B is removed, the rod will freely expand by α TL.

$\therefore$ BB1 = α TL.

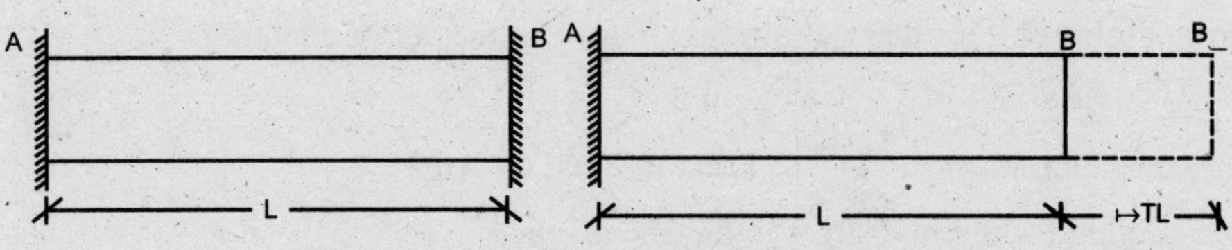

Fig .1.17 Thermal stress and strain

If a compressive load of intensity 'P' is applied at 'B', then the rod will decrease its length from $L + \alpha\, TL$ to L as depicted in Figure 1.18.

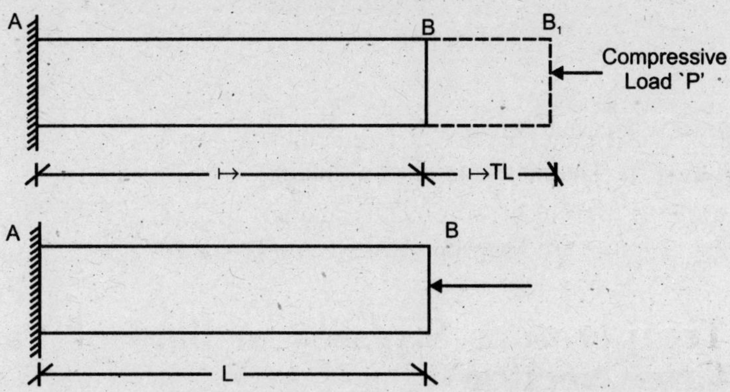

Fig .1.18 Thermal strain under compressive load

We know that

$$\left.\begin{array}{l}\text{Compressive strain i.e.}\\ \text{Thermal strain}\end{array}\right\} \frac{\overset{\text{(Change in length)}}{\text{Decrease in length}}}{\text{Original Length}} = \frac{\alpha TL}{L + \alpha TL} = \frac{\alpha TL}{L} = \alpha T$$

$\therefore$ Thermal strain = α T

Where

α - coefficient of thermal expansion.

T - change in temperature

It is known that Young modulus $E = \dfrac{\text{Stress}}{\text{Strain}} \left(\dfrac{\sigma}{\in}\right)$

i.e, Stress = Strain x E

$$\sigma = \in \times E$$

Now substituting for thermal Strain α T for ϵ in the above equation.

Thermal stress $(\sigma) = \alpha T E$

We know that Stress = $\dfrac{\text{Load}}{\text{Area}}$ $\therefore$ Load = Stress $\times$ Area

Substituting for thermal stress in the above eqn.

Load(P) = $\alpha T E \times A$

Thermal strain and stress with yielding (δ)

$\therefore$ Thermal strain under yielding

$$= \frac{\text{Actual Expansion allowed}}{\text{Original length}}$$

$$= \frac{\alpha T L - \delta}{L}$$

Thermal stress = Thermal Strain $\times$ E

Substituting for Thermal strain under yielding

$$(\sigma)\text{Thermal stress} = \frac{\alpha T L - \delta}{L} \times E$$

1.17 Temperature Stresses in Bars of Varying Cross Section

Let us consider a bar of varying cross section as shown in Figure 1.19. If this bar is fixed at "A" and "C" and undergo change in temperature, the whole assembly will tend to expand or contract.

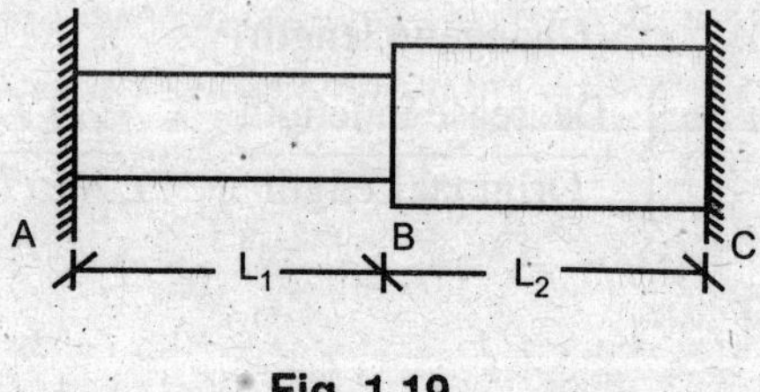

Fig .1.19

If the expansion or contraction is prevented as it is fixed at both ends, temperature stress will be produced in the bar.

Let

L_1	–	Length of portion AB
σ_1	–	Stress induced in portion AB
A_1	–	Cross sectional area of portion AB
L_2	–	Length of portion BC

σ_2 – Stress induced in portion BC
A_2 – Cross sectional area of portion BC
E – Young's modulus of the material
α – Co-efficient of linear expansion
T – Temperature variation

The total force or pull that will be formed in the bar $= P_1 = P_2$

$$\sigma 1 A_1 = \sigma 2 A_2$$

Change in length $\delta L = \dfrac{\sigma_1 L_1}{E} + \dfrac{\sigma_2 L_2}{E} = \dfrac{1}{E}(\sigma_1 L_1 + \sigma_2 L_2)$

If the bar is made up of different materials, then we have to take modulus of elasticity into account

$$\therefore \delta L = \frac{\sigma_1 L_1}{E_1} + \frac{\sigma_2 L_2}{E_2}$$

1.18 Thin Cylinders and Shells

A cylinder is called as thin cylinders when the ratio of wall thickness to the diameters of cylinder is less than 1/20. Thin cylinders and spherical shells are normally used to contain fluids or gas such as in tanks, boilers, compressed air receivers, pipes, pump cylinders, etc.

Stresses in thin cylinder

When the pressure inside the cylinder is developed the cylinder walls are subjected to tensile stress as in Figure 1.20. There are two types of tensile stresses developed according to the direction in which they are acting.

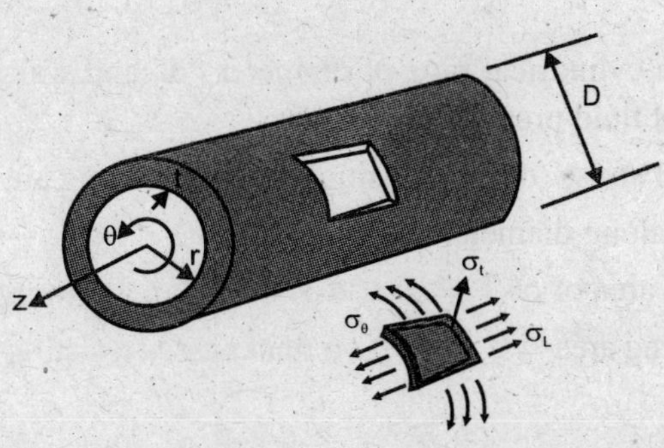

Fig .1.20 Stresses in thin cylinders

1) Circumferential (or) Hoop stress. 2) Longitudinal stress

When these stresses exceed the permissible limit, the cylinder is likely to fail in the following two ways.

1. It may split up into two semi circular halves along the cylinder axis.
2. It may split up into two cylinders.

1.18.1 Circumferential or Hoop Stress

The stress in the circumferential direction due to tendency of bursting the cylinder along the longitudinal axial is called as hoop stress, which is shown in Figure 1.21.

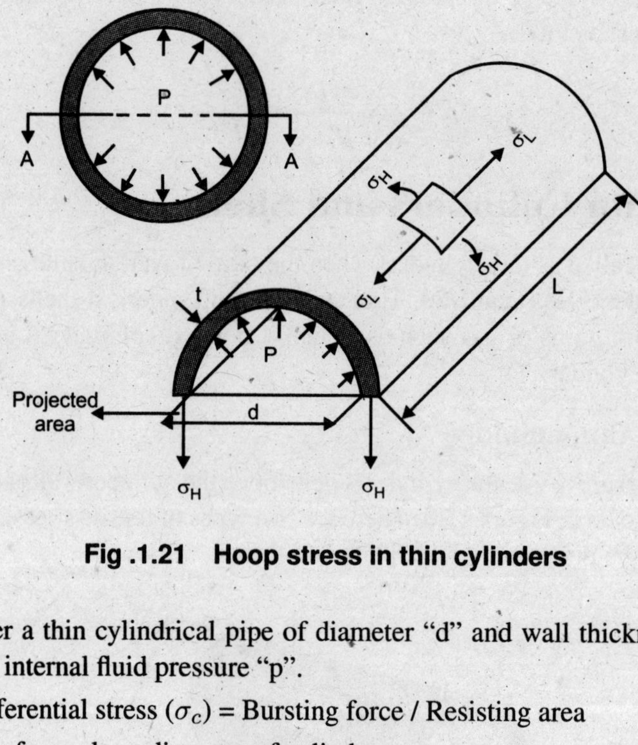

Fig .1.21 Hoop stress in thin cylinders

Consider a thin cylindrical pipe of diameter "d" and wall thickness "t" subjected to an internal fluid pressure "p".

Circumferential stress (σ_c) = Bursting force / Resisting area

Bursting force along diameter of cylinder,

F = Pressure × area of c/s $= P \times d \times \ell$ ($\because$ area of cross section = $d \times \ell$)

The total resisting area $= 2 \times t \times \ell$[two thicknesses resisting the bursting force]

$$\sigma_c = \frac{P \times d \times \ell}{2 \times t \times \ell}$$

$$\sigma_c = \frac{Pd}{2t}$$

Longitudinal Stress

The stress in the longitudinal direction due to tendency of bursting the cylinder along the transverse plane is called longitudinal stress. Consider a thin cylindrical shell of diameter d and wall thickness 't' and length ℓ subjected to an internal pressure P. If the longitudinal stress exceeds the permissible stress limit of the material, the thin cylinder will split longitudinally as shown in Figure 1.22.

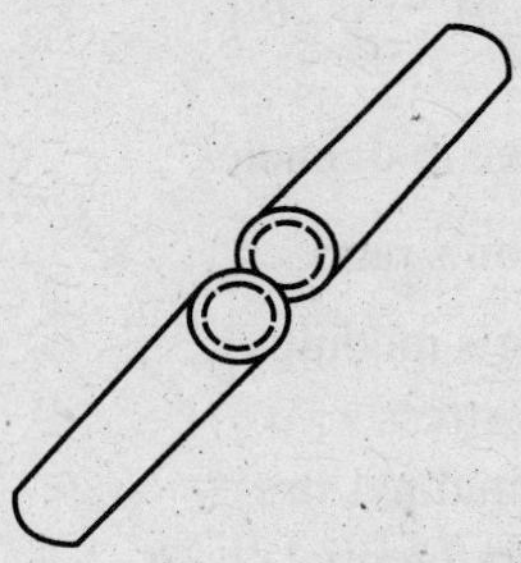

Fig .1.22 Tearing of thin cylinder across the dia due to longitudinal stress

$$\text{Longitudinal stress } \sigma\ell = \frac{\text{Axial bursting force}}{\text{Resiting area}}$$

Axial bursting force (F) = Intensity of stress × area of

$$C/S = P \times \frac{\pi}{4} d^2$$

Total resisting area of cross section = $\pi \times d \times t$

$$\therefore \ \sigma\ell = \frac{P \times \pi d^2/4}{\pi \times d \times t}$$

$$\sigma\ell = \frac{Pd/4}{4t}$$

$$\therefore \sigma_\ell = \frac{1}{2}\sigma_c = \frac{Pd}{4t}$$

Longitudinal stress = Half of the circumferential stress.

Change in dimensions of a thin cylinder due the internal pressure.

Due to internal pressure, the thin cylinders are subjected to lateral and linear strain. Thus the dimensions such as length, diameter may increase.

Circumferential Strain

$$\in_c = \frac{\delta d}{d} = \frac{\sigma_c}{E} = \frac{\sigma_L}{mE};$$

$$\in_c = \frac{Pd}{2tE} - \frac{Pd}{4tmE}$$

$$\in_c = \frac{Pd}{2tE}[1 - 1/2m]$$

Where, $\delta d \Rightarrow$ Change in diameter

$\frac{1}{m} \Rightarrow$ Poisson's ratio

$E \Rightarrow$ Young's modulus

$\sigma_c \Rightarrow$ Hoop stress

$\sigma_\ell \Rightarrow$ Longitudinal stress

$\in_c \Rightarrow$ Circumferential chain.

Change in diameter $\frac{\delta d}{d} = \in_c \times d$

$$= \frac{Pd^2}{2tE}\left(1 - \frac{1}{2m}\right)$$

1.18.2 Longitudinal Strain ($\epsilon\ \ell$)

$$\frac{\delta\ell}{\ell} = \frac{\sigma_\ell}{\in} - \frac{\sigma_c}{mE}$$

$$\Rightarrow \frac{Pd}{4tE} - \frac{pd}{2tmE}$$

$$\varepsilon\ell = \frac{Pd}{2tE}\left(\frac{1}{2} - \frac{1}{m}\right)$$

Change in Length,

$$\delta\ell = \in_\ell \times \ell$$

$$\delta\ell = \frac{pd\ell}{2tE}\left(\frac{1}{2} - \frac{1}{m}\right)$$

Volumetric strain $\epsilon_v = \dfrac{\text{Final volume} - \text{Initial volume}}{\text{Initial volume}}$

$$= \frac{\pi/4\left(d + \delta d^2\right)(\ell + \delta\ell) - \frac{\pi}{4}d^2 \times \ell}{\frac{\pi}{4}d^2 \times \ell}$$

By Neglecting higher order terms δd & $\delta \ell$,

$$\text{Volumetric strain}(\in v) = \frac{2\delta d}{d} + \frac{\sigma \ell}{\ell}$$

$$\in_v = \frac{2\delta d}{d} + \frac{\delta \ell}{\ell} = 2\in_c + \in_\ell$$

$$= \frac{pd}{2tE}\left(1 - \frac{1}{2m}\right) + \frac{2pd}{2tE}\left(\frac{1}{2} - \frac{1}{m}\right)$$

$$\varepsilon_v = \frac{pd}{2tE}\left(2 - \frac{2}{2m} + \frac{1}{2} - \frac{1}{m}\right)$$

$$\frac{\delta_v}{v} = \frac{pd}{2tE}\left(2 + \frac{1}{2} - \frac{2}{m}\right)$$

$$\delta_v = \frac{pdv}{2tE}\left(\frac{5}{2} - \frac{2}{m}\right)$$

Change in volume (δ_v)

$$\delta v = \in_v * v = \frac{Pdv}{2tE}\left(\frac{5}{2} - \frac{2}{m}\right)$$

$$\delta v = v \times \frac{\sigma_c}{E}\left(\frac{5}{2} - \frac{2}{m}\right)$$

1.19 Thin Spherical shells

A thin spherical shell of internal diameter "d" and wall thickness "t" subjected to an internal fluid pressure "P".

The spherical shell will fail be breaking into two hemisphere as shown in the Figure 1.23.

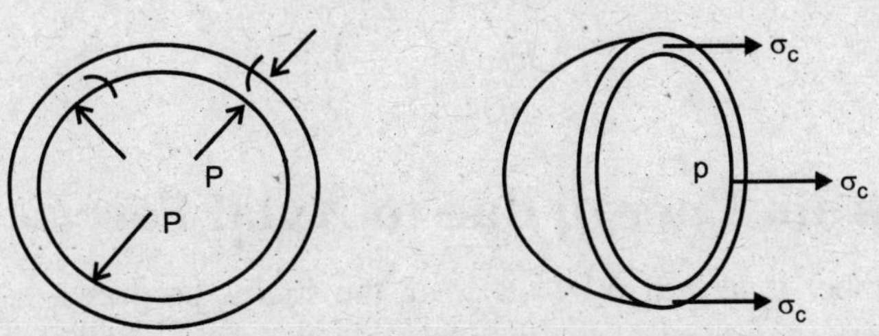

Fig .1.23 Thin spherical shell

The bursting force acting along the centre of the sphere.(F)

= Intensity of stress × Area of cross section

$$\text{Circumferential strain } \in_c = \frac{\delta d}{d} = \frac{\sigma_c}{E} - \frac{\sigma_d}{mE}$$

$$\in_c = \frac{Pd}{4tE} - \frac{Pd}{4tmE}$$

$$\varepsilon_c = \frac{Pd}{4tE}\left[1 - \frac{1}{m}\right]$$

Change in diameter

$$\delta d = \in_c \times d$$

$$(\because \in_c = \frac{\delta d}{d} \text{ and } \delta d = \frac{\rho d^2}{4tE}\left[1 - \frac{1}{m}\right]$$

If δd is the change in diameter, the final volume of shell $= \frac{\pi}{6}(d + \delta d)^3$

$$\therefore \text{ Volumetric strain } \in (\vartheta) = \frac{\text{Final volume} - \text{Initial volume}}{\text{Initial volume}}$$

$$= \frac{\pi/6(d + \delta d)^3 - \pi/6d^3}{\frac{\pi}{6}d^3}$$

On neglecting the higher orders of δd

$$\in_v = 3 \times \frac{\delta d}{d}\in_v = 3 \times \frac{\delta d}{d} = 3 \times \in_c$$

$$\in_v = \frac{3pd}{4tE}\left[1 - \frac{1}{m}\right]$$

Change in volume $(\delta v) = \in_v \times v$

$$= \frac{3pd}{4tE}\left[1 - \frac{1}{m}\right] \times v$$

$$\delta v = \frac{\pi p d^4}{8tE}\left[1 - \frac{1}{m}\right]$$

1.20 Strain Energy due to Axial Force

When a member is stretched by the load, the strains produced in the member increases the energy level of the member. By the principle of conservation of energy, the work done by the load in stretching the member is equal to the energy absorbed or stored by the member. Energy thus stored or absorbed by the member is called strain energy.

Strain Energy stored in a structural member, when applied load is gradual

Let us consider a structural member subjected to a gradual load. If 'p' is the applied load, there will be change in length of the member. Then a load extension diagram as below can be drawn as deposited in the Figure 1.24.

If, x = change in Length
P = Gradually applied load
A = Cross sectional area
L = Length of the body
V = Volume of the body
E = Young's modulus
U = Strain energy stored and the body
σ = Stress induced in the body

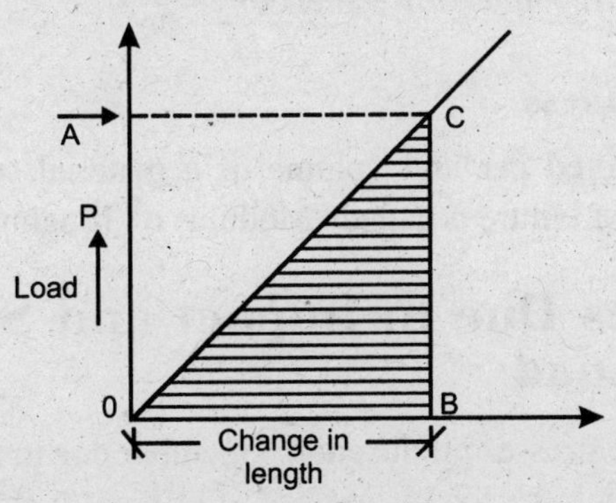

Fig .1.24 Load extension curve

Then, work done by the load (w) = Area of the Load extension curve

$$W = \frac{1}{2}P.x = \frac{1}{2}.\sigma.A.x \qquad \left(\in = \frac{x}{L} \Rightarrow x = \in .L = \frac{\sigma}{E}.L\right)$$

$$= \frac{1}{2}.\sigma.A.\frac{\sigma}{E}.L$$

$$W = \frac{\sigma^2 AL}{2E} = \frac{\sigma^2.v}{2E} \qquad (\because v = A.L)$$

Work done by the Load = Strain energy stored in a body

$$\omega = \frac{\sigma^2}{2E}.v$$

1.21 Resilience

The total strain energy stored in a body is generally known as Resilience.

Strain Energy Density

Strain energy stored in a body per unit volume is known as "Strain Energy Density".

Modulus of Resilience

Strain energy stored in a body per unit volume of a material, when it is stressed to the proportional limit is called as "Modulus of Resilience".

$$\frac{\text{Strain Energy}}{\text{Unit Volume}} = \frac{\frac{\sigma^2.v}{2E}}{v} = \frac{\sigma^2}{2E}$$

$$\text{modulus of Resilience} = \frac{\sigma^2}{2E}$$

Modulus of toughness

The strain energy stored per unit volume of a material, when the material is stressed to the point of failure is called "Modulus of Toughness".

1.22 Stresses Due to Impact and Suddenly Applied Load

The magnitude of the stresses produced in a member due to applied load depends on the manner in which the load is applied. Different types of load may act on a member, Loads may be classified as static or dynamic loads depending upon whether they are constant or vary with time.

Static load is a load which is applied slowly. i.e. gradually, so that there are no vibrations or any dynamics effects on the member. This load is taken to gradually increase from zero to a maximum value, and thereafter remain constant.

A dynamic load may be a suddenly applied load or it may be an impact load or it may be fluctuating load persisting for some period of time continuously varying in its intensity. E.g. Rotating machinery, traffic, earthquake loads. It is already seen that,

Strain Energy stored in a body

$$U = \frac{\sigma^2.v}{2E} \times v = \frac{\sigma^2}{2E}.A.L$$

$$= \frac{\sigma^2}{2\in}.A.L. = Px = P.\frac{\sigma}{E}L = \frac{\sigma.A}{2} = P$$

It is to be understood from the above relation that the maximum stress produced due to the suddenly applied Load is twice the stress produced when the load is applied gradually.

Derivation for stress expression for Impact load

There will be occasions in the industries when the load may come upon a structural member at a high impact effect. If

P = Load applied with impact
L = Length of the rod
A = Cross sectional area
V = Volume of the road = A L
H = Height through which the load will fall before impacting the collar of the bar.
δL = Extension of the rod due to the Load 'P'
E = Modulus of elasticity
σ = Stress induced due to impact loading.

Then,

$$\sigma = \sqrt{\frac{2PE.h}{AL}}$$

Strain Energy stored in a body due to shear stress

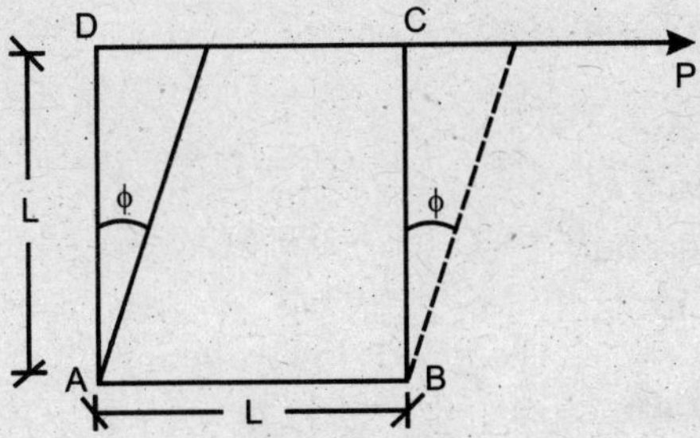

Fig .1.25

Let us consider a side of cube of length 'ℓ' as in diagram 1.25 below

q = shear stress
ϕ = shear strain
c = Modulus of rigidity
p = Force applied tangentially as the face CD.

Then, the strain energy stored in a structural member = $\dfrac{q^2}{2c} \times v$

Part - A

Questions and Answers

1. **Define Stress and Strain**

The internal resisting force per unit area is defined as stress.

$$(\sigma)\text{Stress} = \frac{\text{Load}}{\text{Area}}(N/mm^2)$$

When a body is subjected to an external force, there is some change of dimension in the body. Numerically the strain is equal to the ratio of change in length to the original length of the body.

$$(\epsilon)\text{Strain} = \frac{\text{change in length}}{\text{original length}} \qquad \in = \frac{\sigma L}{L}$$

It has no unit (Dimension less quantity)

2. **What are the different types of strains?**

Different types of strains are

(a) Linear Strain
(b) Lateral Strain
(c) Volumetric strain
(d) Shear strain

3. **State Hooke's Law**

Hooke's law states that when a material is loaded, within its elastic limit, the stress is directly proportional to the strain.

$$\text{i.e., } \quad \text{Strain } \alpha \text{ Strain}$$

$$\text{Stress } \sigma \in;$$

$$\therefore \ \sigma L = E \in$$

$$\boxed{E = \frac{\sigma}{\epsilon}}$$

Where E = Young's modulus, N/mm^2 $\quad \sigma$ = Stress, N/mm^2 $\quad \epsilon$ = Strain

4. **Draw the stress strain behaviour of a ductile material showing all the salient points**

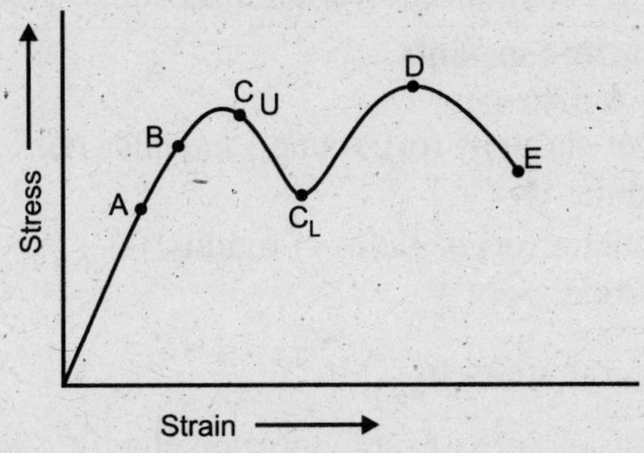

Fig .1.26 Stress - strain curve

A	→	Proportionality limit point
B	→	Elastic limit
C	→	Yield Point
Cu	→	Upper yield point
C_L	→	lower yield point
D	→	Ultimate load point
E	→	Breaking point

5. **Define proportionality limit**

   In the stress - strain diagram of a mild steel, Up to the point of proportional limit, value of stress will be proportional to strain. This is called as proportionality limit.

6. **Define Elastic Limit**

   In the stress - strain diagram of a mild steel, up to the point of elastic limit, the member will come back to its original shape and recover its original shape. This is called a elastic limit.

7. **What is Yield Stress?**

   Beyond the elastic limit, plastic deformation occurs and strains are not totally recoverable. There will be thus permanent deformation or permanent set when load is removed. The stress at the yield point is called the yield strength or Yield stress.

8. **State the meaning of Ultimate Stress**

   The maximum load which the specimen can with stand without failure is called as the ultimate strength or Ultimate stress.

9. **Define factor of safety**

   It is defined as the ratio of ultimate tensile stress to the permissible stress (working stress). Factor of safety = ultimate stress/ Permissible stress

10. **List out the elastic constants**

    - Modulus of elasticity (or) Young's modulus (E)
    - Bulk Modulus (K)
    - Shear Modulus (or) modulus of rigidity(G)
    - Poisson ration μ or $(\frac{1}{m})$

11. **Define Modulus of elasticity**

    The ratio of tensile stress to the corresponding tensile strain is constant within its elastic limit. The ratio is known as young's modulus.

    Young's Modulus (or) = Tensile stress/ Tensile strain Modulus Elasticity

12. **Define Bulk Modulus or Modulus of compression**

    Bulk Modulus = Direct stress / Volumetric strain

    Where, Direct stress= stress which cause the change in volume

    Volumetric strain = sum of linear stress in x, y and z direction.

13. **What is Modulus of Rigidity?**

    It is defined as the ratio of shearing stress to the corresponding shearing strain. It is constant within its elastic limit.

    Modulus of Rigidity (G) = Shearing stress/ Shearing strain

14. **Define Poisson's ratio.**

    When a body is stressed, within its elastic limit, the ratio of lateral strain to the longitudinal strain is constant for a given material. The ratio is called as poisson's ratio.

    Poisson's ratio = Lateral strain / Longitudinal strain.

15. **State the relationship between C and E.**

    $$E = 2c\left(1 + \frac{1}{m}\right)$$

    Where,

    E = Young's Modulus

    C = Rigidity Modulus

    1/m = Poissons ratio

16. **Give the relationship between Bulk Modulus(K) and Young's Modulus(E).**

$$K = \frac{E}{3\left(1-\frac{2}{m}\right)} (or) E = 3k\left[1-\frac{2}{m}\right]$$

Where,

E = young's Modulus

K = bulk modulus

1/m = poisson's ratio

17. **Define Hydrostatic Stress.**

The term hydrostatic stress is used to describe a state of tensile or compressive stress equal in all directions within or external to a body. Hydrostatic stress causes a change in volume of a material, which if expressed per unit of original volume gives volumetric strain denoted by I_v.

18. **Write the expression for change in length of a thin cylindrical shell or diameter 'd' and length "l" when subjected to internal pressure "p".**

Change in diameter, $\delta d = \dfrac{pd^2}{2tE}\left(1-\dfrac{1}{2m}\right)$

Change in Length, $\delta l = \dfrac{pdl}{2tE}\left(\dfrac{1}{2}-\dfrac{1}{m}\right)$

19. **Define resilience or strain energy.**

It is the total strain energy stored in the body.

20. **Define Proof resilience.**

It is the maximum strain energy stored in the body.

21. **Define Modulus of resilience.**

It is the proof resilience per unit volume of the body

22. **Define Thermal Stress.**

It is stress or load induced due to change in temperature. This causes a material to expand or shrink

23. **What is the effect of change in temperature in the composite bar?**

In the composite bar a material with lower - coefficient of expansion will be under tensile stress where as the material with higher co-efficient expansion will be under compressive stress.

24. **Define uniform strength.**

A section is said to be of uniform strength, if the stresses are equal along the length of the section. The self weight of the material is ignored for easier calculation in practice. If self weight is considered then the cross section area of the material will be maximum at the top and minimum at the bottom for the hanging section.

25. **What is meant by resilience of a helical spring and stiffness of a spring?**

(AU Nov's 2013)

The resilience of a helical spring means by amount of energy stored in the helical spring during loading. The stiffness of the spring is defined as the load required to produce unit deflection.

$$\text{Stiffness K} = \frac{W}{\delta}$$

Part - B

Questions and Answers

Problems on stress and strain at a point - Tension, compression and shear stress

1. **Derive the relation for change in length of a bar hanging freely under its own weight.**

Solution

Consider a body subjected to a tensile stress.

We know that Stress, $\sigma = \frac{\text{load}}{\text{Area}} = \frac{P}{A}$

$$\text{Strain, } \varepsilon = \frac{\text{Change in length}}{\text{Original length}} = \frac{\delta L}{L} \qquad (1)$$

From Hooke's law

Young's modulus,

$$E = \frac{\text{Stress}}{\text{Strain}} = \frac{\sigma}{\varepsilon}$$
$$E = \frac{P/A}{\varepsilon}$$
$$E = \frac{P}{A\varepsilon}$$
$$\varepsilon = \frac{P}{AE} \quad (2)$$

Equating (1) and (2)

$$\frac{P}{AE} = \frac{\delta L}{L} \Rightarrow \quad \therefore \ \delta L = \frac{PL}{AE}$$

$$\boxed{\text{Change in length, } \delta L = \frac{PL}{AE}}$$

2. **A rod of 2m long is subjected to an axial pull of 30 KN. Find the minimum diameter of the rod if the permissible stress is limited to 200 × 10⁶N/m^2. Also determine the elongation. Take E = 200KN/mm².**

Given Data

Length (L) = 2m = 2000mm

Axial pull (P) = 30 KN = 30 × 10^3N

$$1kN = 10^3$$
$$1MN = 10^6 N$$
$$1GN = 10^9 N$$

$$\text{Stress, } \sigma = 200 \times 10^6 N/m^2$$
$$= 200 \times 10^6 \frac{N}{10^6 mm^2} = 200 N/mm^2$$
$$E = 200 KN/mm^2 = 200 \times 10^3 N/mm^2$$
$$= 2 \times 10^5 N/mm^2$$

To Find

Minimum diameter (D)

Elongation (δL).

Solution

$$\text{Stress}(\sigma) = \frac{P}{A}$$

$$200 = \frac{30 \times 10^3}{A}$$

$$\text{A} = 50\text{mm}^2$$

$$\frac{\pi}{4}d^2 = 150$$

$$\therefore \boxed{d = 13.8mm}$$

From Hookes Law

$$\Rightarrow \quad E = \frac{\sigma}{\in}$$

$$2 \times 10^5 = \frac{200}{\in}$$

$$\because \quad \in = \frac{200}{2 \times 10^5}$$

$$\in = 1 \times 10^{-3}$$

Substituting for '∈'

$$\in = \frac{\delta L}{L}$$

$$1 \times 10^{-3} = \frac{\delta L}{2000}$$

$$\boxed{\delta L = 2mm}$$

Result

Minimum diameter (D) = 13.8mm

Elongation (δL), = 2mm

3. **A cast iron column has internal diameter of 200mm, what should be the minimum external diameter so that it may carry a load of 1.6 MN and safe stress is 90N/mm^2.**

Given Data

Internal diameter, d = 200 mm

Load, $P = 1.6MN = 1.6 \times 10^6 N$

Stress = 90 N/mm^2

To Find External diameter D

Solution

We know that

Stress, $\sigma = \dfrac{\text{Load}}{\text{Area}}$

Substituting for 'p','A' and σ

$$\sigma = \frac{\text{P}}{\text{A}}$$

$$= \frac{1.6 \times 10^6}{\frac{\pi}{4}(D^2 - d^2)}$$

$$90 = \frac{1.6 \times 10^6}{\frac{\pi}{4}\left(D^2 - (200)^2\right)}$$

$$90 = \frac{1.6 \times 10^6}{\frac{\pi}{4}(D^2 - 200^2)}$$

$$\frac{\pi}{4}(D^2 - 200^2) = \frac{1.6 \times 10^6}{90}$$

$$D^2 - 200^2 = 22635.369$$

$$D^2 = 62635.37$$

$$\boxed{D = 250.27mm}$$

$$\boxed{D = 250\ mm}$$

Result

Minimum external Dia (D) = 250mm

4. **Short hollow case iron cylinder of external diameter 220mm is to carry a compressive load of 600KN. Determine the inner diameter of the cylinder, if the ultimate crushing stress of the material is 540 MN/m^2. Use factor as safety of 6.**

Given Data

External diameter D = 220 mm

P $= 600KN = 600 \times 10^3$ N

Ultimate stress = 540 MN/m^2

$$= 540 \times 10^6 N/m^2$$

$$= \frac{540 \times 10^6}{10^6} = 540 N/mm^2$$

Factor of safety = 6

To Find

Internal diameter = ?

Solution

We know that factor of safety $= \dfrac{\text{Ultimate stress}}{\text{Permissible stress}}$

Permissible stress $= \dfrac{540}{6} = 90 N/mm^2$

$$\text{w.k.t,} \quad \sigma = \frac{P}{A}$$

$$90 = \frac{600 \times 10^3}{\frac{\pi}{4}\left(220^2 - d^2\right)}$$

$$220^2 - d^2 = \left(\frac{600 \times 10^3}{90}\right) \times \frac{4}{\pi} = 8488.26$$

$$\boxed{d = 199.77 mm}$$

$$\boxed{d = 200 \ mm}$$

Result : Internal dia meter (d) = 200mm

5. **A bar of 3m long is 50mm square in section for 120mm of its length, 25mm diameter for 80mm and 40 mm diameter for the remaining length. If a tensile force of 100KN is applied to the bar. Calculate the maximum and minimum stresses produced in it and the total elongation. Take E = 200 GN/m^2 and assume uniform distribution of load over the cross section.** **(AU May 2013)**

Given Data

Load $(P) = 100 kN = 100 \times 10^3 N$

Length of Bar $(\ell) = 0.3 m = 300 mm$

Young's Modulus $(E) = 200 GN/m^2$

$200 \times 10^9/10^6 = 200 \times 10^3 = 2 \times 10^5 N/mm^2$

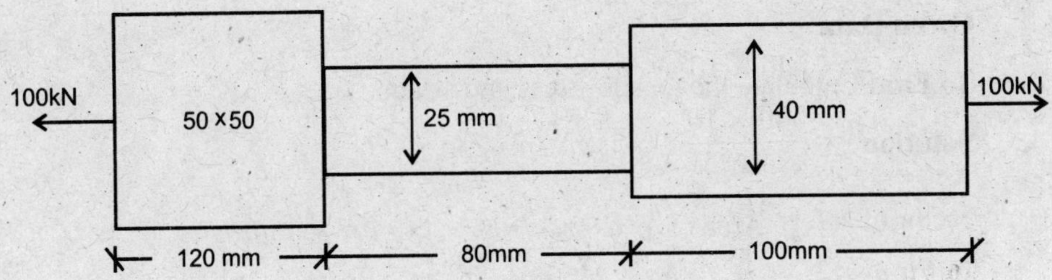

Fig .1.27

Size = 50mm × 50mm (First length) = Square shape

25 mm dia, 80mm length (Second length) = Circular shape

40mm dia, 100mm length (Third length) = Circular shape

Solution

$$\delta l_1 = \frac{Pl_1}{A_1E} = \frac{100 \times 10^3 \times 120}{50 \times 50 \times 2 \times 10^5} = 0.024mm$$

$$\delta l_2 = \frac{Pl_2}{A_2E} = \frac{100 \times 10^3 \times 80}{\pi/4 \times 25^2 \times 2 \times 10^5} = 0.081mm$$

$$\delta l_3 = \frac{Pl_3}{A_3E} = \frac{100 \times 10^3 \times 120}{\pi/4 \times 40^2 \times 2 \times 10^5} = 0.048mm$$

$$\delta l = \delta l_1 + \delta l_2 + \delta_3$$

$$\delta l = 0.024 + 0.081 + 0.048$$

$$\boxed{\delta l = 0.153\ mm}$$

Max and Min stress = ? $\sigma_1 = \frac{P}{A_1}, \sigma_2 = \frac{P}{A_2}, \sigma_3 = \frac{P}{A_3}$

6. **The prismatic bar with rectangular cross section (20 × 40)mm and the length 2.8m is subjected to an axial tensile force of 70 KN. The measured elongation of bar is 1.2mm. Calculate the tensile stress and strain in the bar.**

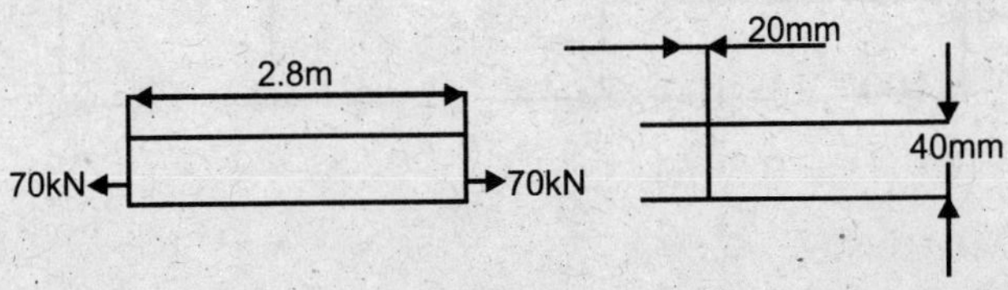

Fig .1.28

Given Data

To Find Calculate the tensile stress and strain

Solution

$$\text{Area of } crosssection = 20mm \times 40mm$$
$$\ell = 2.8m = 2800mm \;\; \delta\ell = 1.2mm$$
$$\text{Load(P)} = 70 \times 10^3 N$$

Result

$$\begin{aligned}\text{Stress}(\sigma) &= \frac{P}{A}\\ &= \frac{70 \times 10^3}{20 \times 40} = 87.5N/mm^2\\ \text{Strain}(\varepsilon) &= \frac{\delta\ell}{\ell}\\ &= \frac{1.2}{2800} = 0.000428\end{aligned}$$

Problems on stress and strain in varying section (Principle of superposition)

7. **A stepped bar is loaded as shown below. Find the stress and corresponding elongation of each part.**

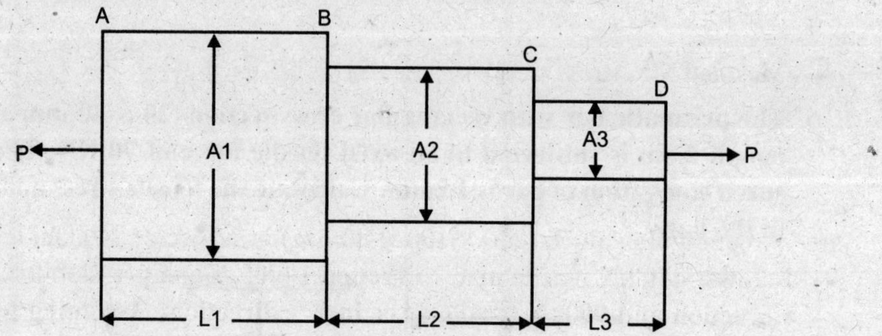

Fig .1.29

P Load applied, A1, A2, A3 are corresponding areas of AB, BC and CD. The free body diagram for each part is

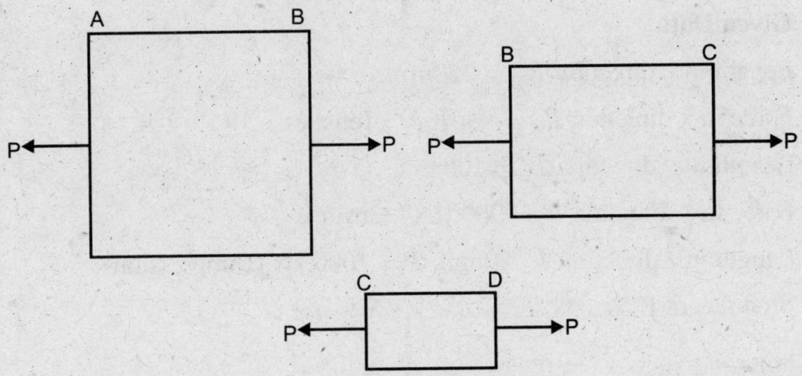

Fig .1.30

The stresses are σ_{AB} = P/Area of AB ; σ_{BC}; = P/Area of BC σ_{CD} = P / Area of CD

The corresponding elongation is,

$$\delta l_1 = \frac{\sigma_{AB} l_1}{E}; \delta l_2 = \frac{\sigma_{Bc} l_2}{;} E \delta l_3 = \frac{\sigma_{CD} l_3}{E}$$

The total elongation in length is

$$\begin{aligned} \delta \ell_1 &= \delta l_1 + \delta l_2 + \delta l_3 \\ &= \frac{1}{E} [\sigma_{AB} l_1 + \sigma_{BC} l_2 + \sigma_{CD} l_3] \end{aligned}$$

Problems on Hookes law and elastic constants

8. **Establish the relation between young's modulus and Bulk modulus.**

   Refer Section 1.10.2

9. **Establish the relation between Young's modulus and rigidity modulus.**

   Refer Section 1.10.1

10. **A rectangular block (250 × 100 × 80mm) is subjected to axial loads as follows: 480 kN tensile in x - direction 1000KN compression force in z -direction and 900kN tensile force in Y - direction. Assuming poisson ratio as 0.25, find the strains in the direction of each force in terms of modulus of elasticity of the material E, If E = 2 × $10^5 N/mm^2$,find the values of the modulus of rigidity and bulk modulus for the material of the block. Also calculate the change in volume of the block due to the loading specified above. (AU Nov/Dec 2012)**

Given Data

Length in x direction $L_x = 250mm,$

Force in x direction $P_x = 480KN,$ (Tensile)

Length in y direction L_y = 100 mm,

Force in y direction P_y = 900 KN (Tensile)

Length in z direction L_z 80mm; P_z= 1000 kN (compression)

Modulus of Elasticity, $E = 2 \times 10^5 N/mm^2$

Poisson's ratio $\frac{1}{m} = 0.25$

To Find

(a) Change in volume of block
(b) Modulus of rigidity (C or G or N)
(c) Bulk modulus,(K)

Solution

Stress in x - direction

$$(\sigma_x) = \frac{\text{Load in } x \text{ direction}}{\text{c/s Area}} = \frac{480 \times 10^3}{100 \times 80} = 60N/mm^2$$

$\therefore$ **Strain along x -direction,**

$$\varepsilon_x = \frac{\text{Stress in x direction}}{\text{Young's Modulus}}$$

$$\varepsilon_x = \frac{60}{2 \times 10^5}$$

$$\boxed{\varepsilon_x = 0.0003}$$

Stress in y - direction

$$(\sigma_y) = \frac{\text{Load in } y \text{ direction}}{\text{c/s Area}} = \frac{900 \times 10^3}{80 \times 250} = 45N/mm^2$$

Strain along y - direction,

$$\varepsilon_y = \frac{\text{Stress in y direction}}{\text{Young's Modulus}}$$

$$= \frac{45}{2 \times 10^5}$$

$$\boxed{\varepsilon_y = 0.000225}$$

Stress in z - direction

$$\sigma_z = \frac{\text{Load on z direction}}{\text{Cross sectional Area}}$$

$$\sigma = \frac{-100 \times 10^3}{100 \times 250} = --4N/mm^2$$

Strain along z -direction

$$\sigma_z = \frac{\text{Stress on z direction}}{E}$$

$$= \frac{-40}{2 \times 10^5}$$

$$\boxed{\varepsilon_z = 0.0002}$$

We known that,

$$\frac{\delta V}{V} = \frac{1}{E}(\sigma_x + \sigma_y + \sigma_z)\left(1 - \frac{2}{m}\right)$$

$$= \frac{1}{2 \times 10^5}(60 + 45 - 40)(1 - 2 \times 0.25)$$

$$\frac{\delta V}{V} = 0.0001625$$

Change in volume $\delta V = 0.0001625 \times$V

$$= 0.0001625 \times (80 \times 100 \times 250)$$

$$\boxed{\delta V = 325mm^3}$$

(ii) We know the relation ship between E, G and $\dfrac{1}{m}$

Substituting for 'E' and $\frac{1}{m}$,

$$E = 2G\left(1 + \frac{1}{m}\right)$$

$$2 \times 10^5 = 2G(1 + 0.25)$$

$$2G = \frac{2 \times 10^5}{(1 + 0.25)} = \frac{2 \times 10^5}{1.25}$$

$$2G = 160 \times 10^3$$

$$G = \frac{160 \times 10^3}{2}$$

$$\boxed{G = 80 \times 10^3 N/mm^2}$$

(iii) We know the relationship between E, k and $\frac{1}{m}$

Substituting for 'E' and $\frac{1}{m}$

$$E = 3K\left(1 - \frac{2}{m}\right)$$

$$2 \times 10^5 = 3k\,(1 - 2\,(0.25))$$

$$3k = \frac{2 \times 10^5}{1 - (2 \times 0.25)}$$

$$3k = \frac{2 \times 10^5}{0.5} = 400 \times 10^3$$

$$k = \frac{400 \times 10^3}{3}$$

$$k = 1.33 \times 10^5 N/mm^2$$

Result

i) Change in volume $\delta v = 325mm^3$
ii) Shear modulus $G = 0.8 \times 10^5 N/mm^2$
iii) Bulk modulus, $k = 1.33 \times 10^5 N/mm^2$

11. **A bar of 25mm diameter is subjected to a pull of 40 KN. The measured extension on gauge length of 200mm is 0.085mm and the change in diameter is 0.003 mm. Calculate the poisson's ratio and the values of the three modulus. (AU Nov' 2013)**

Given Data

Dia of bar (d) = 25mm

Load, p = 40kN = 40 $\times 10^3 N$

Gauge length (ℓ) = 200mm

Change in length $(\delta\ell) = 0.085mm$

Change in dia $(\delta d) = 0.003mm$

To Find μ, E, G, K

Solution

i) Poissions ratio (μ)

$$\mu = \frac{\text{lateral strain}}{\text{Longitudinal strain}} = \frac{\frac{\delta d}{d}}{\frac{\delta\ell}{\ell}} = \frac{\frac{0.003}{25}}{\frac{0.085}{200}} = \frac{0.00012}{0.000425} = 0.28$$

$$\mu = \frac{1}{m} = 0.28$$

ii) Modulus of Elasticity: We know that

$$E = \frac{\sigma}{\in}$$

$$\sigma = \frac{P}{A} = \frac{40 \times 10^3}{\frac{\pi}{4}(25)^2}$$

$$\sigma = 81.49 N/mm^2$$

$$\in = \frac{\delta\ell}{\ell} = \frac{0.085}{200} = 4.25 \times 10^{-4}$$

$$E = \frac{\sigma}{\varepsilon} = \frac{81.49}{4.25 \times 10^{-4}} = 1.91 \times 10^5 N/mm^2$$

iii) Shear Modulus(G), We know that

$$E = 2G[1 + \frac{1}{m}]$$

$$1.91 \times 10^5 = 2G[1 + 0.28]$$

$$2G = \frac{1.91 \times 10^5}{1 + 0.28} = 1.49 \times 10^5$$

$$G = \frac{1.49 \times 10^5}{2} = 7.46 \times 10^4 N/mm^2$$

$$\boxed{G = 7.46 \times 10^4 N/mm^2}$$

iv) Bulk Modulus(k), We know that

$$E = 3K\left[1 - \frac{2}{m}\right]$$

$$1.91 \times 10^5 = 3K[1 - 2(0.28)]$$

$$3k = \frac{1.91 \times 10^5}{1 - 2(0.28)} = 4.34 \times 10^5$$

$$k = \frac{4.34 \times 10^5}{3} = 1.45 \times 10^5$$

$$\therefore \; k = 1.45 \times 10^5 N/mm^2$$

12. **A metallic bar (300 × 100 × 40*mm*) is subjected to a force of 5KN (T), 6KN(C) and 4kN (T) along XY & Z direction. Determine the change in volume of the block $E = 2 \times 10^5 MPA$ and $\frac{1}{m} = 0.25$.**

Given Data

Load in x direction $L_x = 300mm$;

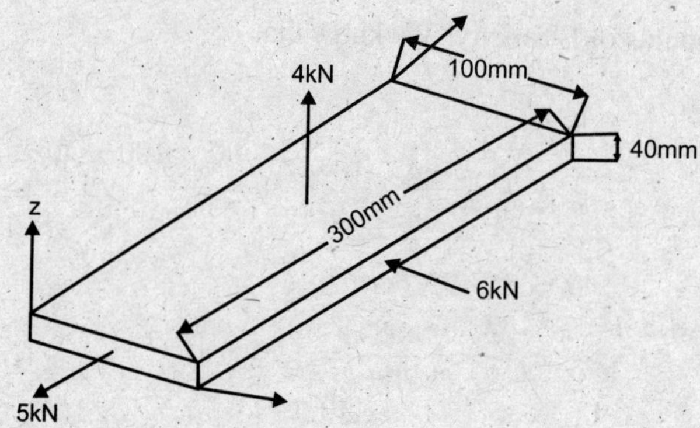

Fig .1.31

Force in x direction $P_x = 5kN(T)$

Load in y direction $L_y = 100mm;$

Force in y direction $P_y = 6kN(c)$

Load in z direction $L_z = 40mm;$

Force in z direction $P_z = 4kN(T)$

Young's Modulus $(E) = 2 \times 10^5 Mpa = 2 \times 10^5 N/mm^2$

Solution

Stress in x direction $\sigma_1 = \dfrac{P_x}{L_y \times L_z} = \dfrac{5 \times 10^3}{100 \times 40} = 1.25N/mm^2$

Stress in y direction $\sigma_y = \dfrac{P_y}{L_x \times L_z} = \dfrac{-6 \times 10^3}{300 \times 40} = -0.5N/mm^2$

Stress in z direction $\sigma_z = \dfrac{P_z}{L_x \times -L_z} = \dfrac{4 \times 10^3}{300 \times 100} = 0.133N/mm^2$

$$\frac{\delta V}{V} = \frac{1}{E}(\sigma_x + \sigma_V + \sigma_z)\left(1 - \frac{2}{m}\right)$$

$$= \frac{1}{2 \times 10^5}(1.25 - 0.5 + 0.133)(1 - 2(0.25))$$

$$= \frac{1}{2 \times 10^5}(0.883)(0.5)$$

$$\therefore \quad \frac{\delta V}{V} = 2.21 \times 10^{6}$$

$$\delta V = 2.21 \times 10^{-6} \times V$$

$$= 2.21 \times 10^{-6} \times 300 \times 100 \times 40$$

$$\boxed{\delta v = 2.65 mm^3}$$

13. **A concrete cube of 150mm side is subjected to three pairs of axial forces on the three pairs of face along the three mutually perpendicular axes xy,and z axis. The forces along x and z directions are tensile while the force along the y direction is compressive. If $Px = 54kN$, $Py = 72kN, Pz = 36kN, E = 125N/mm^2$ and $\mu = 0.25$, compute the change in volume of the cube given that $E = 12.5\times 10^3 N/mm^2$**

Given Data

Load in x Direction $(Px) = 54kN$ (Tensile)

Load in y Direction $(Py) = 72kN$ (Compressive)

Load in z Direction $(Pz) = 36kN$ Tensile

Area of cube will 150 $\times$ 150mm

Solution

Stress in X direction,

$$\sigma_x = \frac{\text{Load in X direction}}{\text{Area}}$$

$$= \frac{54 \times 10^3 N}{150 \times 150}$$

$$\boxed{\sigma_x = 2.4 N/mm^2}$$

Stress in Y direction,

$$\sigma_y = \frac{\text{Load in Y direction}}{\text{Area}}$$

$$\sigma_y = \frac{-72 \times 10^3 N}{150 \times 150 mm^2}$$

$$\boxed{\sigma_y = -3.2 N/mm^2}$$

Stress in 'Z' direction,

$$\sigma_z = \frac{\text{Load in Z direction}}{\text{Area}}$$

$$\sigma_z = \frac{36 \times 10^3 N}{150 \times 150 mm^2}$$

$$\boxed{\sigma_z = 1.6 N/mm^2}$$

We know that,

$$\frac{dV}{V} = \frac{1}{E}(\sigma_x + \sigma_y + \sigma_z\left[1 - \frac{2}{m}\right])$$

$$\Rightarrow \frac{dV}{V} = \frac{1}{12.5 \times 10^3}(2.4 - 3.2 + 1.6)[1 - \frac{2}{m}]$$

$$\frac{dV}{V} = \frac{1}{12.5 \times 10^3}(0.8)\,(1 - 2\,(0.25))$$

$$= 3.2 \times 10^{-5}$$

$$dV = 3.2 \times 10^{-5} \times V$$

$$= 3.2 \times 10^{-5} \times 150 \times 150 \times 150$$

$$\boxed{dV = 108 mm^3}$$

Problems on Stress - Strain curve of mild steel, Tor steel, concrete, ultimate stress, yield stress - factor of safety etc.,

14. **Draw stress strain curve for a mild steel rod subjected to tension and explain the salient points on it.**

Figure - 1.17 Refer page no. 17 of the book ?

(a) **Limit of Proportionality (A):** It is the limiting value of the stress up to which stress is proportional to the strain.

(b) **Elastic limit :** This is the limiting values of stress up to which if the materials, is stressed and then released, strain disappears completely and the original length is regained. This point is slightly beyond the limit of proportionality.

(c) **Upper yield point (B):** This is the stress at which the load starts reducing and the extension increases. This phenomenon is called yielding of material. At this stage strain is about 0.125 percent and stress is about $250 N/mm^2$.

(d) **Lower yield point (C):** At this stage the stress remains the same but strain increase for sometimes.

(e) **Ultimate Stress (D):** This is the maximum stress the material can resist. This stress is about $370 - 400N/mm^2$. At this stage, the cross sectional area at a particular section starts reducing very fast. This is called neck formation. After this stage load is resisted and hence the stress developed starts reducing.

(f) **Breaking point (E):** The stress at which finally the specimen fails is called breaking point. At this stage strain is about 20 to 25 percent.

15. **The following results were obtained in a tensile test of mild steel specimen of original diameter 20mm and a gauge length of 40mm. At the limit of proportionality the load was 80kN and the maximum load with stand was 150kN. When the two broken parts were fitted together the length between the gauge workings was found to be 55.6mm and the minimum diameter at the neck was 15.8mm. Calculate the (i) modulus of elasticity(ii) stress at the limit of proportionality, (iii) Yield stress, (iv) ultimate stress (v)% elongation is length, (vi)% reduction in area.**

Given Data

Length at specimen : 40mm

Specimen Dia : 20 mm

Load at Proportional Limit : 80 KN

Ultimate Load = 150KN

Final Length of Bar : 55.6mm

Dia at necking : 15.8mm

Solution

Area of mild steel,

$$A = \frac{\pi}{4}d^2 = \frac{\pi}{4} \times (20)^2$$

$$\boxed{A = 314.5mm^3}$$

Stress at Proportional limit, $\sigma_\rho = \frac{\text{Load}}{\text{Area}}$

$$= \frac{80kN}{314.15\ mm^2}$$

$$= \frac{80 \times 10^3}{314.15\ N/mm^2}$$

$$\boxed{\sigma_\rho = 254.645\ N/mm^2}$$

$$\text{Strain}, \in = \frac{\text{Change in length}}{\text{Original length}}$$

$$= \frac{0.048mm}{40mm}$$

$$\boxed{\epsilon = 0.0012}$$

Young's modulus or modulus of elasticity

$$E = \frac{\text{Stress}}{\text{Strain}} = \frac{254.65}{0.0012}$$

$$\boxed{E = 2.12 \times 10^5 N/mm^2}$$

$$\text{Yield stress}, (\sigma_y) = \frac{\text{Yield load}}{Area} = \frac{85kN}{314.15} N/mm^2$$

$$= \frac{85 \times 10^3}{314.15} N/mm^2$$

$$\boxed{\sigma y = 270.57N/mm^2}$$

Ultimate stress, σ_u

$$\sigma_u = \frac{\text{Ultimate load}}{\text{Area}}$$

$$\sigma_u = \frac{150kN}{314.15mm^2} = \frac{150 \times 10^3}{314.15} N/mm^2$$

$$\boxed{\sigma_u = 477.48\ N/mm^2}$$

$$\%\text{of elongation} = \frac{\text{Final Length} - \text{Initial Length}}{\text{Initial Length}} \times 100$$

$$= \frac{55.6mm - 40mm}{40mm} \times 100 = 39\%$$

$$\%\text{of reduction in area} = \frac{\text{Initial Dia} - \text{Final Dia}}{\text{Initial Dia}} \times 100$$

$$= \frac{\frac{\pi}{4}(20^2) - \frac{\pi}{4}(15.8)^2}{\frac{\pi}{4}(20^2)} \times 100 = 37.59\%$$

16. **A tensile test was conducted on a mild steel bar. The following data was obtained from the test.**

(i) Diameter of the steel bar = 3cm
(ii) Gauge length of the bar = 20cm
(iii) Load at Elastic limit = 250 kN
(iv) Extension at a load at 150 KN = 0.21mm
(v) Maximum load = 380kN
(vi) Total extension = 60 mm
(vii) Diameter of rod at failure = 2.25cm

Determine the following properties

(i) The young's modulus
(ii) The stress at elastic unit
(iii) The percentage of elongation
(iv) The percentage decrease in area.

Given Data

(i) Dia of the steel bar = 3cm
(ii) Gauge length of the bar = 20cm = 200mm
(iii) Load st elastic limit = $250kN = 250 \times 10^3 N$
(iv) Extension at a load at 150kN = 0.21 mm
(v) Maximum load = $380kN = 380 \times 10^3$ N
(vi) Total extension = 60mm
(vii) Diameter of rod at failure = 2.25cm = 22.5mm

To Find

(i) The Young's modulus
(ii) The stress at elastic limit
(iii) The percentage of elongation

(iv) The percentage of decrease in area.

Solution

(i) **Young's modulus**

$$\text{Young's modulus}(E) = \frac{\text{stress}}{\text{strain}}$$

$$\text{Stress, } \sigma = \frac{\text{load at elastic limit}}{\text{Area}}$$

$$= \frac{250 \times 10^3}{\frac{\pi}{4} \times 30^2} = 353.68\ N/mm^2$$

$$\text{Strain, } \in = \frac{\text{Change in length}}{\text{Original length}} = \frac{0.21}{200} = 1.05 \times 10^{-3}$$

$$E = \frac{353.68}{1.05 \times 10^{-3}} = 336.84 \times 10^3 N/mm^2$$

$$\boxed{E = 336.84 \times 10^3 N/mm^2}$$

(ii) **Percentage elongation**

$$= \frac{\text{Final extension}}{\text{Original length}} = \frac{60}{200} \times 100\% = 30\%$$

(iii) **Percentage decrease in area**

$$= \frac{\text{initial area} - \text{final area}}{\text{initial area}} \times 100$$

$$= \frac{(\pi/4 \times 30^2) - (\pi/4 \times 22.5^2)}{\pi/4 \times 30^2} \times 100$$

$$= \frac{706.85 - 397.6}{706.8} \times 100$$

$$\boxed{\text{Percentage decrease in Area} = 43.75\%}$$

Result

(a) Young's modulus $E = 336.84 \times 10^3 N/mm^2$

(b) Stress at elastic limit $= 353.68 N/mm^2$

(c) Percentage elongation = 30%

(d) Percentage decrease in Area = 43.75%

Problems on Stresses in Compound Bars

17. **A steel rod of cross - sectional area 2000mm^2 and two brass each of cross - sectional area of 1200 mm^2 together supports a load of 60 KN as in fig.1.30**

(i) Find the stresses in the rods. Take E for steel as 2×10^5 N/mm^2 and E for brass as $1 \times 10^5 N/mm^2$. **(AU Nov 2013)**

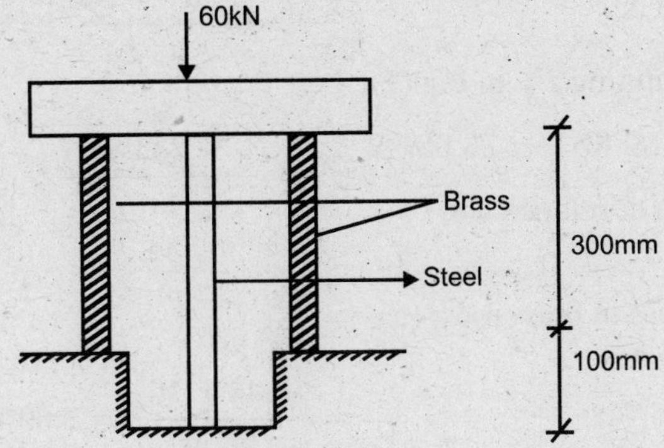

Fig .1.32

Given Data

Cross sectional Area of steel $A_s = 2000mm^2$

Cross sectional Area of Brass $A_b = 2 \times 1200 = 2400mm^2$

Load = 60 kN

Solution

Let p_b be the force in a each brass rods and P_s be the force in the steel rod,

For equilibrium, $\sum v = 0$

$$P_b + P_s + P_b = 60$$

$$2P_b + P_s = 60 \qquad (1)$$

The compatibility condition $\delta_s = \delta b$

$$\frac{P_s L_s}{A_s E_s} = \frac{P_b L_b}{A_b E_b}$$

$$\frac{P_s \times 400}{2000 \times 2 \times 10^5} = \frac{P_b \times 300}{2400 \times 1 \times 10^5}$$

$$1 \times 10^{-6} P_s = 1.25 \times 10^{-6} P_b$$

$$P_s = 1.25 P_b \qquad (2)$$

Substituting (2)in (1)

$$(1) \Rightarrow 2P_b + 1.25P_b = 60$$
$$P_b(3.25) = 60$$
$$\therefore \quad P_b = \frac{60}{3.25} = 18.46kN$$
$$\boxed{P_b = 18.46kN}$$

Now, Substituting P_B in Eqn(2)

$P_s = 1.25(18.46) \quad = 23.08kN \quad \therefore \quad \boxed{P_s = 23.08kN}$

The stresses developed are,

$$\text{Stress in brass rods } (\sigma_b) = \frac{P_b}{A_b}$$
$$= \frac{18.46 \times 10^3}{2400} = 7.69N/mm^2$$
$$\text{Stress in steel rods}(\sigma_s) = \frac{Ps}{A_s}$$
$$= \frac{23.08 \times 10^3}{2000} = 11.54N/mm^2$$

We know that $\varepsilon v = 0$

$$\boxed{P_s + 2P_b = 60KN}$$

For Compatibility, $\delta L_s = \delta L_B$

$$i.e., \quad \frac{P_s L_s}{A_s E_s} = \frac{P_b L_b}{A_b E_b}$$
$$\sigma_s \cdot \frac{L_s}{E_s} = \sigma_b \cdot \frac{L_b}{E_b}$$
$$\sigma_s \cdot \frac{400}{2 \times 10^5} = \sigma_b \cdot \frac{300}{1 \times 10^5}$$
$$2 \times 10^{-3}\sigma_s = 3 \times 10^{-3}\sigma_b$$
$$(\therefore \quad \frac{P_s}{A_s} = \sigma_s \text{ and } \frac{P_b}{A_b} = \sigma_b)$$
$$i.e., \sigma_s = \frac{3 \times 10^-3}{2 \times 10^{-3}}.\sigma_b = 1.5\sigma_b$$

We know that from(1)

$$P_s = \sigma_s.\ \text{As and}\ \ P_b = \sigma_b.A_b$$

$$\sigma_b = 1.5\sigma_b$$

$$P_s + 2P_b = 60 \times 10^3$$

$$\sigma_s \cdot A_s + 2\sigma_b \cdot A_b = 60 \times 10^3$$

$$(1.5\sigma_b)\,(2000) + 2\,(\sigma_b)\,(2400) = 60 \times 10^3$$

$$3000\sigma_b + 4800\sigma_b = 60 \times 10^3$$

$$\boxed{\sigma_b = 7.69N/mm^2}$$

Sub σ_b in

$$\sigma_s = 1.5\ \sigma_b$$

$$= 1.5(7.69)$$

$$\boxed{\sigma_s = 11.54N/mm^2}$$

18. **A compound bar of length 500mm consist of a strip of aluminium 40mm wide × 15mm thick and a strip of steel 40mm wide × 10mm thick rigidly joined at ends. If the bar is subjected to a load of 50kN find the stress developed in each material and the extension of the bar. Take modulus of elasticity of aluminium and steels as 1.1 × 10^5 N/mm^2 and 2.1 × 10^5 N/mm^2**

Given Data

Length of compressed bar $(L) = 500$

Area of Aluminium Strip $A_a = 40 \times 15 = 600mm^2$

Area of steal Strip $A_s = 40 \times 10 = 400mm^2$

Load, P = 50KN = 50,000N

Modulus of Elasticity for Aluminium $E_a = 1.1 \times 10^5 N/mm^2$

Modulus of Elasticity for Steel $E_s = 2.1 \times 10^5 N/mm^2$

To Find

1. Stress Developed in each Material
2. Extension in each Material

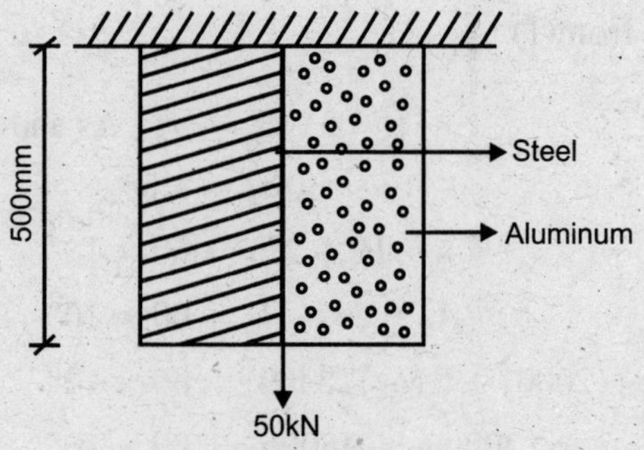

Fig .1.33

Solution

$$\text{Total load}(P) = P_a + P_s$$
$$P = 50 \times 10^3 N$$
$$50 \times 10^3 = P_a + P_s \qquad (1)$$

For compatibility conditions

Change in length of Aluminium = Change in length of steel

$$i.e., \quad \delta L_a = \delta L_s$$
$$\frac{P_a\, L_a}{A_a\, E_a} = \frac{P_s.L_s}{A_s.E_s} = [L_a = L_s]$$
$$\therefore \quad P_a = \frac{A_a E_a}{A_s E_s} \times P_s$$
$$\text{Load on aluminium } (P_a) = \frac{600 \times 1.1 \times 10^5}{400 \times 2.10 \times 10^5} \times P_s$$
$$P_a = 0.785 P_s \qquad (2)$$

We know that,

$$P_a + P_s = 50000$$
$$P_a = 50000 - P_a$$

Substituting P_a = 0.785 Ps in equation (1)

$$50000 - P_s = 0.785 P_s$$
$$\therefore \quad \boxed{P_s = 28011.2N}$$

Substituting P_s = 28011.2N in equation (2), we get

$$\boxed{P_a = 21988.7N}$$

$$\text{Stress Developed in Aluminium v } (\sigma_a) = \frac{\text{Load on Aluminium}}{\text{Area of Aluminium}}$$

$$\sigma_a = \frac{P_a}{A_a} = \frac{21988.7}{600}$$

$$\sigma_a = 36.65\ N/mm^2$$

$$\text{Stress Developed in steel} = \frac{\text{Load on Steel}}{Area of Steel}$$

$$i.e.\ \sigma_s = \frac{P_s}{A_s} = \frac{28011.2}{400}$$

$$\sigma_s = 70.03N/mm^2$$

Extension in each Material we know that for compatibility,

Change in length of Aluminium = Change in length of steel

$$i.e.\ \delta\ell_a = \delta\ell_s$$

$$i.e.\ \frac{P_a\ell_a}{A_aE_a} = \frac{P_s\ell_s}{A_sE_s}$$

$$\therefore\ \delta L = \frac{P_a\ell_a}{A_aE_a} = \frac{21988.7 \times 500}{600 \times 1.1 \times 10^5} = 0.166$$

$$\text{i.e. } \boxed{\delta L = 0.166\ mm}$$

19. **A short steel tube of external diameter 70mm and internal diameter 40mm is surrounded by a brass tube of same length and having external diameter 90mm and Internal diameter 80mm. The tubes are rigidly fired and an axial load of 20KN is placed on the tubes. Find the load carried in each tube and shortening of each tube. Take L = 30cm, Modulus of elasticity of steel = 2.1 × $10^5 N/mm^2$, $Brass = 1.1 \times 10^5 N/mm^2$**

Given Data

Steel Tube	Brass Tube
External Dia (D) = 70mm	External Dia (D) = 90mm
Internal Dia (d) = 40mm	Internal Dia (d) = 80mm

Length(L) = 30 cm = 300 mm

Modulus of Elasticity of steel $(E_S) = 2.1 \times 10^5 N/mm^2$

Modulus of Elasticity of Brass $(E_b) = 1.1 \times 10^5 N/mm^2$

Axial Load = 20kN

To Find

The Load carries by each tube and shortening of each tube

Solution

$$\text{Cross sectional area steel tube } A_s = \frac{\pi}{4}\left(70^2 + 40^2\right) = 2591.8mm^2$$

$$\text{Cross sectional area Brass tube } A_b = \frac{\pi}{4}\left(90^2 + 80^2\right) = 1335.1mm^2$$

Now,

i) Loads as each tube for compatibility conditions,

Now Substituting for values

$$\delta L_s = \delta \ell_b$$

$$\frac{P_s \ell_s}{A_s E_s} = \frac{P_b \ell_b}{A_b E_b} \qquad (\because\ l_s = l_b)$$

$$\frac{P_s}{2591.8 \times 2.1 \times 10^5} = \frac{P_b}{1335.1 \times 1.1 \times 10^5}$$

$$\therefore \boxed{P_s = 3.7 P_b} \qquad (1)$$

We know that substituting for values,

$$P = P_s + P_b$$

$$20 \times 10^3 = 3.7P_b + P_b$$

$$\therefore \boxed{P_b = 4255.3N} \text{ Substituting in(1)}$$

$$\therefore P_s = 3.7 \times P_b$$

$$\boxed{P_s = 15{,}744N}$$

ii) Shortening of tube

$$\therefore \delta L_s = \delta \ell_b$$

$$\therefore \delta L = \frac{P_s L_s}{A_s E_s} = \frac{15744 \times 300}{2591.8 \times 2.1}$$

$$\boxed{\delta L = 0.0087mm}$$

Result

$$P_s = 15.74kN$$
$$P_b = 4.26kN$$
$$\delta\ell = 0.0087mm$$

20. **A steel rod of c/s area 2000mm^2 and two brass roads each of cross section of 1200mm^2 together support a load of 500kN. Find the stresses in the rods. Take E for steel 2 × 10^5 N/mm^2 E for brass = 1 × $10^5 N/mm^2$ (AU Nov'2012)**

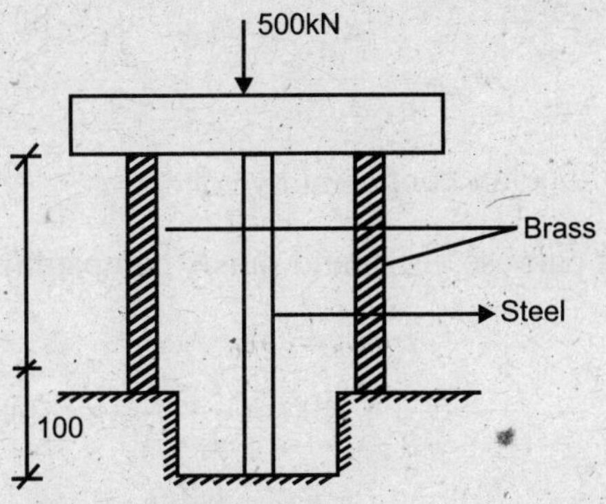

Fig .1.34

Given Data

Length of Steel rod $L_s = 300 + 100 = 400mm$

Length of brass rod $L_b = 300mm$

$$\text{Elasticity of steel } E_s = 2 \times 10^5 N/mm^2$$
$$\text{Elasticity of brass} E_B = 1 \times 105N/mm^2$$
$$A_s = 2000mm^2$$
$$A_b = 1200mm^2$$
$$P = 500kN = 500^2 \times 10^3 N$$

To Find

i) Stress in steel rod, σ_s

ii) Stress in brass rod, σ_b

Solution

For equilibrium condition, decrease in length on the steel rod is equal to the decrease in length on brass rods.

Now, decrease in length on steel rod,

$$\delta L_s = \frac{P_s L_s}{A_s E_s}$$

$$\text{We know that } \frac{P_s}{A_s} = \sigma_s$$

$$\therefore \delta L_s = \frac{\sigma_s}{E_s} \times L_s = \frac{\sigma_s}{2 \times 10^5} \times 400$$

Similarly, decrease in length on brass rod, δL_b

$$\delta L_b = \frac{P_b L_b}{A_b E_b} \qquad \left[\because \frac{P_b}{A_b} \times \sigma_b\right]$$

$$\delta L_b = \frac{\sigma_b}{E_b} \times L_b = \frac{\sigma_t}{1 \times 10^5} \times 300$$

For equilibrium purpose, its should satisfy compatibility condition

$$\delta L_s = \delta L_b$$

$$\frac{\sigma_s}{2 \times 10^5} \times 400 = \frac{\sigma_s}{1 \times 10^5} \times 300$$

$$\sigma_s = 1.5\sigma_b \tag{1}$$

Now,Total load acting = Load on Steel rod + Load on brass rod

We know that Load = Stress $\times$ Area

$$P = P_s + P_b$$

$$500 \times 10^3 = \sigma_s \times A_s + \sigma_b \times A_b$$

$$500 \times 10^3 = (1.5\sigma_b)\, A_s + \sigma_b \times A_b$$

$$\Rightarrow 500 \times 10^3 = (1.5 \times 2000\sigma_b) + (\sigma_b \times 1200) \therefore \sigma_b = 119 N/mm^2$$

$\boxed{\sigma_b = 119 N/mm^2}$; Now, substituting σ_b ineqn.

$$\therefore \sigma_s = 1.5\sigma_b$$

$$= 1.5 \times 119$$

$$\boxed{\sigma_s = 178.5 N/mm^2}$$

Result

Stress in steel rod, $\sigma_s = 178.5 N/mm^2$

Stress in brass rod, $\sigma_b = 119 N/mm^2$

21. **Two rods one of steel and other of copper are of same length and each rod have a diameter.of 10mm. A horizontal bar of negligible weight is attached to the lower end as shown. Find where a 20KN load is to be placed on the bar so that it remains horizontal E_S/E_C = 13/7.m**

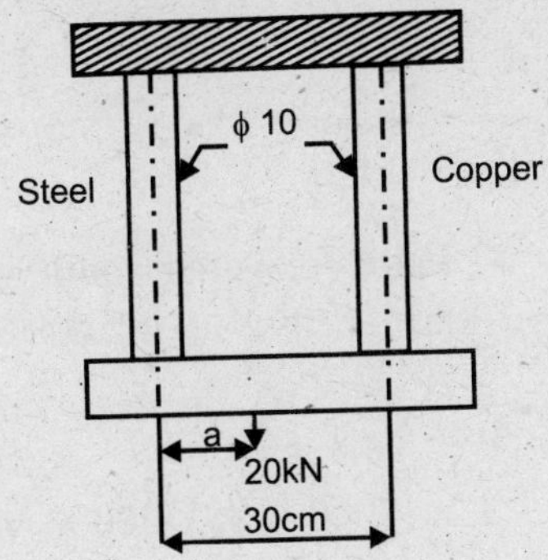

Fig .1.35

If the load of 20KN is placed at the middle of the bar then both the rods will share the load 20KN equally. As the copper has low value of Modulus of elasticity when compared to steel, the copper bar will elongate more and the connecting bar no more remains horizontal.

So the load must be placed near the steel rod so that the steel rod and then making the deformation of steel and Copper bar equal share more load. Let a be the distance between steel rod and the load.

P_s = load shared by steel bar

P_c = load shared by Copper bar

Since 20KN is shared by steel and Copper rods

We have Ps + Pc = 20KN

For compatibility conditions i.e.,

$$\frac{P_s L_s}{A_s E_s} = \frac{P_c L_c}{A_c E_c}$$

Since $L_s = L_c$

and $A_s = A_c$

$$\frac{P_s}{E_s} = \frac{P_c}{E_c}$$

$$\therefore \quad P_s = \frac{E_s}{E_c}.P_c = \frac{13}{7}.P_c \qquad (2)$$

Solving (1) and 2 we get

$$\begin{cases} Ps + Pc = 20kN \\ Ps = \frac{13}{7}.Pc \end{cases}$$

Pc = 7000 N

Ps = 13000 N

To find the distance 'a', Let us consider equilibrium steel at equilibrium, Substituting for $P_c = 7000N,\ \ 20kN = 20,000N$

$$Pc \times 30 = 20 \times a$$
$$\therefore \quad a = (7000 \times 30)/20000 = 105mm$$
$$a = 10.5cm$$

Solution 10.5cm from Left

22. **Three bars made of copper; zinc and aluminium are of equal length and have cross section 500, 700 and 1000 sq.mm respectively. They are rigidly connected at their ends. If this compound member is subjected to a longitudinal pull of 250 kN, estimate the proportion of the load carried on each rod and the induced stresses. Take the value of E for copper = 1.3 × 10$^5 N/mm^2$, for zinc = 1 × 10$^5 N/mm^2$ and for aluminium = 0.8 × 10$^5 N/mm^2$.**

Given Data

Area of cross section of Copper $(A_c) = 500mm^2$

Area of cross section of Zinc $(A_z) = 700mm^2$

Area of cross section of Aluminium $(A_l) = 1000mm^2$

$$E_{\text{copper}} = 1.3 \times 10^5 N/mm^2$$
$$E_{\text{zinc}} = 1 \times 10^5 N/mm^2$$
$$E_{\text{aluminium}} = 0.8 \times 10^5 N/mm^2$$

To Find

Proportion of the load carried on each rod and the induced stresses

Solution

1. Stress on copper,

$$\sigma_c = \frac{P}{A_c \times E_z}$$
$$= \frac{250 \times 10^3}{500 \times 1.3 \times 10^5}$$
$$\boxed{\sigma_c = 3.84 \times 10^{-3} N/mm^2}$$

2. Stress on zinc

$$= \frac{P}{A_Z \times E_z} = \frac{250 \times 10^3}{700 \times 1 \times 10^5 N/mm^2}$$
$$\boxed{\sigma_z = 3.57 \times 10^{-3} N/mm^2}$$

3. Stress on aluminium

$$= \frac{P}{A_A \times E_A} = \frac{250 \times 10^3}{1000 \times 0.8 \times 10^5}$$
$$\boxed{\sigma_a = 3.125 \times 10^{-3} N/mm^2}$$

Problems on Thermal Stresses in Simple and Compounds bars

23. **A rod of 1.5m long, 10mm diameter is fixed at the ends and subjected to axial pull of 8KN. Find the residual stress due to increase in temperature of 25°c take $E = 2.1 \times 10^5 N/mm^2, \alpha = 15 \times 10^{-6}/°C$**

Given Data

Stress due Load (σ_1) = $\frac{\text{load}}{\text{area}} = \frac{8000}{\pi/4 \times 10^2} = 101.85 N/mm^2$

Length of rod (L) = 15m

Dia of Rod = 10mm

Arial load = 8 kN = 8000 N

Temperature increase = 25°

Young's modulus (E) = $2.1 \times 10^5 N/mm^2$

Coefficient of thermal expansion (α) = 15×10^{-6}°C.

Solution

$$\sigma_1 = \frac{\text{load}}{\text{area}} = \frac{8000}{\frac{\pi}{4} \times 10^2} = 101.85 N/mm^2$$

Initial Stress

Increase in length of the rod due to the Increase in temp

Thermal stress $(\sigma_2) = \alpha$ TE $=15\times10^{-6}\times25\times2.1\times10^5 = 78.75N/mm^2$.

Residual Stress = $\sigma_1 - \sigma_2 = 101.85 - 78.75 = 23.1N/mm^2$

24. **A rod of 3m long is initially at a temperature of 15° C and it is raised to 90° C find the expansion of the rod and if the expansion is prevented find the stress in the Material. Take 'E' as 2 × 10^5 N/mm^2, α = 12 × 10^6 /°C**

Length of Rod = 3m = 300mm

Temperature increase = 90°

Given Data

$$E = 2 \times 10^5 N/mm^2$$

$$\alpha = 12 \times 10^{-6\circ}C \quad \ell = 3m = 3000mm$$

$$Temp, (T) = 90^\circ C - 15^\circ c = 75^\circ C$$

To Find

1. Expansion of the rod
2. Stress in the Material

Solution

Expansion of rod = $\alpha TL = 12 \times 10^{-6} \times 75 \times 3000 = 2.7mm,$

Thermal Stress =α TE = 12 × 10^{-6} × 75 × 2 × 10^5

$$= 180N/mm^2$$

25. **A rod is made of steel, copper and aluminium between two rigid supports at A and D. Calculate the stresses developed in each material when the temperature of the system is raised by 40°C. Values of young's modulus and coefficient of thermal expansion are as below.**

$$E_s = 2.1 \times 10^5 N/mm^2, \qquad \alpha_s = 12 \times 10^{-6}/\circ C$$

$$E_c = 1.1 \times 10^5 N/mm^2, \qquad \alpha_c = 17 \times 10^{-6}/\circ C$$

$$E_a = 0.7 \times 10^5 N/mm^2, \qquad \alpha_a = 21 \times 10^{-6}/\circ C$$

Given Data

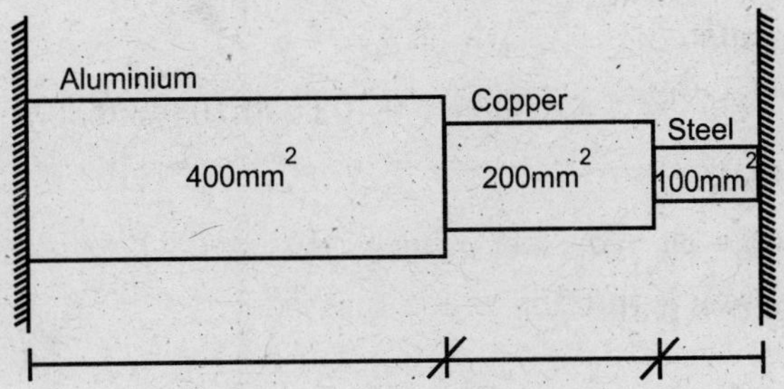

Fig .1.36

Temperature raise = 40°

$$E_s = 2.1 \times 10^5 N/mm^2$$
$$E_c = 1.1 \times 10^5 N/mm^2$$
$$E_a = 0.7 \times 10^5 N/mm^2$$
$$\alpha_s = 12 \times 10^{-6}/°C$$
$$\alpha_c = 17 \times 10^{-6}/°C$$
$$\alpha_a = 21 \times 10^{-6}/°C$$

Solution

Thermal stress (σ) = α.TE

We know that,

$$\because \; E = \frac{\sigma}{\epsilon}$$
$$\epsilon = \frac{\sigma}{E}$$

Substituting for E and Thermal stress (σ), we get

$$\therefore \; \frac{\delta\ell}{\ell} = \frac{\alpha TE}{E} \quad \therefore \; \delta\ell = \alpha TL$$
$$\delta\text{steel} = \alpha_{\text{steel}} \; T_{\text{steel}} \; L_{\text{steel}}$$
$$= 12 \times 10^{-6} \times 40 \times 70 = 0.0336mm$$
$$\delta\text{copper} = \alpha_{\text{copper}} \; T_{\text{copper}} \; L_{\text{copper}}$$
$$= 17 \times 10^{-6} \times 40 \times 150 = 0.102mm$$
$$\delta\text{aluminium} = \alpha_{\text{aluminium}} \; T_{\text{aluminium}} \; L_{\text{aluminium}}$$
$$= 21 \times 10^{-6} \times 40 \times 200 = 0.168mm$$

Total elongation $(\delta) = \delta_s + \delta_c + \delta_a$

$$= 0.0336 + 0.102 + 0.168 = 0.30m$$

We know that,

Load P = σ_s; $A_s = \sigma_c$; $A_c = \sigma_A$ A_A

Taking first two parts.

$$\boxed{\sigma_s A_s = \sigma_c A_c}$$

$$\therefore \frac{\sigma_s}{\sigma_c} = \frac{A_c}{A_s} = \frac{200}{100} = 2$$

$$\therefore \sigma_s = 2.\sigma_c$$

Also

$$\boxed{\sigma_c A_c = \sigma_a A_a}$$

$$\therefore \frac{\sigma_a}{\sigma_c} = \frac{A_c}{A_a} = 0.5$$

$$\therefore \sigma_A = 0.5\sigma_c$$

We known that,

$$\delta_L = \frac{\sigma_s L_s}{E_s} + \frac{\sigma_c L_c}{E_c} + \frac{\sigma_a L_a}{E_a}$$

$$0.30 = \frac{2\sigma_c \times 75}{2.17 \times 10^5} + \frac{\sigma_c \times 150}{1.1 \times 10^5} + \frac{0.5\sigma_c \times 200}{0.7 \times 10^5}$$

$$0.30 = 3.48 \times 10^3 \sigma_c$$

$$\therefore \boxed{\sigma_c = 86.12 \ N/mm^2}$$

26. **A weight of 200KN is supported by the pillars of equal height, each $75mm^2$ in section. The central pillar of steel and the other ones are of copper. The pillar are so adjusted at a temperature of 20° each carries equal loads. Find the stresses in each pillar at 120°. Take**

$$E_s = 2 \times 10^5 N/mm^2, \ E_c = 0.8 \times 10^5 N/mm^2$$

$$\alpha_s = 12 \times 10^{-6}/°c; \ \ \alpha_c = 18.5 \times 10^{-6}/°c;$$

Given Data

Load, $P = 200kN = 200^3 N$

Area of each pillar, $A = 75mm^2$

Rise in temperature, $(T) = 120°C - 20°C = 100°C$

Young's modulus of steel, $(E_s) = 2 \times 10^5 N/mm^2$

Young's modulus of copper, $(E_c) = 0.8 \times 10^5 N/mm^2$

Coefficient of linear expansion of steel, $(\alpha_s) = 12 \times 10^6/°C$

Coefficient of linear

Expansion of Copper, $(\alpha_c) = 18.5 \times 10^{-6}/°C$

To Find

Stress developed in steel and copper (σ_s and σ_c)

We know that, Load carried by each pillar $= \dfrac{200 \times 10^3}{3}$

Solution

Stress in each pillar,

$$\sigma = \frac{\text{Load}}{\text{Area}} = \frac{\frac{200 \times 10^3}{3}}{75}$$

$$\boxed{\sigma = 888.89 N/mm^2}$$

Under equilibrium condition,

Compression in the copper bar is equal to tension in the steel tube.

i.e Load on the copper = Load on the steel

$$\sigma_c A_c = \sigma_s A_s$$

$$\sigma_s = \sigma_c \times \frac{150}{75}$$

$$\boxed{\sigma_s = 2\sigma_c}$$

$$\left\{\begin{array}{l}\text{Actual expansion}\\ \text{of steel}(\delta Ls)\end{array}\right\} = \left\{\begin{array}{l}\text{Free expansion}\\ \text{of steel}\end{array}\right\} + \left\{\begin{array}{l}\text{Expansion due to}\\ \text{tensile stress in steel}\end{array}\right\}$$

$$dL_s = \sigma_s T L_s + \frac{\sigma_s}{E_s} \times L_s \qquad (2)$$

$$\left\{\begin{array}{l}\text{Actual expansion}\\ \text{of steel }(\delta Lc)\end{array}\right\} = \left\{\begin{array}{l}\text{Free expansion}\\ \text{of copper}\end{array}\right\} + \left\{\begin{array}{l}\text{Contraction due to compressive}\\ \text{stress induced in copper}\end{array}\right\}$$

$$\sigma_c.T.L_c - \frac{\sigma_c}{E_c} \times L_c$$

Actual expansion of steel = Actual expansion of copper

$$\Rightarrow \sigma_s.T.L_s + \frac{\sigma_s}{E_s} \times L_s = \sigma_c.T.L_c - \frac{\sigma_c}{E_c} \times L_c$$

$$\Rightarrow L_s\left[\sigma_s T + \frac{\sigma_s}{E_s}\right] = L_c\left[\alpha_c T - \frac{\sigma_c}{E_c}\right]$$

$$\Rightarrow \sigma_s.T + \frac{\sigma_s}{E_s} = \sigma_c.T - \frac{\sigma_c}{E_c} \qquad [\because\ L_s = I]$$

$$\Rightarrow 12 \times 10^{-6} \times 100 + \frac{2\sigma_c}{2 \times 10^5} = 18.5 \times 10^{-6} \times 100 - \frac{\sigma_c}{0.8 \times 10^5}$$

$$\Rightarrow 0.0012 + \frac{\sigma_c}{10^5} = 0.00185 - \frac{\sigma_c}{0.8 \times 10^5}$$

$$\frac{\sigma_c}{10^5} + \frac{\sigma_c}{0.8 \times 10^5} = 0.00065$$

$$\boxed{\sigma_c = 28.88N/mm^2}$$

We know that,

$$\sigma_s = 2\sigma_c$$

$$= 2 \times 28.88$$

$$\sigma_s = 57.78N/mm^2$$

Final stress in copper $= 888.89 + 28.89 = 917.78N/mm^2$

Final stress in steel $= 888.89 - 57.78 = 831.11N/mm^2$

Problems on Thin Cylinders and Shells

27. **Derive relations for change in length, thickness and volume of a thin cylinder subjected to an internal pressure. Also explain the failure of thin cylinders. (AU May 2012).**

Solution

Due to internal pressure, the cylindrical shells are subjected to lateral and linear strain. Thus the change in dimensions such as length,diameter may increase. Now consider a thin cylindrical shell of same dimensions as in previous case for the following analysis.

We know that

Circumferential Strain,

$$e_c = \frac{\delta d}{d} = \frac{\sigma_c}{E} - \frac{\sigma_a}{mE}$$

$$\delta d = \text{Change in diameter}$$

$$\frac{1}{\delta} = \text{Poisson's ratio}$$

E = Young's modulus

$$e_c = \frac{pd}{2tE} - \frac{pd}{4tmE}$$

$$\boxed{e_c = \frac{pd}{2tE}\left(1 - \frac{1}{2m}\right)}$$

Change in diameter, $\delta d = e_c \times d$

$$\delta d = \frac{pd^2}{2tE}\left(1 - \frac{1}{2m}\right)$$

$$e_a = \frac{\delta l}{l} = \frac{\sigma_a}{E} = \frac{\sigma_c}{mE}$$

Longitudinal Strain,

$$= \frac{pd}{4tE} - \frac{pd}{2tmE}$$

$$\boxed{e_a = \frac{pd}{dtE}\left(\frac{1}{2} - \frac{1}{m}\right)}$$

Change in length, $\delta l = e_a \times l$

$$\boxed{\delta l = \frac{pdl}{dtE}\left(\frac{1}{2} - \frac{1}{m}\right)}$$

Volumetric strain,

$$e_v = \frac{\text{Final volume - Initial volume}}{\frac{\pi}{4}d^2 l}$$

$$= \frac{\frac{\pi}{4}(d + \delta d)^2 (l + dl) - \frac{\pi}{2} \times d^2 l}{\frac{\pi}{4}d^2 l}$$

By neglecting higher order terms of δd and δl

$$\begin{aligned} e_v &= \frac{2\delta d}{d} + \frac{\delta l}{l} \\ &= 2e_c + e_a \\ &= \frac{2pd}{2tE}\left(1 - \frac{1}{2m}\right) + \frac{pd}{2tE}\left[\frac{1}{2} - \frac{1}{m}\right] \\ &= \frac{pd}{2tE}\left(2 - \frac{2}{m} + \frac{1}{2} - \frac{1}{2}\right) \\ &= \frac{pd}{2tE}\left(2 + \frac{1}{2} - \frac{1}{m}\right) \end{aligned}$$

$$\boxed{e_v = \frac{pd}{2tE}\left(\frac{E}{2} - \frac{2}{m}\right)}$$

Change in volume,

$$\begin{aligned} \delta v &= e_v \times v \\ &= \frac{pdv}{2tE}\left[\frac{5}{2} - \frac{2}{m}\right] \end{aligned}$$

$$\boxed{\delta v = v \times \frac{\sigma_c}{E}\left[\frac{5}{2} - \frac{2}{m}\right]}$$

28. **A cylindrical air drum is 2.25 m in diameter with plates 1.2cm thick. The efficiencies of the longitudinal and circumferential joints are respectively 75% and 40%.If the tensile stress in the plating is to be limited to 120 MN/m^2, Find the maximum safe air pressure (AU Nov'2013)**

Given Data

Diagram of air drum (d) $= 2.25m = 2250mm$

Thickness(ℓ) $= 1.2cm = 1.2 \times 10^{-2}m = 12mm$

Tensile stress = 120MN/m^2

Solution

Converting to N/mm^2

$$\sigma = 120 \times \frac{10^6 N}{10^6 mm^2} = 120 N/mm^2$$

The tensile stress mentioned on the problem is the Hoop stress which is otherwise called circumferential stress $(\sigma t) = 120N/mm^2$

We know that,

$$\text{Hoop stress}(\sigma t) = \frac{Pd}{2t_\eta}$$

Substituting the value of $\sigma t, d, t, ad$, efficiency η

$$120 = \frac{P \times 2250}{2 \times 120 \times 0.75} \qquad (\therefore\ d = Diameter = 2250mm\ \ t = 120mm)$$

$$\therefore \text{ The pressure} = \frac{120 \times 2 \times 120 \times 0.75}{2250}$$

$$P = 9.6N/mm^2$$

$\therefore$ The safe air pressure, $P = 9.6N/mm^2$

29. **A cylindrical shell 3m long which is closed the ends has an internal diameter of 1m and a wall thickness of 15mm. Calculate the circumferential and longitudinal stresses induced and also change in the dimensions of the dimensions of the shell if it is subjected to an internal pressure of 1.5MN/m^2, Take $E = 2 \times 10^5 N/mm^2$ and $\frac{1}{m} = 0.28$ (AU May 2013)**

Given Data

$$\text{Length of Shell} \quad (\ell) = 3m$$

$$\text{Length and Shell Dia of Shell} \quad (d) = 1m = 1000mm$$

$$\text{To wall thickness} \quad (t) = 15m$$

$$\text{Internal Pressure} \quad (P) = 1.5MN/m^2 = 1.5N/mm^2$$

$$\text{Young's modulus} \quad (E) = 2 \times 10^5 N/mm^2; 1/m = 0.28$$

$$\text{Poissons ratio}(\frac{1}{m}) = 0.28$$

Solution

We know that, circumlental stress (σ_c)

$$\sigma c = \frac{pd}{2t}$$

$$\text{Now, substituting for} \quad = \frac{1.5 \times 1000}{2 \times 15} = 50N/mm^2$$

$$\text{Longitudinal stress}(\sigma\ell) = \frac{pd}{4t}$$

$$\text{value} \quad \sigma_c = \frac{1.5 \times 1000}{4 \times 15} = 25N/mm^2$$

Substituting for value

Length of shell $(l) = 3m = 3000mm$

To Find change in dimension

We know that Lateral strain

$$e_t = \frac{\delta d}{d}$$

$$\therefore \delta d = e_t \times d \rightarrow \boxed{1}$$

Also we know that, actual strain take into acute elastic modulus

$$e_t = \frac{\sigma_c}{E} - \frac{\sigma_\ell}{mE}$$

$$e_t = \frac{50}{2 \times 10^5} - \frac{25}{2 \times 10^5} \times 0.28$$

$$e_c = 2.5 \times 10^{-4} - 1.25 \times 10^{-4} \times 0.28$$

$$e_t = 2.5 \times 10^{-4} - 3.5 \times 10^{-5}$$

$$\boxed{e_t = 2.15 \times 10^{-4}} \rightarrow \boxed{2}$$

Now substituting (2) in (1) will given δd

$$\delta d = e_t \times d$$

$$\delta d = 2.15 \times 10^{-4} \times 1000$$

$$\boxed{\delta d = 0.215 \ mm}$$

We know that

Longitudinal Strain

$$e_\ell = \frac{\sigma_1}{E} - \frac{\sigma_t}{mE}$$

$$e_\ell = \frac{25}{2 \times 10^5} - \frac{50}{2 \times 10^5} \times 0.28$$

$$= 1.25 \times 10^{-4} - 2.5 \times 10^{-4} \times 0.28$$

$$= 1.25 \times 10^{-4} - 7 \times 10^{-5}$$

$$e_\ell = 5.5 \times 10^{-5}$$

Also we know that, Longitudinal strain its terms of dimension of specimen

$$e_\ell = \frac{\delta l}{l}$$

$$\delta l = e_\ell \times l$$

Now substitutes the vale of $e\ell = 5.5 \times 10^{-5}$

$$\delta l = 5.5 \times 10^{-5} \times 3000$$

$$\boxed{\delta\ell = 0.165\ mm}$$

Volumetric Strain (ev)

Volumetric Strain $= 2e_1 + e_1$

Now substituting for e_t and de_1

$$e_v = 2 \times 2.15 \times 10^{-4} + 5.5 \times 10^{-5}$$

$$e_v = 4.3 \times 10^{-4} + 5.5 \times 10^{-5}$$

$$e_v = 4.85 \times 10^{-4}$$

Also we know that

$$e_v = \frac{\delta v}{v}$$

$$\delta_v = e_v \times v$$

$$\delta_v = 4.85 \times 10^{-4} \times \frac{\pi}{4} \times 1000^2 \times 3000$$

$$\boxed{\delta v = 1.14 \times 10^6 mm^3}$$

30. **A cylindrical vessel, whose ends are closed by means of rigid flange plate, is made up of steel plate 3mm thick. The length and internal diameter of the vessel are 50 cm and 25cm respectively. Determine the longitudinal and hoop stress in the cylindrical shell due to an internal fluid pressure of $3N/mm^2$. Also calculate the increase in length, diameter and volume of vessel. Take $E = 2 \times 10^5$ N/mm^2 and $\mu = 0.3$ (AU May 2010)**

Given Data

Wall thickness of cylindrical vessel = 3mm

Length = 50cm = 500mm
Diameter = 25cm = 250mm
Fluid pressure = 3 N/mm^2
young's modulus (E) = $2 \times 10^5 N/mm^2$
Poissons ratio$(\mu) = 0.3$

To Find

i) Longitudinal stress
ii) Hoop stress
iii) Increase in length
iv) Increase in diameter
v) Increase in volume

Solution

i) Longitudinal stress, (σ_ℓ)

$$(\sigma_\ell) = \frac{Pd}{4t} = \frac{3 \times 250}{4 \times 3}$$

$$\boxed{\sigma_a = 62.5N/mm^2}$$

ii) Hoop stress, (σ_c)

$$\text{Circumferential stress} \quad (\sigma_c) = \frac{Pd}{2t} = \frac{3 \times 250}{2 \times 3}$$

$$\boxed{\sigma_c = 125N/mm^2}$$

iii) Increase in length,

$$\delta\ell = \frac{Pd\ell}{2tE}\left(\frac{1}{2} - \frac{1}{m}\right) = \frac{3 \times 250 \times 500}{2 \times 3 \times 2 \times 10^5}\left(\frac{1}{2} - 0.3\right)$$

$$\boxed{\delta l = 0.0625mm}$$

iv) Increase in diameter

$$\delta d = \frac{Pd^2}{2tE}\left(1 - \frac{1}{2m}\right) = \frac{3 \times 250^2}{2 \times 3 \times 2 \times 10^5}\left[1 - \frac{1}{2} \times 0.3\right]$$

$$\boxed{\delta d = 0.132mm}$$

v) Increase in volume

$$\delta v = V \times \frac{Pd}{2tE}\left[\frac{5}{2} - \frac{2}{m}\right]$$

$$V = \frac{\pi}{4} \times d^2 \times l = \frac{\pi}{4} \times 250^2 \times 500$$

$$= 24.54 \times 10^6 mm^3$$

$$\delta v = 24.54 \times 10^6 \times \frac{3 \times 250}{2 \times 3 \times 2 \times 10^5}\left[\frac{5}{2} - 2 \times 0.3\right]$$

$$= 15337.5[1.9]$$

$$\boxed{\delta v = 29145.63mm^3}$$

Result

i) Longitudinal stress, $\sigma_a = 62.5N/mm^2$
ii) Hoop Stress, $\sigma_c = 125N/mm^2$
iii) Increase in length, $\delta l = 0.0625mm$
iv) Increase in diameter, $\delta d = 0.132mm$
v) Increase in volume $\delta v = 29145.63mm^3$

31. **A cylinder shell 3m long has 1m internal diameter and 1.5cm metal thickness. Calculate the maximum intensity of shear stress induced and also the changes in the dimension of the shell if it is subjected to an internal pressure of 15 kg/cm^2. Take $E = 2.04 \times 10^5 N/mm^2$ and** $\mu = 0.3$.

Given Data

Length of shell, $(l) = 3m = 3000mm$

Diameter of shell, $(d) = 1m = 1000mm$

Thickness of shell, $(t) = 1.5cm = 15mm$

Internal pressure, $(p) = 15kg/cm^2$

Converting internal pressure in N/cm^2

$$= 15 \times 9.81N/cm^2$$
$$= 147.15N/cm^2$$
$$= 147.15N/10^2mm^2$$
$$= 147.15 \times 10^{-2}N/mm^2$$
$$P = 1.47N/mm^2$$

Young's modulus, $E = 2.04 \times 10^5 N/mm^2$

Poisson' ratio, $\frac{1}{m}$ (or) $\mu = 0.3$

To Find

i) Maximum intensity of shear stress, $\tau\ max$
ii) Change in dimensions of the shell $(\delta d, \delta l, \delta v)$

Solution

Circumferential stress,

$$\sigma_c = \frac{pd}{2t} = \frac{1.47 \times 1000}{2 \times 15}$$

$$\boxed{\sigma_c = 49N/mm^2}$$

Longitudinal stress,

$$\sigma_\ell = \frac{pd}{4t} = \frac{1.47 \times 1000}{4 \times 15}$$

$$\boxed{\sigma_\ell = 24.5 N/mm^2}$$

Maximum shear Stress,

$$\tau_{\max} = \frac{\sigma_c - \sigma_\ell}{2} = \frac{49 - 24.5}{2}$$

$$\tau_{\max} = 12.25 N/mm^2$$

Circumferential strain,

$$e_c = \frac{\sigma_c}{E} - \frac{\sigma_a}{mE} = \frac{49}{2.04 \times 10^5} - 0.3 \frac{24.5}{2.04 \times 10^5}$$

$$\boxed{e_c = 2.04 \times 10^{-4}}$$

We know that,

$$e_c = \frac{\delta d}{d}$$

$$\therefore \quad \delta d = e_c \times d = 2.04 \times 10^{-4} \times 1000$$

Change in diameter, $(\delta d) = 0.204 mm$

Longitudinal strain,

$$e_\ell = \frac{\sigma_\ell}{E} - \frac{\sigma_c}{mE} = \frac{24.5}{2.4 \times 10^5} - (0.3) \times \frac{49}{2.04 \times 10^5}$$

$$\boxed{e_\ell = 3 \times 10^{-5}}$$

$$e_\ell = \frac{\delta \ell}{l} \quad \Rightarrow \quad \therefore \quad \delta \ell \Rightarrow e_\ell \times \ell = 3 \times 10^{-5} \times 3000$$

$\therefore$ Change in length $\quad \delta \ell = 0.09 mm$

We know that,

Volumetric strain,

$$e_v = e_\ell + e_c$$

$$= 3 \times 10^{-5} + 2\left(2.04 \times 10^{-4}\right)$$

$$\boxed{e_v = 4.38 \times 10^{-4}}$$

We know that,

$$e_v = \frac{\delta v}{v}$$

$$v = \frac{\pi d^2}{4}.\ell = \frac{\pi\left(1000^2\right)}{4} \times 3000$$

$$= 2.36 \times 10^9 mm^3$$

$$\delta v = e_v \times v$$

$$= 4.38 \times 10^{-4} \times 2.36 \times 10^9$$

$$= 1.03 \times 10^6 \ mm^3$$

32. **A seamless spherical shell is 0.8m internal diameter and 4 mm thickness. It is filled with fluid under pressure until its volume increase by 50cm^3. Determine the find pressure taking $E = 2 \times 10^5 N/mm^2$ and poisson's ratio = 0.2.** **(AU Nov'2012)**

Given Data

Diameter of vessel, (d) = 0.8mm

Thickness of vessel, (t) = 4mm

Additional liquid pumped = 50cm^3

Young's modulus, $(E) = 2 \times 10^5 N/mm^2$

Poisson's ratio, $(\frac{1}{m}) = 0.2$

Solution

Change in volume, $(\delta v) = \dfrac{\pi p d^4}{8tE}\left(1 - \dfrac{1}{m}\right)$

Now, substituting for values

$$50 \times 10^3 = \frac{\pi \times p \times (800)^4}{8 \times 4 \times 2 \times 10^5}(1 - 0.3)$$

$$P = 0.355 N/mm^2$$

Result Pressure exerted by the fluid, p = 0.355 N/mm^2

Review Questions

Part - A

1. Define composite bar.
2. Define factor of safety.
3. Define modulus of rigidity.
4. Give the relationship between stress and strain for a biaxial state of stress.
5. What are elastic constants? Given the relationship between them.
6. What are the effects of thermal stresses in composite bars?
7. What is strain hardening?
8. What is the principle of superposition?
9. A gas cylinder of internal diameter 1.2m metres is 25mm thick. Find the allowable pressure of the gas inside the cylinder if the tensile stress in the material is not to exceed 100MN/m^2.
10. A cylindrical shell 1000mm long, 150mm internal diameter, 8 mm thick is filled with a fluid at atmospheric pressure. If an additional 20000mm^3 of fluid is pumped into the cylinder find the pressure exerted by the fluid on the cylinder and the hoop stress induced. $E = 2 \times 10^5 N/mm^2$ and $\gamma = 0.3$.
11. A cylindrical shell 1000mm long, 200mm internal diameter having thickness of metal as 8mm is filled with fluid at atmospheric pressure. If an additional 2cm^3 of fluid is pumped into the cylinder, find (i) the pressure exerted by the fluid on the cylinder and (ii) the hoop stress induced. Take $E = 2 \times 10^5 N/mm^2$and $\gamma = 0.3$.
12. A spherical shell of 800mm is subjected to an internal pressure of $2N/mm^2$. Find the thickness of the shell if the allowable stress in the material of the shell is 100 N/mm^2.
13. A thin cylinder is subjected to internal fluid pressure. what are the stresses developed in the material of the cylinder?
14. Determine the Poisson's ratio of a material for which young's modulus is $1.2 \times 10^5 N/mm^2$ and modulus of rigidity is $4.8 \times 10^4 N/mm^2$.
15. Find the diameter of a circular bar which is subjected to a axial pull of150kN, if the maximum allowable shear stress on any section is 60N/mm^2.

Part - B

1. A compound tube consist of a steel tube 150mm internal diameter and 170mm external diameter and an outer brass tube 160mm internal diameter and 180mm external diameter. The two tubes are of the same length. The compound tube carries and axial load of 1000 KN. Find the stresses and load carried by each tube and the amount it shortens. Length of each tube is 140mm.Take E for steel as $2 \times 10^5 N/mm^2$ and for brass as $1 \times 10^5 N/mm^2$.
2. A compound tube consist of a steel tube 170mm external diameter and 150mm thickness and an outer brass tube 190mm extended diameter and 10mm thickness. The two tubes are of the same length. The compound tube carries and axial tensile load of 1500 KN. Find the stresses and loads carried by each tube and the amount by which it shortens. Length of each tube is 150mm. Assume the modulus of elasticity of steel and that of brass as $200kN/mm^2$ and $100kN/mm^2$ respectively.
3. A metallic bar $300mm \times 10mm \times 40mm$ is subjected to a force of 5 kN (tensile), 6KN (tensile) and 4kN(tensile) along x, y and z directions respectively. Determine the change in the volume of the block. Take $E = 2 \times 10^5 N/mm^2$ and Poisson's ratio = 0.25.
4. A shell 3.25m long, 1m dia, is under an internal pressure of 1Mpa. If the thickness of the shell is 10mm, find(i) hoop and longitudinal stresses (ii) maximum shear stress and (iii) Change in the dimensions. $E = 2 \times 10^5 Mpa$ and Poisson's ratio 0.3.

Fig .1.37

Chapter 2

SHEAR AND BENDING OF BEAMS

Beams and Bending - Types of loads, supports - Shear Force and Bending Moment Diagrams for statically determinate beam with concentrated load, UDL, Uniformly varying load. Theory of simple Bending - Analysis of Beams for Stresses - Stress Distribution at a cross Section due to bending moment and shear force for Cantilever, simply supported and overhanging beams with different loading conditions - Flitched Beams.

2.1 Introduction

Structural members like beams, columns etc, in a building or in any other structures have to resist forces that are coming upon them either laterally or transversely. Such a structural and laterally loaded member, whose cross sectional dimensions are small as compared to its length is called as a beam. Alternatively a beam can also be defined as a bar which is subjected to forces or couples that lies in a plane containing the longitudinal axis of the bar.

2.1.1 Types of beams

Beams in general can be classified based on following factors.

1. Geometry
2. Materials
3. Cross sections
4. Supports

Based on geometry : On the basis of geometry, beam can be classified on below:

1. Straight beams
2. Curved beams

Based on materials : On the basis of materials, beam can be classified on below:

1. Steel beam
2. Concrete beam
3. Plastic beam
4. Composite beam

Based on cross section

Based on cross section, the beam can be classified as below.

1. Rectangular section
2. T - section
3. I - section
4. Triangular section
5. Circular section
6. Channel section
7. Angle section
8. Built up sections etc.,

Figure 2.1. deposits few of the beam cross - sections.

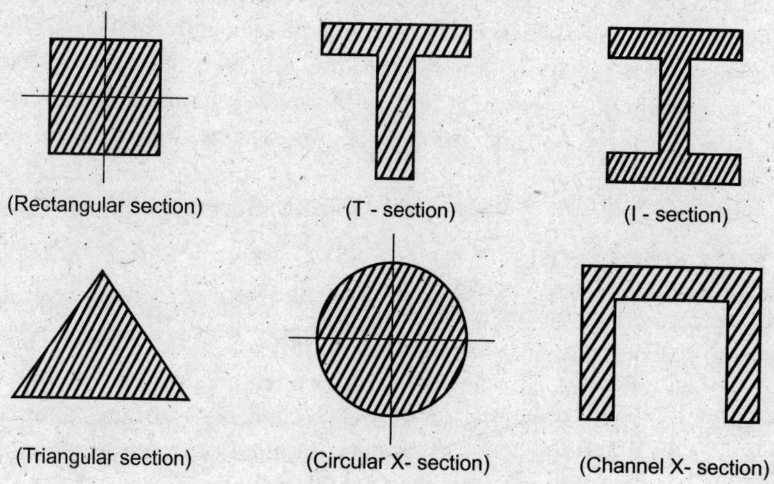

Fig .2.1 Cross section of Beams

Based on supports

On the basis of the support, the beams may be classified as below:

1. Cantilever beam
2. Simply supported beam
3. Over hanging beam
4. Propped cantilever beam
5. Fixed beam
6. Continuous beam

1. **Cantilever beam :** A beam which is fixed or built in at one end and free at other end is called as cantilever beam. A typical cantilever beam with a point load at the free end is shown in Figure 2.2.

2. **Simply supported beam :** If the ends of a beam are made to freely rest on supports, the beam is called as freely or simply supported beam, Figure 2.3 shows a typical simply supported beam with a point load at the mid span is shown in Figure 2.3.

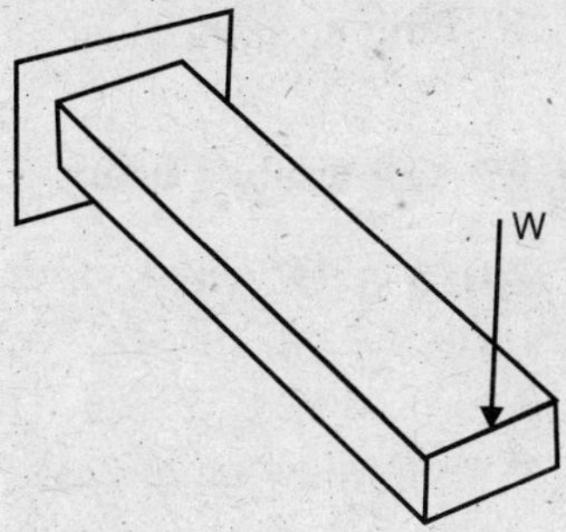

Fig.2.2 Cantilever beam

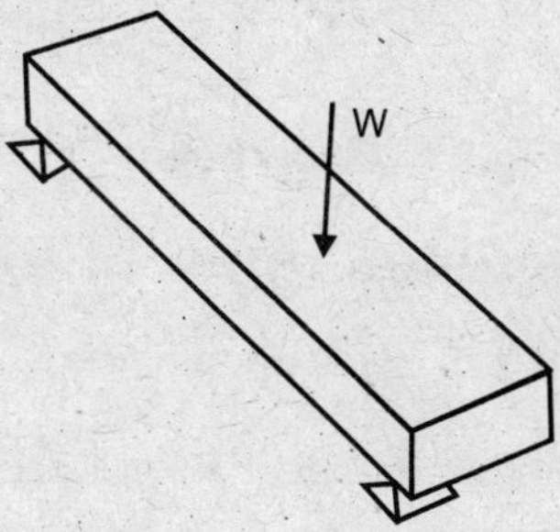

Fig.2.3 Simply supported beam

3. **Over hanging beam :** If one or both the end portions are extended beyond the support, then it is called as over hanging beam. Figure 2.4 shows a typical over hanging beam.
4. **Propped cantilever beam :** A beam which is fixed or built in at one end and simply supported at the other end, is called propped cantilever beam. Figure 2.5 shows a typical propped cantilever beam.

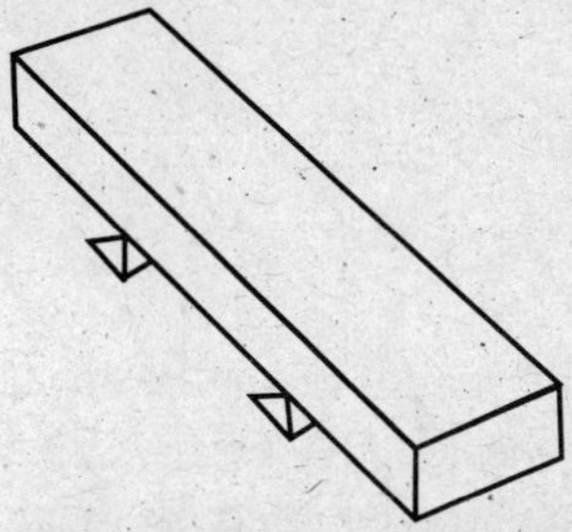

Fig.2.4 Over hanging beam

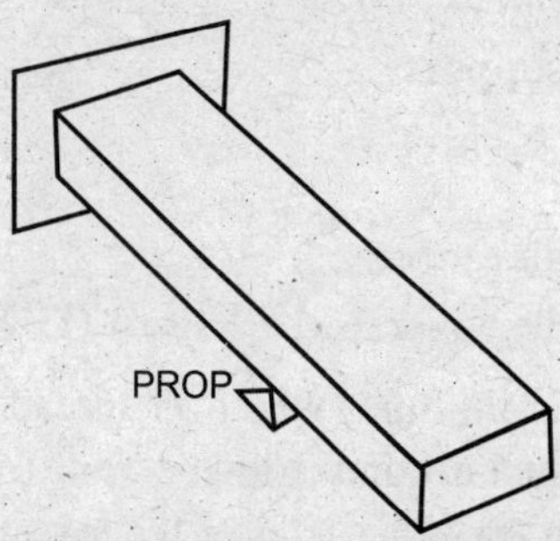

Fig.2.5 Propped Cantilever beam

5. **Fixed beam :** A beam which is fixed or built in at both the ends is called as fixed or encastred or built-in beams Figure 2.6 shows a typical fixed beam.
6. **Continuous beam :** A beam which is provided with more than two supports is called as a continuous beam Figure 2.7 depicts a typical continuous beam.

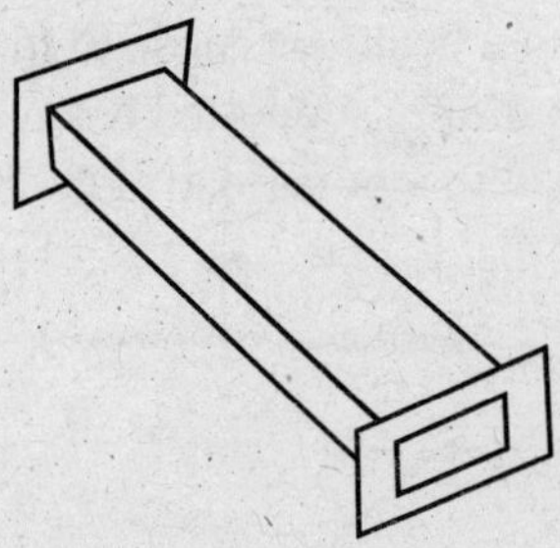

Fig.2.6 Fixed beam

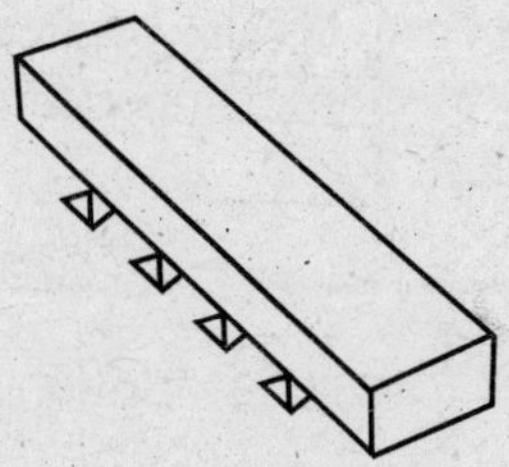

Fig.2.7 Continuous beam

2.1.2 Common types of loads upon a beam

Common types of loads upon a beam are as below

1. Concentrated load or Point Load
2. Uniformly Distributed load
3. Uniformly Varying load
4. Couple

Concentrated load or Point load

A concentrated load is also called as point load which is a load applied over a very small area and is regarded as a single load. Figure 2.8 shows a point load 'W' acting upon the centre of the beam.

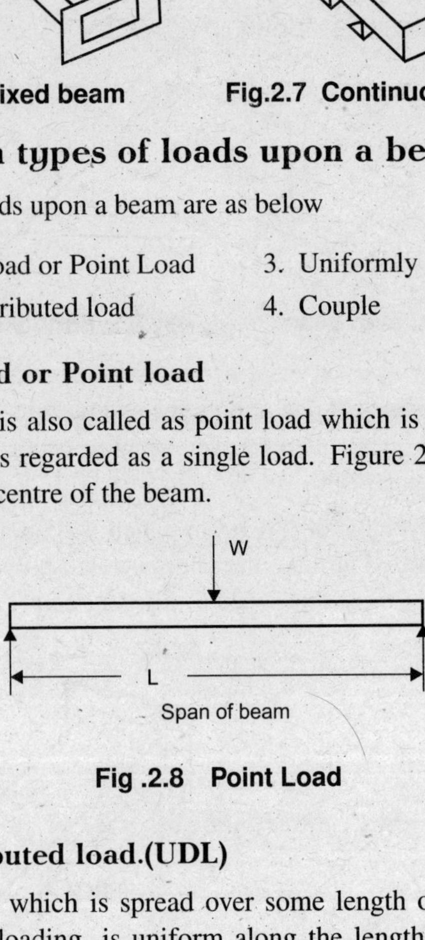

Fig .2.8 Point Load

Uniformly distributed load.(UDL)

An UDL is the one, which is spread over some length of the beam in such a manner that rate of loading, is uniform along the length. Figure 2.9 shows a typical uniformly distributed load spanned to the entire length of beam at the rate of W/m.

Uniformly varying load.(UVL)

An UVL is the one, which is spread over some length of the beam in such a manner that rate of loading, is uniformly varying along the length of the beam. In Figure 2.10, load is uniformly varying from 'O' at the left end to 'W' at the end of the span 'L' i.e., right end of the beam.

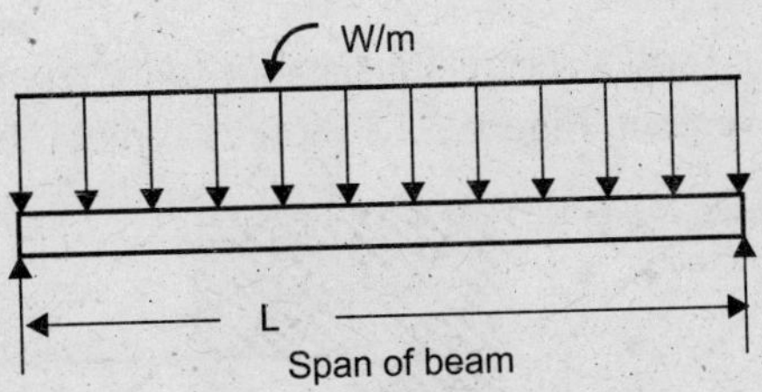

Fig .2.9 Uniformly Distributed Load

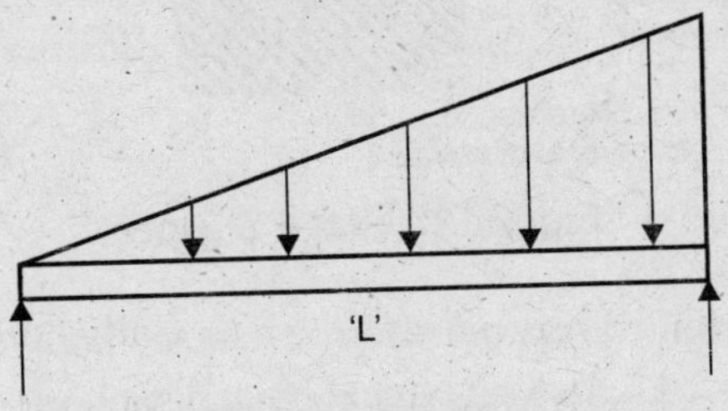

Fig .2.10 Uniformly Varying Load

Couple or concentrated moment(Fig.2.11)

A couple is the one, which is the action of two parallel, equal and opposite forces. The magnitude of the couple is given as the product of the force 'F' and the perpendicular distance between the two parallel forces as below.

Fig .2.11 Couple

Where 'M' (couple) = $F \times d$

A typical couple at a point 'A' of the beam is shown in Fig. 2.11.

2.1.3 Common types of beam supports

Common types of beam supports are as below.

1. Fixed support 2. Hinged support 3. Roller support

Fixed supports : Fixed support offers resistance to the moment, horizontal movement and vertical movement, Figure 2.12 shows a typical fixed support of a beam.

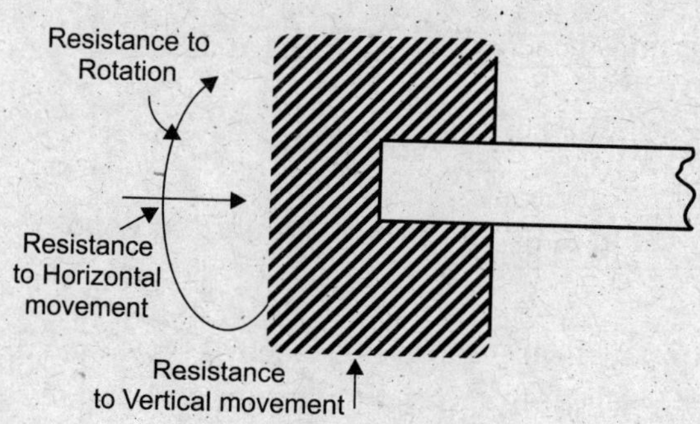

Fig .2.12 Fixed Support

Hinged supports : Hinged support offers resistance both to horizontal and vertical movement Figure 2.13 shows a typical hinged support.

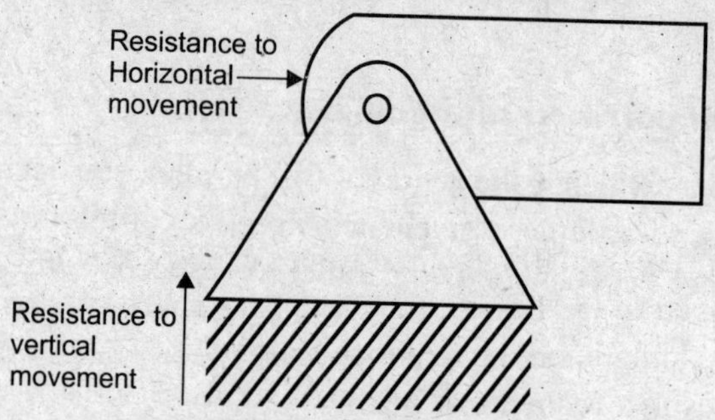

Fig .2.13 Hinged Support

Roller supports : Roller offers resistance to vertical movement only. Figure 2.14 shows a typical Roller support.

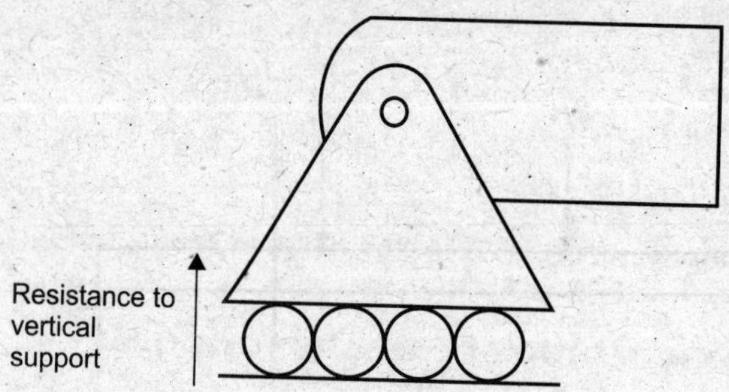

Fig .2.14 Roller Support

2.1.4 Determinately of a beam

Determinately of a beam is generally referred by two types.

They are:

1. Statically Determinate beam
2. Statically indeterminate beam

▶ **Statically determinate beams :** If all the external forces and moments acting on a beam can be determined from the equilibrium conditions, such a beam is called as statically determinate beam.

▶ **Statically in determinate :** If all the external forces and moments acting on a beam can't be determined from the equilibrium conditions alone such a beam is called as statically indeterminate beam, where deflection equations are also taken into account to solve such problems.

2.2 Internal Forces in Beams (Shear Force and Bending moment)

When a beam is loaded externally, internal forces called shear force and bending moments are developed in the beams. The intensity and nature of such internal forces will vary depending upon the intensity and type of loads. Commonly considered internal forces in a beam are as below.

1. Shear Force
2. Bending moment

Shear Force

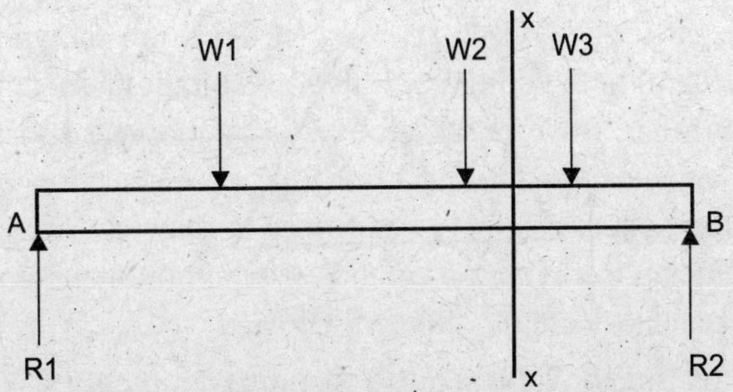

Fig .2.15 Shearing action of a Loaded beam

Now let us consider a beam as shown in Figure 2.15 which is simply supported at both the ends. Let W_1, W_2, W_3 be the loads and reactions at the supports be R_1 and R_2 respectively.

If we cut a section at x, and diagnose for internal forces at that section due to the loads and reactions, It can be seen that there is a tendency for the portion of the beam on one side of the section to slide or shear laterally relative to the other portion due to a resultant force to the left or right of the section called shear force (SF).

In the light of the above statement, Shear force (SF) can be defined as the algebraic sum of all the lateral components of the forces acting on either side of the section $x - x$.

2.2.1 Sign Convention for Shear Force

The shear force at a section is considered positive when the resultant of the forces to the left to the section is upwards or to the right of the section down wards. Similarly the shear force at a section is negative if the resultant of the forces to the left of the section is down wards or to the right of the section upwards. Refer Figure 2.16.

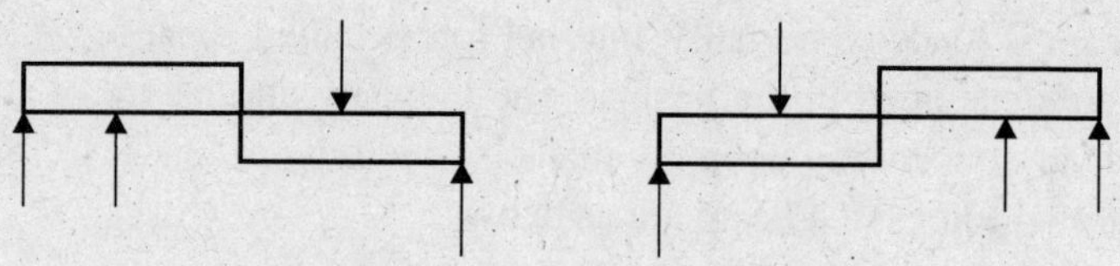

(+ ve) Shear Force (– ve) Shear Force

Fig .2.16 Sign convention for SF

Bending Moment

If we further diagnose the section at $x - x$ of the beam shown in fig.2.15, it will be quite interesting to note that there is also a resultant moment about that section besides the shear force. If we say that the resultant moment about the section $x-x$ of all the loads and reactions to the left of the $x - x$ section at is in clock wise direction, then the resultant moment of forces to the right of $x - x$ section must be "in counter clock wise direction. such a resultant moment to the left or right of the section is termed as Bending moment (B.M.).

In the light of the above statement, Bending moment (B.M.) can be shortly defined as the algebraic sum of the moments of all the forces acting on either side of the section, $x - x$.

2.2.2 Sign Conventions for the Bending Moment

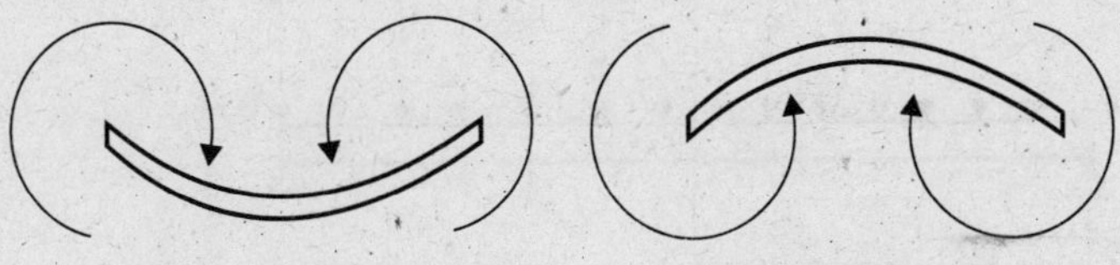

a. (+ ve) BM Force
Sagging BM

b. (- ve) BM Force
Hogging BM

Fig .2.17 Sign convention for BM

The bending moment at a section is considered positive if the bending moment at that section is such that it tends to bend the beam to a curvature having concavity at the top as shown in Figure 2.17.(a) The positive BM is called as sagging BM.

Similarly the bending moment at a section is negative if the bending moment at a section is such that it tends to bend the beam to a curvature having convexity at the top as shown in Figure 2.17.(b) Negative BM is called as Hogging BM.

Shear Force and Bending Moment Diagrams

The intensity of shear force and bending moment will vary along the length of the beam depending upon the intensity and nature of loads.

For analysing and designing a beam, the intensity and nature of the shear force and bending moment along the length of the beam for any fixed loading conditions should be known. Once such variations internal forces, are known, graphical plots can be generated, called shear force and bending moment diagram.

Shear force diagram is a graphical plot, showing the variation of shear force 'F' along the length of beam. If x refers to the length of the beam, then F is function of x i.e., $F(x)$.

Bending moment diagram is a graphical plot, showing the variation of bending moment long the length of the beam. If x refers to the length of the beam, then M is a function x i.e. $M(x)$.

2.2.3 Relationship between the Rate of loading shear force and bending moment

Let us consider a simply supported beam 'AB' carrying a uniformly distributed load w/length and let us imagine to cut a short slice of length 'dx' from this loaded beam at distance 'X' from the origin 'θ' as shown below in Figure 2.18.

Let us cut remove the such a slice from the beam at section (1) - (1) and (2)-(2) and if we draw the free body diagram of such a slice, it will be as shown in Figure 2.19.

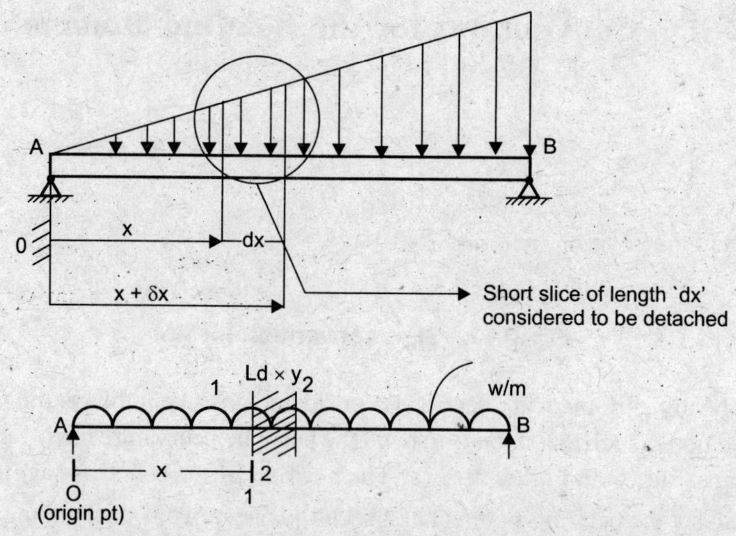

Fig .2.18 Rate of loading shear force and bending moment

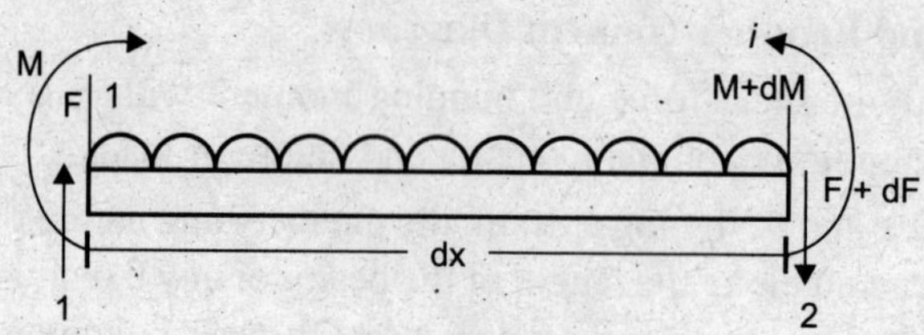

Fig .2.19

Let F = Shear force at section(1) - (1) and

$F + d_F$ = Shear force at section (2) - (2) .

M = Bending moment at section (1) - (1)

$M + d_M$ = Bending moment at section (2) - (2)

Various forces and moments acting on the slice 'dx' of the beam are as below.

1. Shear force'F' acting vertically up at section(1) - (1)

2. Shear force($F + df$ acting vertically down at section (2)) - (2)

3. The load $w \times dx$ acting down wards.

4. The moments 'M' and $(M + dM)$ acting at section (1) - (1) and (2) - (2) respectively.

Now,if we resolve the forces acting on this slice 'dx' particularly we get

$$F - W.dx + -(F + dF) = 0$$

$$- dF = \text{W.dx} \therefore \quad \boxed{\frac{dF}{dx} = -\omega}$$

From the above equation it is to be understood that rate of change of shear force with respect to the position (length) of the beam is equal to the rate of loading(w).

Now let us take the moments of the forces and couples, and resolve, we get

$$M - \omega.dx.\frac{dx}{2} + F.dx = M + dM$$

$$M - W.\frac{dx^2}{2} + F.dx = M + dM$$

Neglecting the higher powers of little quantities, i.e., $\frac{dx^2}{2} = 0$

$$\text{F.dx} = dM$$

$$\therefore \quad \boxed{\frac{dM}{\text{dx}} = F}$$

From the above equation, it is understand that rate of change of bending moment is equal to the shear force of the section.

Bending and Shear stresses in beams

When a beam is subjected to external loads, internal forces such as the Shear force and Bending moments are set up at all sections. Due to such internal forces, beam will tend to undergo deformations, under such situations, the beam will develop, resistive forces or stresses against shearing force and bending moment. The stresses induced in the beam because of bending movement are called as bending stresses and similarly by shear forces are called are shear stresses.

2.3 Theory of Simple Bending

If a segment of beam is subjected to a constant bending moment and no shear force, then the stresses will be set up along the length of the beam due to bending moment only and is interesting to note that, this particular length of the beam is said to be under simple bending or pure bending. The stresses set up in that stretch of beam are known a bending stresses.

Let us consider a beam simply supported at A and B and over hanging both side as shown in Figure 2.20 and having point load at both the ends. Under such condition portion 'AB' is subjected to only pure bending.

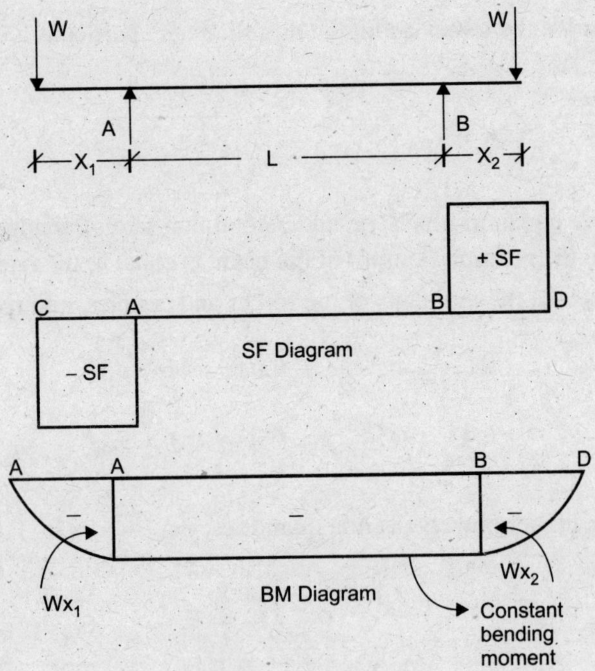

Fig .2.20

2.3.1 Assumptions in theory of simple bending

Assumptions in general considered in theory of simple bending are as below.

1. The material of the beam is homogeneous and isotropic.
2. Value of modulus of elasticity is same in tension and compression.
3. Transverse sections which were plane before bending remain plane after bending also.
4. The beam is initially straight and all longitudinal filaments bend into circular arcs with a common curvature centre.
5. Radius of curvature is large compared to the dimension of the cross - section of beam.
6. Each layer of the beam is free to expand or contract independent of the layer, above or below it.

2.3.2 Bending equation

Let us consider a small segment of Length 'dx' of a beam subjected to pure bending and Let AB and CD be the two sections of beam which are normal to the axis of the beam N - N. Due to the bending action the beam will undergo deformation as shown in Figure 2.21.

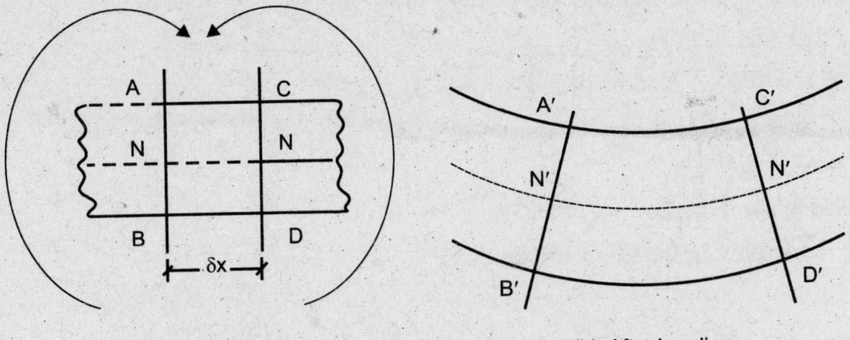

(a) Before bending (b) After bending

Fig .2.21

Because of the bending action the top layer AC become A'C' (shortened) and the bottom layer BD became B'D' (extended)'. The neutral layer N-N become N'N' but did not undergo any change in length hence it is neutral axis of the beam.

In the Figure 2.22 (a) a beam layer 'EF' is shown whose behaviour can be diagnosed for stress and strain of that layer EF.

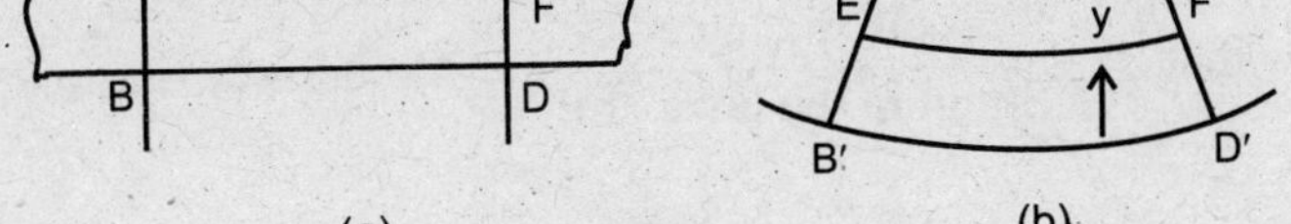

(a) (b)

Fig .2.22

The deformation of the layer EF due to bending with radius of curvature 'R' at the centre of curvature 'O' can be drawn as Figure 2.22.(b).

Strain variation along the depth of the beam

The strain variation along the depth of the beam can be derived as below.

Let

EF = dx = original length of the beam layer diagonsed for.

NN = dx = Neutral layer length

(As the NA will not undergo change in length NN = N'N' = dx).

From Fig 2.22(b)

N'N = RQ ($\because$ Radius of E'F' = R + y)

$\therefore$ E'F' = (R + y)θ

N'N' = NN = dx

We know that, dx = R θ

$\therefore$ Increase in length of the layer EF

$$= E'F' - EF = (R + y)\theta - R\theta$$
$$= R\theta + y\theta - R\theta = y\theta$$

$\therefore$ Strain in the layer EF

$$= \frac{\text{Increase in Length}}{\text{Original length}} = \frac{y\theta}{EF} = \frac{y\theta}{R\theta}$$
$$= \boxed{\frac{y}{R}}$$

As 'R' is constant, it is to be inferred that strain in a layer is proportional to its distance from the neutral axis.

Stress variation along the depth of the beam

Let 'σ' be the stress of the layer EF

E = modulus of elasticity of the beam

We know that, $\therefore$ $\dfrac{\text{Stress in the layer 'EF'}}{\text{Strain in the layer 'EF'}}$ = Modulus of Elasticity the beam

$$\therefore \frac{\sigma}{\frac{y}{R}} = E$$
$$\therefore \frac{\sigma}{y} = \frac{E}{R}$$
$$\therefore \sigma = \frac{E}{R}.y$$

From the above equation, it can be understood that stress in the layer is proportional to the distance of the layer from the neutral axis.

Moment of Area

Let us compute the moment offered by the area of the section against bending.

Figure 2.23 shows the cross - section of a beam. Let N.A. be the neutral axis of the section. Let us consider a small layer at a distance 'y' from NA.

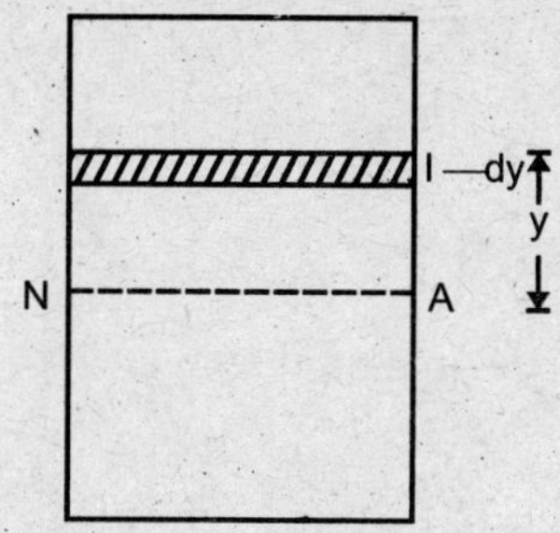

Fig .2.23 Cross - section of a beam

Let dA = area of the layer.

$\therefore$ Force on the that layer = stress in that layer × area of that layer = $\sigma \times dA$.

Substituting for stress 'σ'

$$= \frac{E}{R}.y.dA \quad (\because \quad \sigma = \frac{E}{R}.y)$$

Total force on the beam can be obtained by integrating above equation.

$\therefore$ Total force an the beam section

$$= \int \frac{E}{R}.y.dA = \frac{E}{R}\int y.dA. \left(\because \frac{E}{R} \text{ is constant}\right)$$

We know that under pure bending, there is no force on the section of the beam (ie force is zero)

$$\therefore \quad \frac{E}{R}\int .y.dA = 0$$

$$\text{or} \quad \int y.\ dA = 0 \qquad (\text{As} \frac{E}{R} \text{ Cannot be zero})$$

Now $y.dA$ represent the moment of area dA about NA.

$\int y.\ dA$ represents the moment of the entire area of section about NA.

Moment of resistance

Due to pure bending, the layers above 'NA' (Neutral axis) are subjected to compressive stresses and the layers below are subjected to tensile stresses. Due to these stresses, there will be forces acting. These forces will have moment about the NA. The total moment of these forces about NA for a section is given known as moment of resistance 'M'.

The force on the layer at a distance 'y' from neutral axis as in Figure 2.23 is given by the following equation.

$$\frac{E}{R}.y.dA$$

∴ Moment of their force about NA.

$$= \text{Force on the layer} \times y$$

$$= \frac{E}{R}.y.dA.y = \frac{E}{R}y^2.dA$$

Total moment of the forces on the section of the beam (Moment of resistance).

$$= \int \frac{E}{R}.y^2 dA = \frac{E}{R}\int y^2.dA$$

If 'M' is the external moment applied an the beam section, for equilibrium contribution,

$$M = \frac{E}{R}\int y^2.dA$$

$$M = \frac{E}{R}.I$$

(∵ $\int y^2.dA =$ moment of inertia of the area of the section about NA)

$$\Rightarrow \frac{M}{I} = \frac{E}{R}$$

We have already derived that

Therefore a new equation can be written as

$$\frac{\sigma}{y} = \frac{E}{R}$$

$$\therefore \boxed{\frac{M}{I} = \frac{\sigma}{y} = \frac{E}{R}},$$

Which is called as bending equation.

Where 'M' is expressed in N.mm

I (moment of inertia)in mm^4

σ Bending stress is $N/mm^2, y$ in 'mm'

E Young's modulus in N/mm^2 and R in 'mm'.

The above bending equation is applicable to a member which is subjected to a constant bending moment and is free from shear force.

2.4 Section Modulus

Section modulus is defined as the ratio of moment of inertia of a section about the neutral axis to the distance of the outermost layer from the neutral axis. It is denoted by the symbol 'Z'. hence it is mathematically written as

$$z = \frac{I}{y_{max}}$$

Where

I = moment of inertia about NA.

y_{max} = Distance of the outer most layer from the NA.

We know from sending equation,

$$\frac{M}{I} = \frac{\sigma}{y}$$

We can deduce that stress σ will be maximum, when 'y' is maximum.

Hence above equation can be written as with σ maximum and y maximum

$$\frac{M}{I} = \frac{\sigma_{\text{max}}}{y_{\text{max}}}$$

$$\therefore \; M = \sigma_{max} \frac{I}{y_{max}}$$

Now substituting for in the above equation

$$\frac{I}{y_{max}} = z$$

We get, $\quad M = \sigma_{max}.Z$

Hence moment of resistance offered by the section is maximum, when section modulus 'Z'' is maximum. Hence section modulus represents the strength of the section.

Section modulus of various shapes and sections of beams

1. Rectangular section

$$z = \frac{I}{y_{\max}} = \frac{\frac{bd^2}{12}}{d/2} = \frac{bd^3}{12} \times \frac{2}{a} = \frac{bd^2}{6}$$

2. Hollow Rectangular section.

$$Z = \frac{I}{y_{\max}} = \frac{\frac{Bd^3 - bd^3}{12}}{D/2} = \frac{1}{6D}\left(BD^3 - bd^3\right)$$

3. Circular section

$$I = \frac{\pi}{64}.d^4$$

$$y_{\max} = \frac{d}{2}$$

$$z = \frac{I}{y_{\max}} = \frac{\frac{\pi}{64}.d^4}{d/2} = \frac{\pi}{3^2}d^3$$

4. Hollow circular section

$$z = \frac{I}{y_{\max}} = \frac{\frac{\pi}{64}\left(D^4 - d^4\right)}{D/2} = \frac{\pi}{32D}\left(D^4 - d^4\right)$$

2.5 Flitched beams (Composite beams)

A beam which is made up of two or more material is called as flitched or composite beam. A typical flitched beam is shown in Figure 2.24.

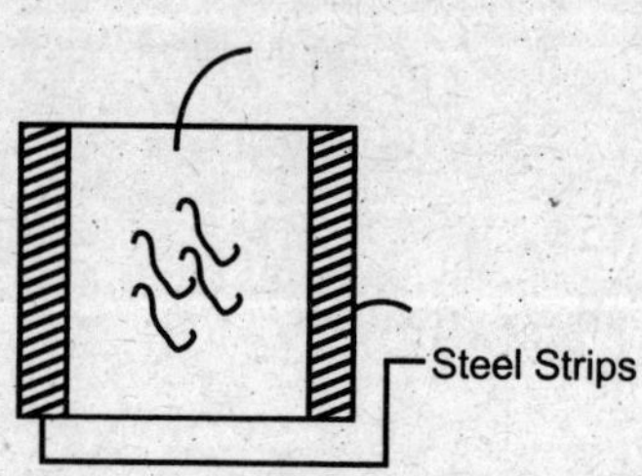

Fig .2.24 Flitched beam

Such a beam will behave as a single unit. Therefore strain in wood and steel are same.

$$i.e. \frac{\sigma_\omega}{E_\omega} = \frac{\sigma_s}{E_s}$$

Where

E_ω and E_s are modulus of elasticity of wood and steel respectively.

$$\sigma_s = \frac{E_s}{E_w}.\sigma_w$$

$$\therefore \sigma_s = m.\sigma_w$$

Where $m = \dfrac{E_s}{E_\omega}$ = modulus ratio.

Moment of resistance of the flitched beam section

Let M_f be the moment of resistance of the flitched beam.

M_w	=	Moment of resistance offered by wood.
M_s	=	Moment of resistance offered by steel.
b	=	breath of wooden section
d	=	depth of wooden section
t	=	thickness of steel strip
d	=	depth of the steel section.

Then $M_F = M_w + M_s$

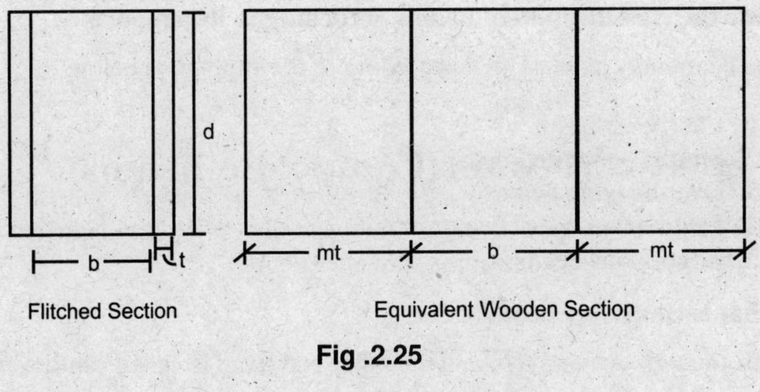

Fig .2.25

If σ_w and σ_s are stresses in wood and steel and Z_w and Z_s be the section modulus of wood and steel, the

$$M_F = \sigma_w.z_w + \sigma_s.z_s$$

$$= \sigma_w.\frac{bd^2}{6} + \sigma_s\frac{2.td^2}{6} \quad (\because \ 2 \text{ steel strips of thickness 't' each})$$

We know that, $\sigma_s = m.\sigma_w$

$$\therefore \ M_F = \sigma_w.\frac{bd^2}{6} + m.\sigma_w.\frac{2td^2}{6} = \frac{\sigma_w.bd^2}{6}(b + m2t)$$

From the above equation, it is inferred that, Moment of resistance of flitched section M_F is moment of resistance of wooden section if width $(b + m2t)$.

This wooden section of width $(b + m2t)$ and depth 'd'is known as equivalent wooden section equivalent wooden reaction is shown an Figure 2.25.

Part - A

Question and Answer

2 Marks

1. **Define beam**

   Beam is a structural member, which is supported along the length and subjected to external loads acting perpendicular to the centre line of the beam.

2. **What is meant by transverse loading an beams?**

   If a load is acting on the beam perpendicular to the center line of it then it is called transverse loading.

3. **State the classification of beams according to its supports?**

   The beam may be classified according to the support as below.

   1. Cantilever beam
   2. Simply supported beam
   3. Over hanging beam
   4. Fixed beam and
   5. Continuous beam

4. **What is cantilever beam?**

   A beam with one end free and the other end fixed is called cantilever beam.

5. **What is simply supported beam?**

   A beam supported or resting freely on the supports at its both ends which is called as simply supported beam.

6. **What is meant by overhanging beam?**

   If one or both of the end portions are extended beyond the support then it is called overhanging beam.

7. **What are the types of Transverse load?**

   Different types of transverse loads, are as below.

   1. Point or Concentrated load.

2. Uniformly Distributed load.
3. Uniformly Varying loads.

8. **What is meant by point or concentrated loads?**

   A load (w) which is acting at a point upon a beam is called point load.

9. **What is uniformly distributed load?**

   If a load which is spread over a beam in such a manner, so that rate of loading 'w' is uniform through out the length, is called as Uniformly distributed load.

10. **Define shear force and Bending Moment at a section.**

    - **Shear force:** Shear force at any cross section is defined as the algebraic sum of all the forces acting either side of the section of the beam.
    - **Bending Moment:** Bending moment at any cross section is the algebraic sum of the moments of all the forces which are acting either side the section of the beam.

11. **What is meant by Positive or Sagging BM?**

    Bending moment is said to be positive or sagging if the moment of the left side of the beam is clockwise and in the right side of the beam is counter clockwise. Otherwise the beam tends to bend like concave manner.

12. **What is meant by negative or Hogging BM ?**

    Bending moment is said to be negative or hogging if the moment of the forces in the left side of the beam is counter clockwise or right side of the beam is clockwise. Other wise beam tends to bend like convex manner.

13. **Draw the rough sketch of SF and BM for the beam given below.**

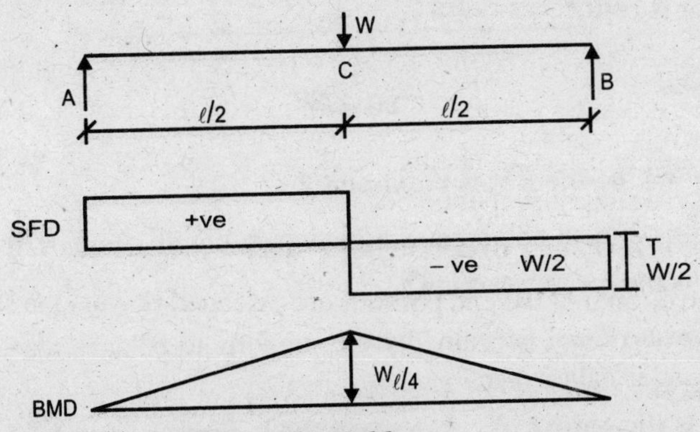

Fig .2.26

14. **Draw the rough sketch of SF and BM for the beam given below.**

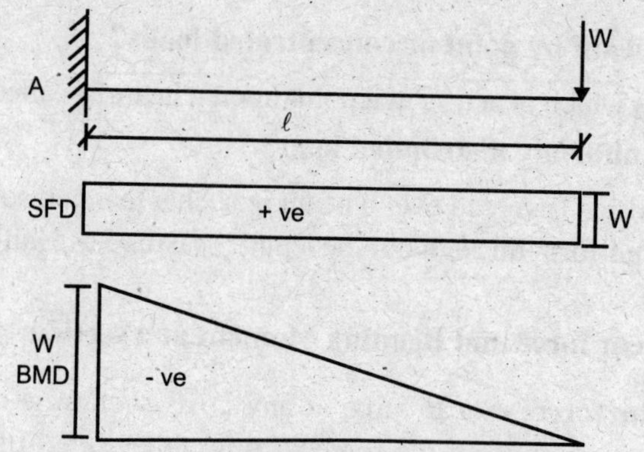

Fig .2.27

15. **What will be the SF and BM diagrams for the simply supported beam of length 'L' subjected to UDL of w/m length through out its length.**

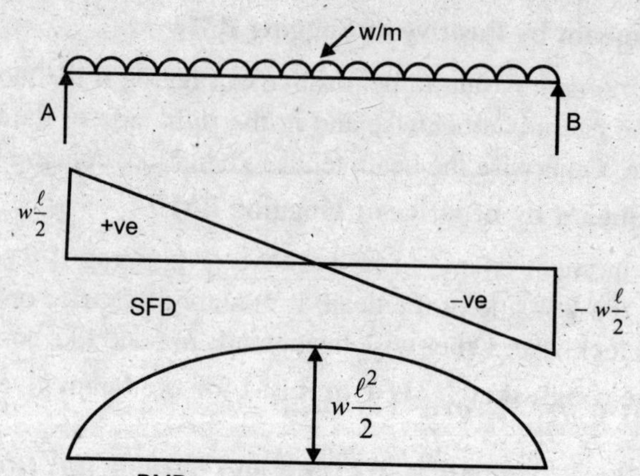

Fig .2.28

16. **When will be the BM is maximum?**

Bending moment will be maximum where the SF changes its Sign.

17. **What is meant by reaction?**

The upward force provided by the supports to balance the vertical downward load is called reaction.

18. **What is the nature of curve when the beam carrying UDL?**

The curve follows parabolic relation (or) parabolic curve.

19. **What are the SF and BM diagram for the beam given below?**

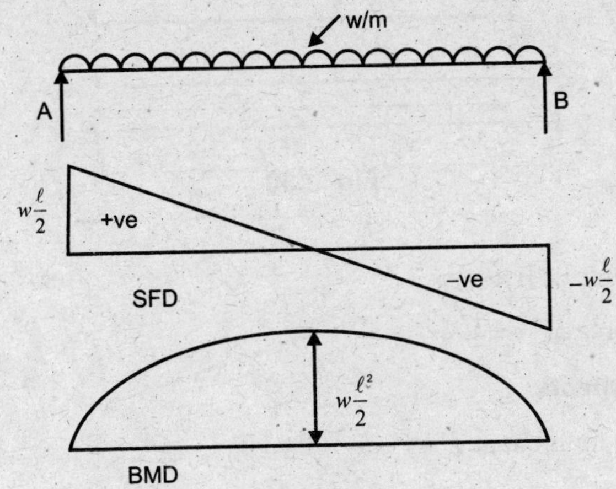

Fig .2.29

20. **Define the term point of contra flexure.**

The point where the Bending moment changes its sign or point at which Bending moment is zero is called the point of contra flexure.

Part - B

Question and Answer

Problems On Cantilever beams

1. **What are the types of supports in beams and Explain any two types?**

Refer Section 2.1.3 of this book.

2. **Draw the shear force and Bending moment diagrams for the following cases.**

   (a) Cantilever with a point load at free end.
   (b) Cantilever with a Uniformly distributed Load (UDL) throughout the span.
   (c) Cantilever with a Uniformly Varying Load(UVL).

(a) **Cantilever with a point load at free end.**

Shear Force

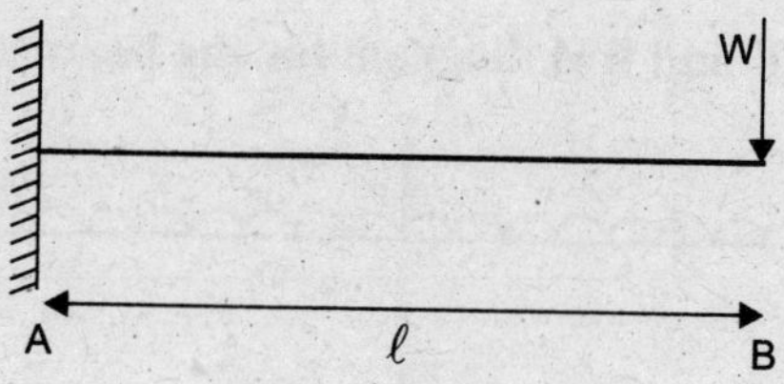

Fig .2.30

Shear force at B = $+w$

Shear force at A = $+w$

Bending Moments

Bending moment at $B = -w \times 0 = 0$

Bending moment at $A = -w \times \ell = -w\ell = -w\ell$

Shear force and Bending moment diagram.

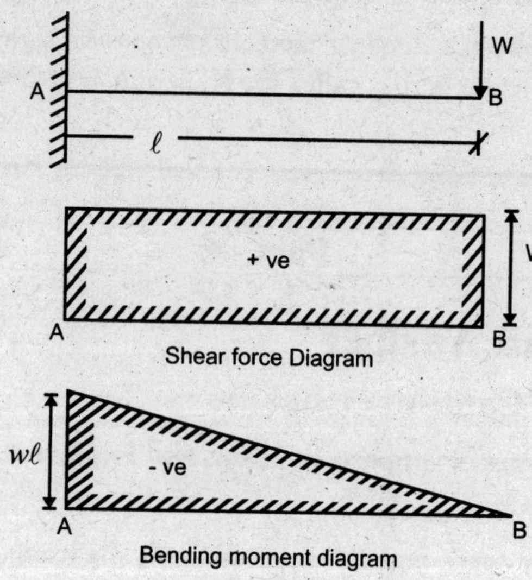

Fig .2.31

(b) **Cantilever with a Uniformly Distributed Load throughout the span**

Shear Force

When $x = 0$, Shear force at $B = 0$.

When $x = \ell$, Shear force at $A = w \times L = wL$

Bending moment

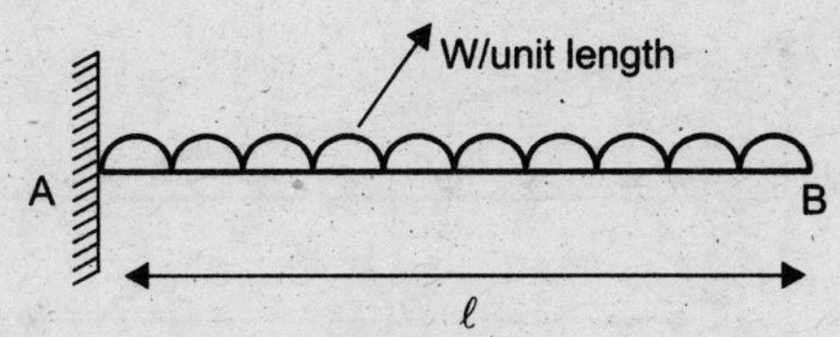

Fig .2.32

When $x = x$ $M_x = -w.x \times \frac{x}{2} = \frac{-wx^2}{2}$

When $x = 0.$ Bending moment at $B = 0$

When $x = 0$ Bending moment at $B = 0$

When $x = \ell$ Bending moment at $A = \frac{-wL^2}{2}$

Result

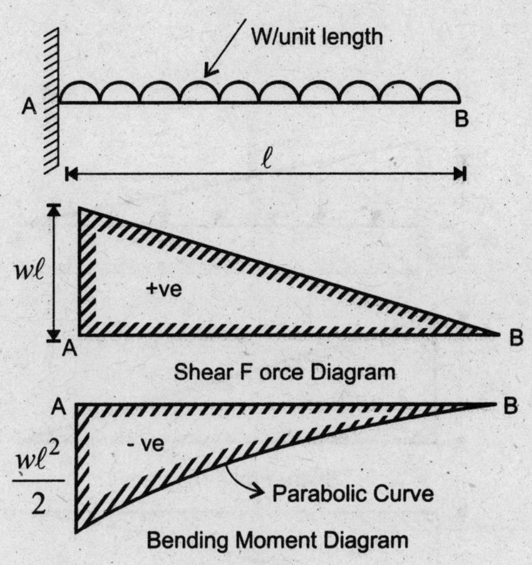

Fig .2.33

(c) **Cantilever with a Uniformly Varying Load (UVL) throughout the span.**

Shear Force

When shear force at $x = \frac{wx}{\ell} \times \frac{x}{2} = \frac{wx^2}{2\ell}$

When $x = 0$ Shear force at $B = 0$

When $x = \ell$ Shear force at $A = \frac{w\ell}{2} = \frac{w\ell}{2}$

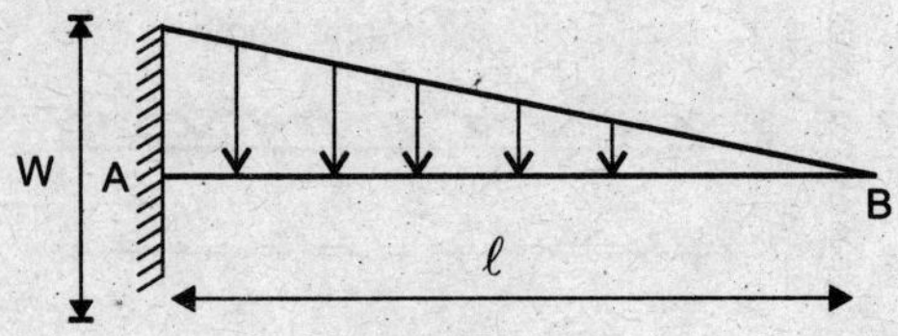

Fig .2.34

Bending moment

Bending moment @ any section x-x → at a distance 'x' from the right end.

$x, M_x = \frac{-wx}{\ell}.\frac{x}{2}.\frac{x}{3}$

When $x = 0$, BM @ $B = 0$

When $x = \ell$, Bending Moment @ A

$$M_A = -\frac{w\ell^3}{6\ell} = -\frac{w\ell^2}{6}$$

Result

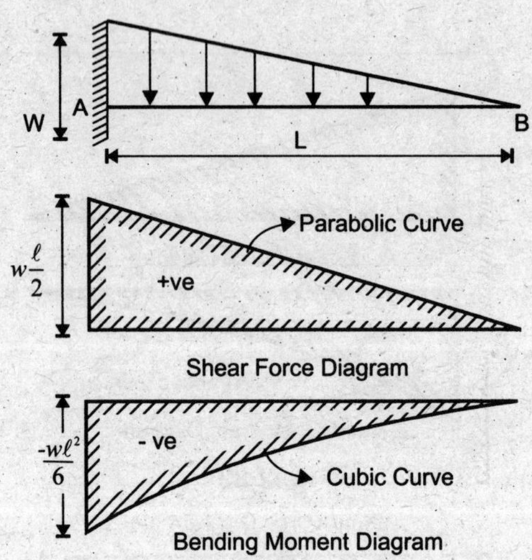

Fig .2.35

3. **A cantilever beam of length 2m carries a uniformly distributed load of 3kN/m Draw shear force and Bending moment Diagrams.**

Solution

Given L = 2m w = 3kN/m

We know that

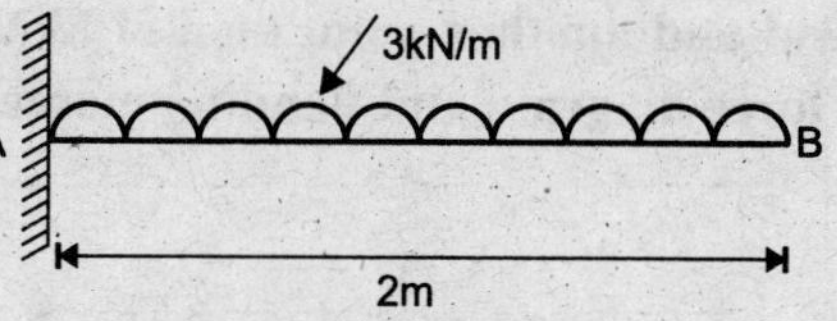

Fig .2.36

Shear force $= wL$

When $x = 0$

Shear force @ $B = 3 \times 0 = 0$

When $x = 2$

Shear force @ $A = 3 \times 2 = 6kN$

Bending moment $= -\dfrac{wx^2}{2}$

When $x = 0$

Bending moment @ $B = \dfrac{-3 \times 0^2}{2} = 0$

When $x = 2$

Bending moment @ $A = \dfrac{-3 \times 2^2}{2} = -6$

Result

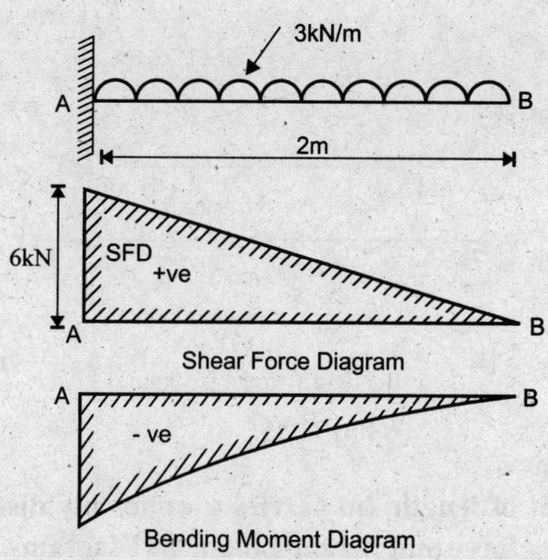

Fig .2.37

4. **A cantilever of 3m long carries a uniformly distributed load of 12kN/m is spread over a length of 1.5m from the free end. It also carries a point load**

of 15kN at free end and another point load of 8kN at 1m from the fixed end. Draw shear force diagram and Bending moments diagrams.

Solution

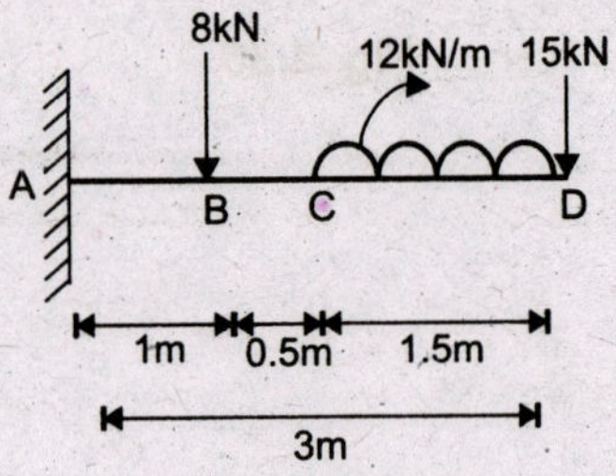

Fig .2.38

Shear force

Shear force at D = 15kN

Shear force at C = $15 + 12 \times 1.5 = 33kN$

Shear force at B = $15 + 12 \times 1.5 + 8 = 41kN$

Shear force at A = 41 kN

Bending Moment

BM @ D = 0

Bending moment @ C

$$= -(15 \times 1.5) - (12 \times 1.5 \times \frac{1.5}{2}) = -36kN.m$$

Bending moment @ B

$$= -15 \times 2 - 12 \times 1.5 \left\{ \frac{1.5}{2} + 0.5 \right\} = -52.5kNm$$

Bending moment @ A

$$= -15 \times 3 - 12 \times 1.5 \left\{ \frac{1.5}{2} + 1.5 \right\} - 8 \times 1 = -93.5kNm$$

Result

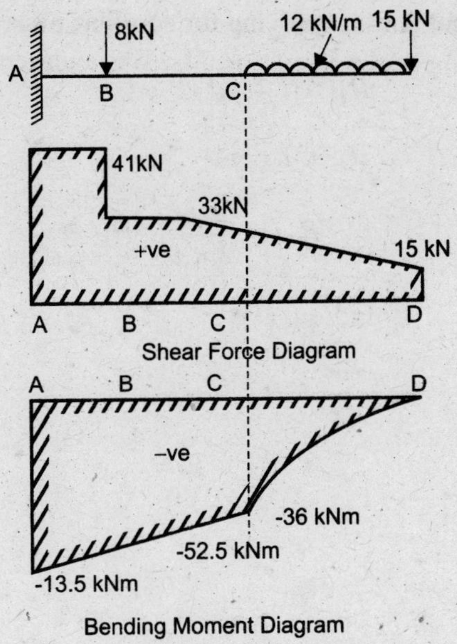

Fig .2.39

Problems on Simply Supported Beams

5. **Draw the shear force and Bending moment diagrams for simply supported Beams for following cases.**

   (a) Simply supported beam with Central Point Load.
   (b) Simply supported beam with Uniformly Distributed Load.
   (c) Simply supported beam with Uniformly Varying Load.

Solution

(a) **Simply supported beam with central point load.**

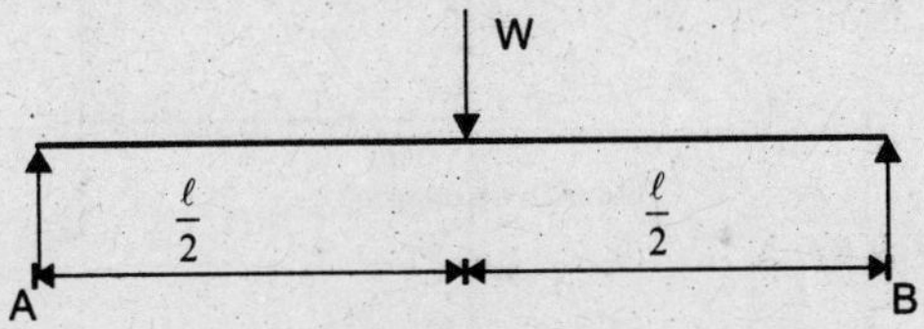

Fig .2.40

Applying equilibrium conditions, $\sum V = 0$

$$R_A + R_B = w \tag{1}$$

Taking moment about A (i.e. applying for equilibrium conditions. $\sum M = 0$)

$$R_B \times \ell - w\ell/2 = 0$$

$$R_B \times \ell = w\frac{\ell}{2}$$

$$\therefore \quad R_B = \frac{w\ell}{2 \times \ell} \quad R_B = \frac{w}{2}$$

Substituting for $R_B = \frac{W}{2}$ in Eqn. (1)

$$\frac{w}{2} + R_A = w$$

$$\therefore \quad R_A = w - \frac{w}{2} = \frac{w}{2}$$

Shear force

Shear force @ $B = -w/2 = R_B$
Shear force @ $C = -w/2 + w - w/2$
Shear force @ $A = w/2 = R_A$

Bending moment

Bending moment @ B = 0
Bending moment @ $c = \frac{w}{2} \times \frac{\ell}{2} = \frac{w\ell}{4}$
Bending moment @ $A = 0$

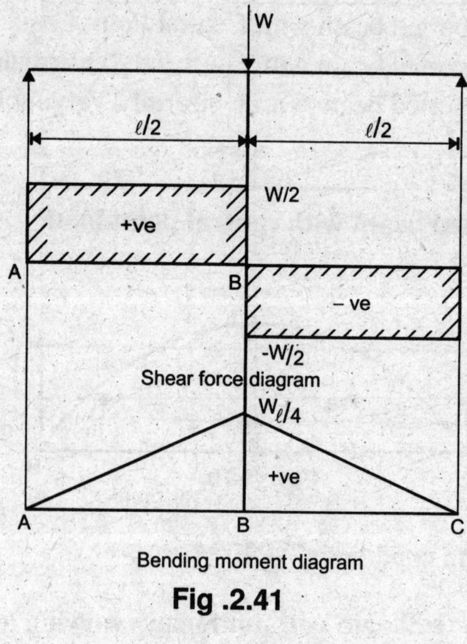

Fig .2.41

(b) **Simply supported beam with Uniformly Distributed load.**

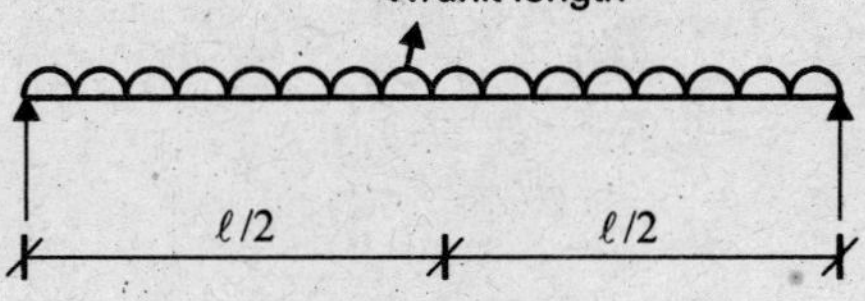

Fig .2.42

Shear force

Shear force @ $B = -R_B = -w\dfrac{L}{2}$

Shear force @ $A = +R_B = w\dfrac{L}{2}$

Bending Moment

Bending moment @ $B = 0$

Bending moment @ C,

$$= \frac{w\ell}{2} \times \frac{\ell}{2} - \frac{w\ell}{2} \times \frac{\ell}{2 \times 2} = \frac{w\ell^2}{4} - \frac{w\ell^2}{8} = \frac{w\ell^2}{8}$$

Bending moment@ A = 0

Result

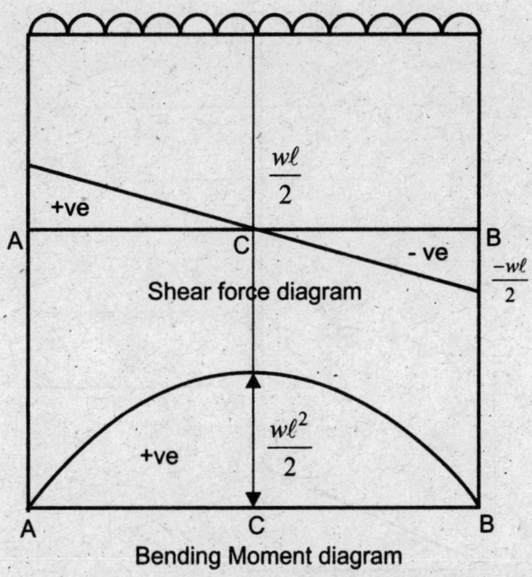

Fig .2.43

(c) **Simply supported beam with uniformly varying load.**

Applying for equilibrium condition $\sum M = 0$ and $\sum V = 0$

$$R_B \times \ell = \frac{1}{2} \times \ell \times w \times \frac{2}{3} \times \ell$$
$$\therefore \quad R_B = \frac{w\ell}{3}$$
$$\Rightarrow \quad R_A + R_B = \frac{1}{2} \times \ell \times w$$

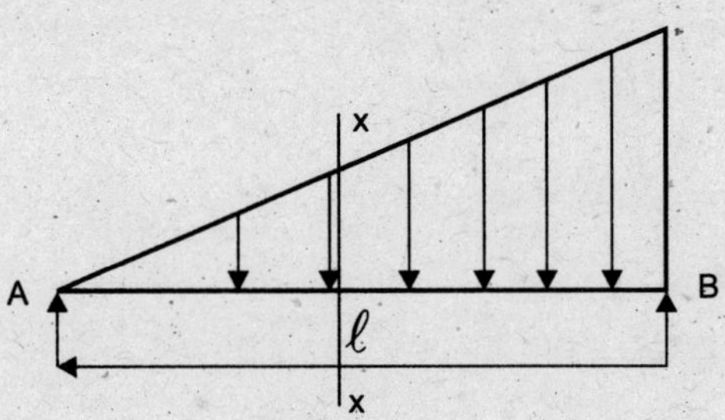

Fig .2.44

Substituting for R_B

$$R_A + \frac{w\ell}{3} = \frac{1}{2} \times \ell \times w$$
$$\therefore \quad R_A = \frac{w\ell}{2} - \frac{w\ell}{3} = \frac{w\ell}{6}$$

Shear Force

Shear force @ $B = -\dfrac{w\ell}{3}$

Shear force @ $A = R_A = \dfrac{w\ell}{6}$

$$S.F.@\ell\times = \frac{w\ell}{6} - \frac{1}{2}.\frac{wx}{x} - x$$
$$= \frac{w\ell}{6} - \frac{wx^2}{2\ell}$$

Bending moments

BM @ B = 0 Bending moment @ A = 0

Maximum bending moment

BM at xx

M_x = R_A.X - Load on $Ax = \dfrac{1}{3x}$

$$= \frac{w\ell}{6}x - \frac{1}{2}.\frac{wx}{\ell}.x$$
$$= x\frac{1}{3}.x$$
$$= \frac{w\ell x}{6} - \frac{wx^3}{6\ell}$$

Maximum BM, will occur where shear force changes it segment.

$\therefore$ Equality shear force equation to zero will get,

$$S_{Fx} = \frac{w\ell}{6} - \frac{wx^2}{2\ell} = 0$$
$$\therefore \frac{w\ell}{6} = \frac{wx^2}{2\ell}$$
$$\therefore x^2 = \frac{w\ell}{6}$$
$$= \frac{2\ell}{w} = \frac{\ell^2}{3}$$
$$\therefore x = \frac{\ell}{\sqrt{3}} = 0.577\ell$$

Form A

Now we know that

$$Mx = \frac{w\ell^x}{6} - \frac{wx^3}{6\ell}$$

Substituting for $x = \frac{\ell}{\sqrt{3}}$

$$\therefore mx = \frac{w\ell}{6}.\frac{\ell}{\sqrt{3}} - \frac{w}{6\ell}.\left(\frac{\ell}{\sqrt{3}}\right)^3$$
$$= \frac{w\ell^2}{6\ell} - \frac{w\ell^3}{6 \times 3\sqrt{3}} = \frac{3w\ell^2 - w\ell^2}{18\sqrt{3}}$$
$$= \frac{w\ell^3}{9\sqrt{3}}$$

Result

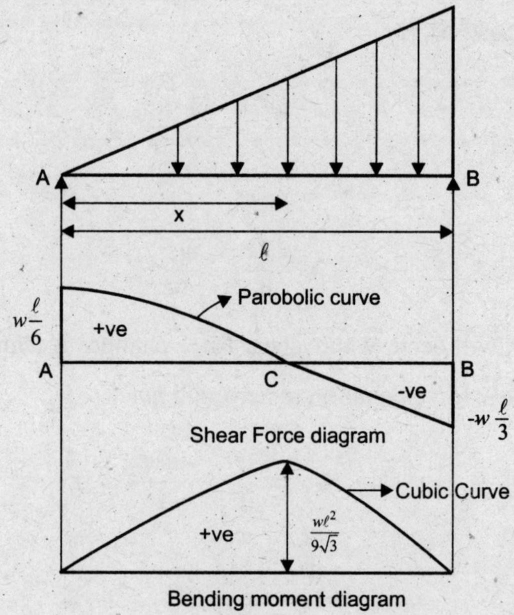

Fig .2.45

6. **A beam freely supported over an effective span of 5m carries point 3kN, 4.5kN and 7kN at 1,2.5 and 3.5m respectively from the left hand support construct the shear force and Bending moment diagrams**

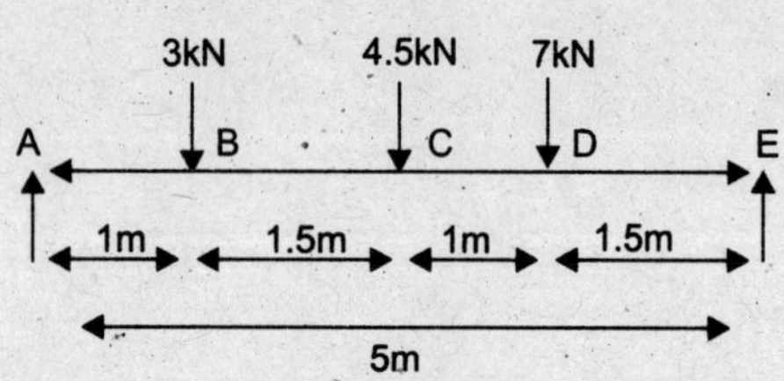

Fig .2.46

Applying Equilibrium conditions $\sum V = 0$

$$R_A + R_E = \text{(Total Down Wards Load)}$$
$$R_A + R_E = 3 + 4.5 + 7$$
$$R_A + R_E = 14.5kN \tag{1}$$

Applying Equilibrium conditions $\sum M = 0$

Taking moment about A

$$R_E \times 5 = 3 \times 1 + 4.5 \times 2.5 + 7 \times 3.5$$
$$R_E \times 5 - 7 \times 3.5 - 4.5 \times 2.5 - 3 \times 1 = 0$$
$$R_E \times 5 = 38.75kN$$
$$R_E = 7.75kN$$

We know that $R_A + R_E = 14.5kN$ substituting for RE

$$R_A = 6.75kN$$

We get,

Shear force

Shear force at $E = -R_E = -7.75kN$

Shear force at $D = -7.75 + 7 = -0.75kN$

Shear force at $C = -R_E + 7kN + 4.5kN = 3.75kN$

Shear force at $B = 3.75 + 3 = 6.75kN$

Shear force at $A = R_A = 6.75kN$

Bending moment

Bending moment at $E = 0$

Bending moment at $D = 7.75 \times 1.5 = 11.625kNm$

Bending moment at $C = 7.75 \times 2.5 - 7 = 12.375kNm$

Bending moment at $B = R_A \times 1 = 6.75kNm$

Bending moment at $A = 0$.

Result

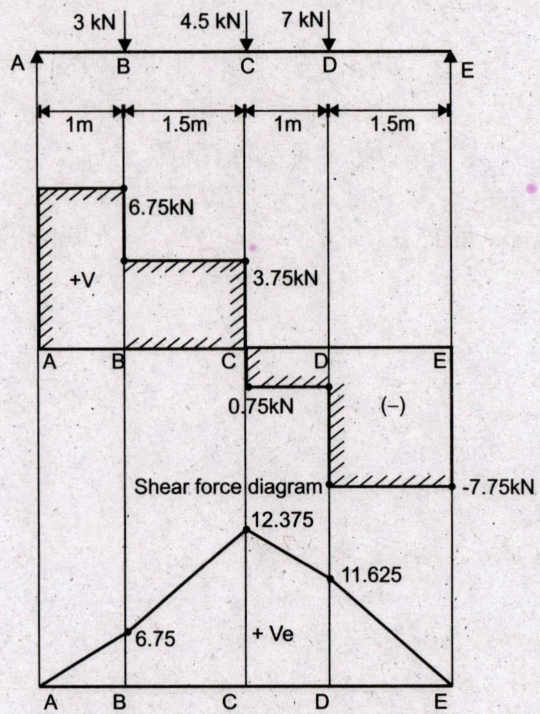

Fig .2.47 Bending Moment

7. **A simply supported Beam of span 5m carries a point Load of 30kN, 40kN and 50kN at 1m, 2.5m, 3.5m from the left end support. Draw SF and BM diagram.**

Solution Apply or equilibrium conditions $\sum V = 0$

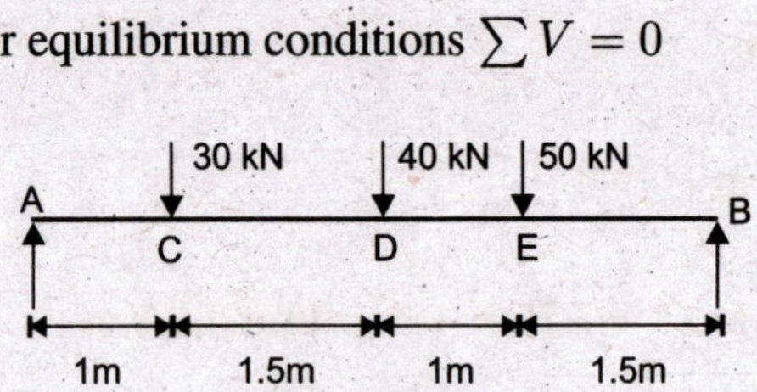

Fig .2.48

$$R_A + R_B = 30 + 40 + 50$$

$$R_A + R_B = 120kN \tag{1}$$

Equilibrium condition $\sum M = 0$, Taking moment about 'A'

$$(30 \times 1) + (40 \times 2.5) + (50 \times 3.5) = 5R_B$$
$$30 + 100 + 175 = 5R_B$$
$$\therefore \quad \boxed{R_B = 61kN}$$

Substituting $R_B = 61$in equation (1)

$$R_A + 61 = 120$$
$$\therefore \quad \boxed{R_A = 59kN}$$

Shear force

SF @ B = −61 kN

SF @ E = $-61 + 50 = -11$ kN

SF @ D = $-11 + 40$ = 29kN

SF @ C = 29 + 30 = 59 kN

SF @ A = 59 kN

Bending moment

BM @ B = 0

BM @ E = 61 × 1.5= 91.5 kNm

BM @ D = (61 × 2.5) – (50 × 1) = 102.5kNm

BM @ C = 59 × 1 = 59 kN

BM @ A = 0.

Result

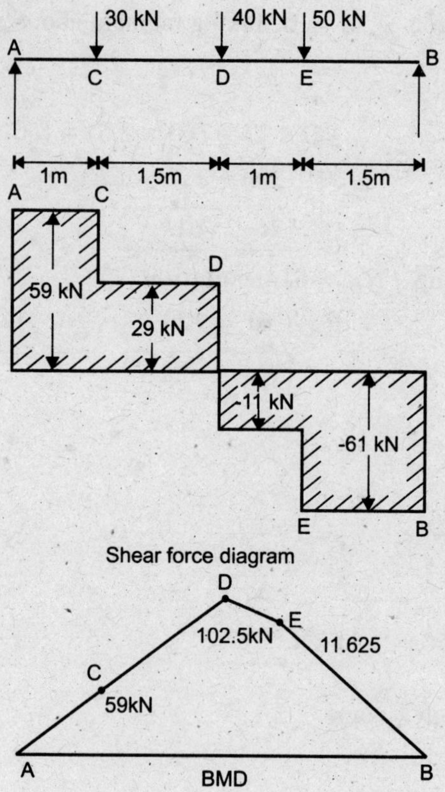

8. **A simply supported beam of 16m effective span carries the concentrated load 4kN, 5kN and 3kN at distance 3m, 7m and 10m respectively from he left support. Calculate the maximum shearing force and bending moment. Draw SFD and BMD.**

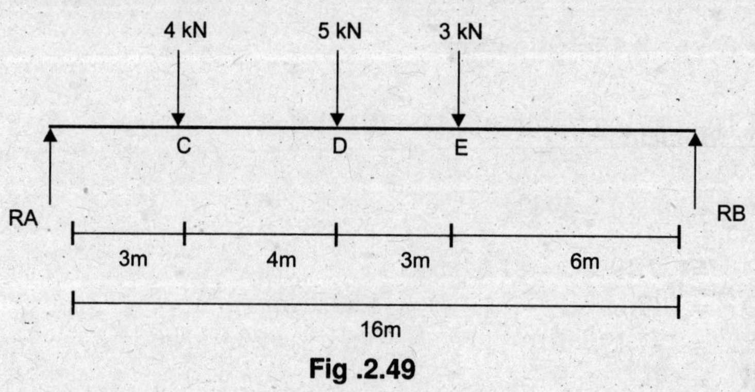

Fig .2.49

Solution

Applying equilibrium condition

$$\sum V = 0$$

$$R_A + R_B = 4 + 5 + 3 = 12kN \qquad (1)$$

Applying equilibrium condition

$$\sum M_A = 0$$

$$(4 \times 3) + (5 \times 7) + (3 \times 10) = R_B \times 16$$

$$(4 \times 3) + (5 \times 7) + (3 \times 10) = V_B \times 16$$

$$12 + 35 + 30 = 16V_B$$

$$\therefore \quad \boxed{R_B = 4.81kN}$$

Substituting for R_B.

We get $\boxed{R_A \text{ as } 7.19\text{kN}}$

Shear Force

@ A = 7.19kN
@ C = 7.19 – 4 = 3.19kN
@ D = 3.19 - 5 = -1.81kN
@ E = $-1.81 - 3 = -4.81kN$
@ B = - 4.18kN

Bending Moment

@ A = 0
@ C = 7.19 × 3 = 21.57kNm
@ D = (7.19 ×7) – (4× 4) = 34.33 kNm
@ E = (7.19 ×10) – (4 × 7) – (5× 3)= 28.9 kNm
@ B = 0

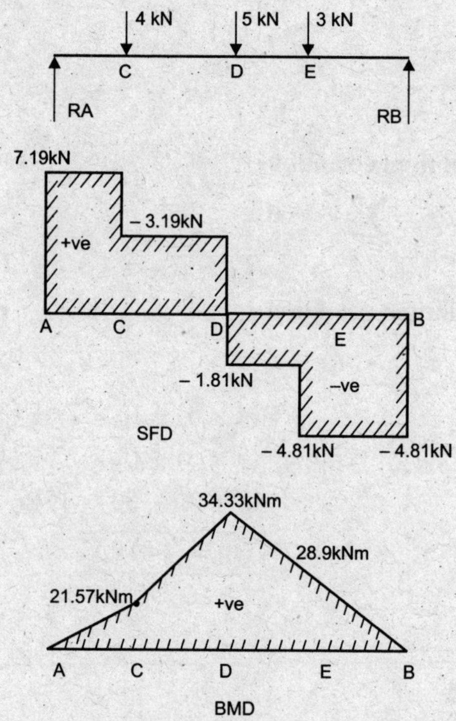

Fig .2.50

9. A simply supported beam of length 10m carries UDL and two point loads as shown in fig shown below. Draw SFD and BMD and Calculate the maximum bending moment.

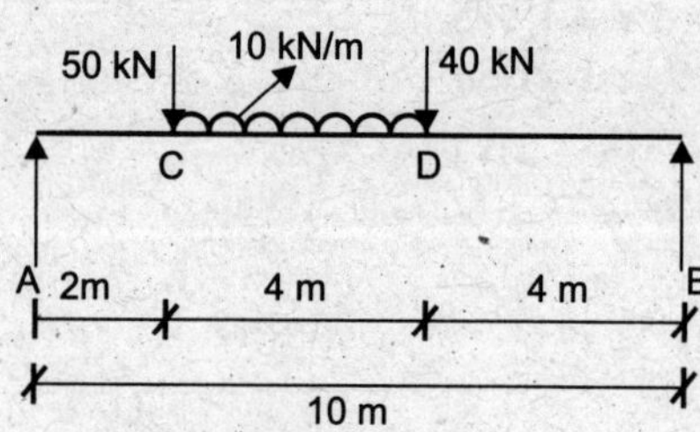

Fig .2.51

Solution

Applying equilibrium conditions $\sum M = 0$

Taking moment about 'A'

$$\sum M_A = 0$$

$$(R_B \times 10) - (40 \times 6) - (50 \times 2) - (10 \times 4 \times 4) = 0$$

$$R_B = \frac{500}{10} = 50kN$$

Applying equilibrium conditions

$$\sum V = 0$$

$$\therefore \quad R_A = R_A + R_B = 50 + 40 + (10 \times 4)$$

$$R_A + 50 = 130 \qquad \text{Substituting for} R_B = 50$$

$$R_A = 80kN$$

Shear Force

Shear Force @	A	=	80 kN
Shear Force @	C	=	80 – 50 = 30kN
Shear Force @	D_1	=	$30 - 40 = -10$ kN
Shear Force @	D_2	=	$-10 - 40 = -50$ kN
Shear Force @	B	=	$-50 = -50$ kN

Bending Moment

Bending Moment @	A	=	0
Bending Moment@	B	=	0
Bending Moment @	C	=	80 ×2 = 160kNm
Bending Moment @	D	=	50 ×4 = 200kNm

Max BM occurs at where SF is zero.

$$\frac{50}{x} = \frac{10}{4 - x}$$

$$\Rightarrow \quad 5 = \frac{x}{4 - x} \quad x = 3.33m \quad \text{from 'C'}$$

∴ Max BM occurs at 5.33m From 'A'

Taking moment about that point from 'A'

$$(R_A \times 5.33) - (50 \times 3.33) - \left(10 \times \frac{3.33^2}{2}\right)$$

Substituting for R_A

$$80 \times 5.33 - 50 \times 3.33 - (10 \times \frac{3.33^2}{2})$$

$$426.4 - 166.5 - 55.44$$

$$= 204.46kNm$$

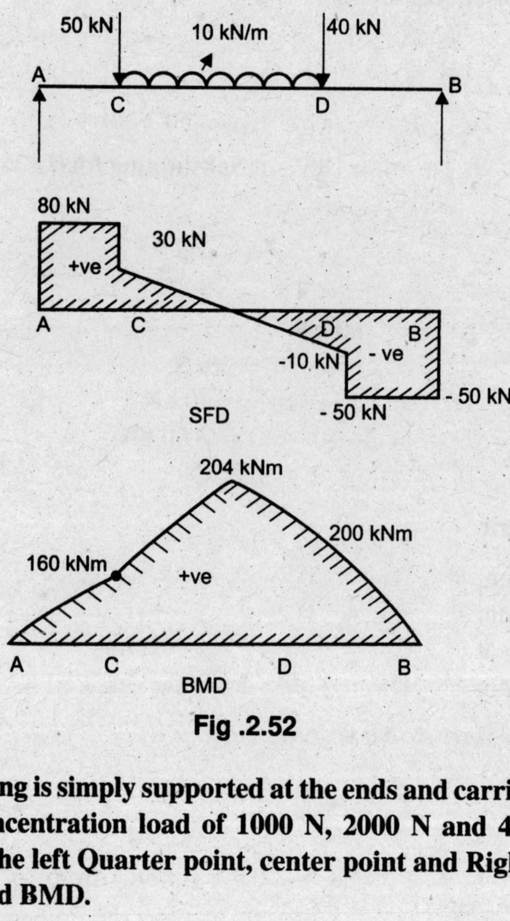

Fig .2.52

10. **A Beam 8m long is simply supported at the ends and carries a UDL 1500N/m and three concentration load of 1000 N, 2000 N and 4000 N acting respectively at the left Quarter point, center point and Right Quarter point. Draw SFD and BMD.**

Solution

Applying equilibrium condition $\sum V = 0 \quad R_A + R_E$= Total downward load

$$\text{i.e.} \quad R_A + R_E = 1500 \times 8 + 1000 + 2000 + 4000$$

$$R_A + R_E = 19000N \qquad (1)$$

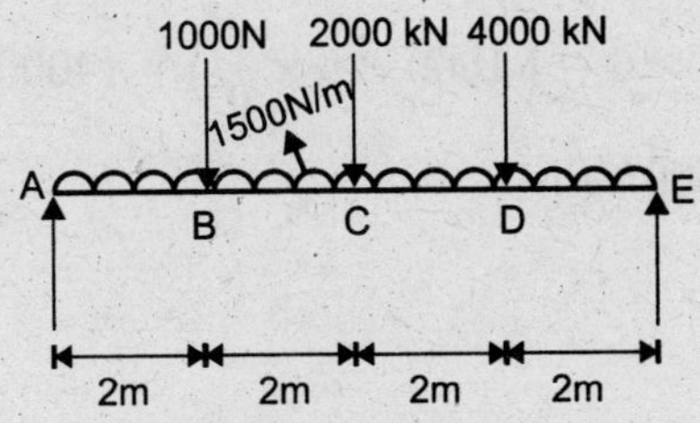

Fig .2.53

Applying equilibrium condition $\sum M = 0$, Taking moment about 'A'.

$$R_E \times 8 = \left(1500 \times 8 \times \frac{8}{2}\right) + (1000 \times 2)$$

$$+ (2000 \times 4) + (4000 \times 6) = 82000N$$

$$\therefore \quad R_E = \frac{82000}{8} = 10,250N$$

Now substituting for R_E in eqn, (1) R_A = 19000 - 10250

$$\boxed{R_A = 8750N}$$

Shear Forces,(SF) SF @ E = – RE = –10, 250N

Without Point Load	**With Point Load**
SF @ D = $-10250 + 1500 \times 2$	S.F @ D = $-10250 + 1500 \times 2$
$= -7250$ N	+ 4000 = -3250 N
SF @ C = $-3250 + 1500 \times 2$	S.F @ C = $-3250 + 2000 + 1500 \times 2$
$= -250$ N	= 1750 N
SF @ B = $-1750 + 1500 \times 2$	S.F @ B = $1750 + 1500 \times 2 + 1000$
= 4750 N	= 5750 N
	SA @ A = 5750 + (1500 × 2) = 8750 N

Bending Moments (BM)

BM@ $E = 0$

BM@ $D = R_E \times 2 - \left(1500 \times 2 \times \frac{2}{2}\right) = 17500Nm.$

BM@ $C = R_E \times 4 - \left(1500 \times 4 \times \frac{4}{2}\right) - 4000 \times 2 = 21000Nm.$

$$\text{BM@ } B = R_E \times 6 - \left(1500 \times 6 \times \frac{6}{2}\right) - (4000 \times 4) - (2000 \times 2)$$

$$= 14500Nm.$$

$$\text{BM@ } A = 0$$

Result

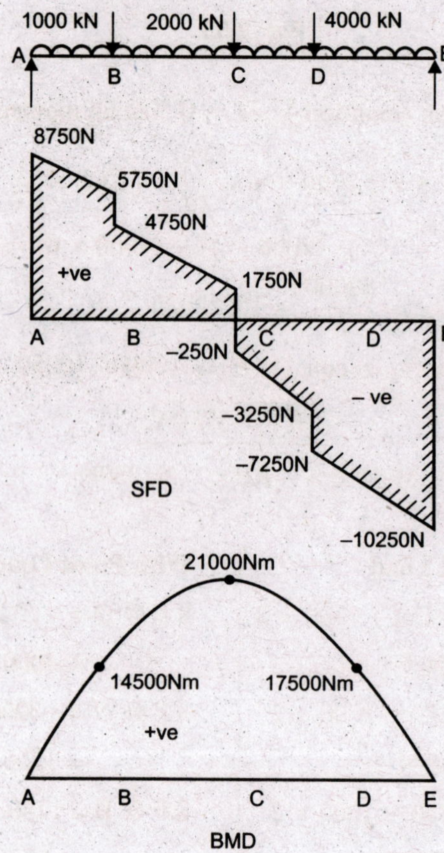

Fig .2.54

11. **A Beam of span 8m is supported at its ends. It is loaded with a gradually varying load and of 1kN/m from the left hand support to 2 kN/m to the Right hand Support. Construct the SFD, BMD.**

Solution

Applying or equilibrium condition $\sum V = 0$

$$R_A + R_B = 1 \times 8 + \left(\frac{1}{2} \times 8 \times 1\right) \Rightarrow \boxed{R_A + R_B = 12} \qquad (1)$$

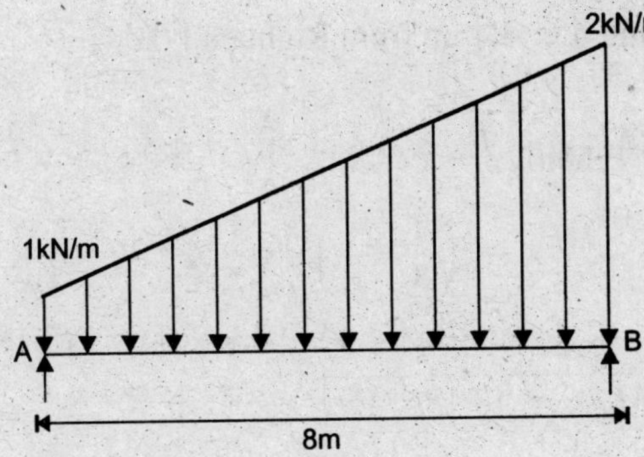

Applying or equilibrium condition $\sum M = 0$

Taking moment about 'A'

$$R_B \times 8 = \left(1 \times 8 \times \frac{8}{2}\right) + \left(\frac{2}{3} \times 8 \times \frac{1}{2} \times 8 \times 1\right)$$

$$R_B = 53.33/8$$

$$\boxed{R_B = 6.67kN}$$

Substituting for R_B in equation (1), we get

$$R_A = 12 - 6.67$$

$$\boxed{R_A = 5.33kN}$$

Shear Forces (SF)

@ B = $-R_B = -6.67$ kN

@ A = $+R_A = 5.33$ kN

Bending Moment (BM)

@ B = 0

@ A = 0

BM is maximum at SF = 0

$$SF_x \Rightarrow -R_B + (1 \times x) + \left(\frac{1}{2} \times x \times 1\right) = 0 = -6.67 + x + \frac{1}{2}x = 0$$

$$1.5x = 6.67$$

$$\boxed{x = 4.45m}$$

The BM is maximum @ 4.45m from Right end 'B'.

$$BM_{max} = R_B \times 4.45 - \left(1 \times 4.45 \times \frac{4.45}{2}\right)$$
$$- \left(\frac{2}{3} \times 4.45 \times \frac{1}{2} \times 4.45 \times 1\right)$$
$$= 6.67 \times 4.45 - 9.9 - 6.6 = 13.18 kN.m$$

$$\boxed{BM_{max} = 13.18 kNm}$$

Result

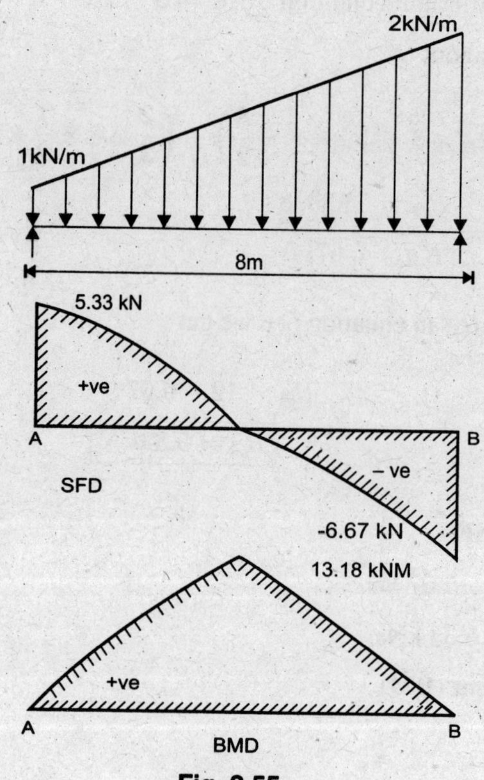

Fig .2.55

Problems on Overhanging Beam

12. **A Overhanging beam ABC of length 6m is supported at A and B. The Overhanging portion BC is of 2m length. A UDL of intensity 2kN/m acts over the entire span. Two point loads of intensity 5kN and 3kN are acting**

at a distance of 2m from the left support A and at the end C respectively. Draw SFD, BMD and also locate the point of contra flexure

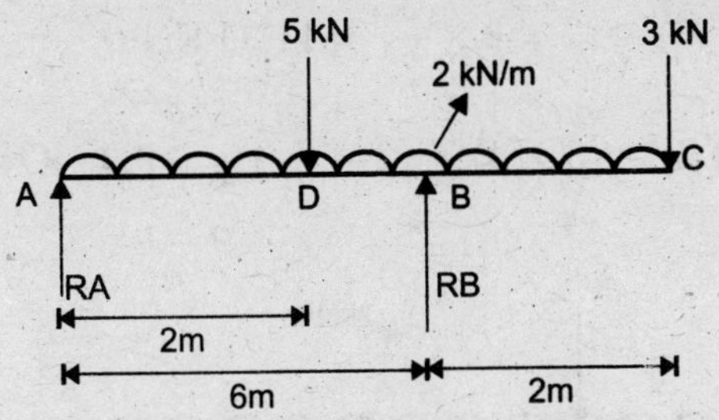

Fig .2.56

Solution

Taking Moment about A

Applying equilibrium condition $\sum V = 0$

$$R_A + R_B = (2 \times 8) + 5 + 3 = 24 \quad (1)$$

Applying equilibrium condition $\sum M = 0$

$$(R_B \times 6) - \left(2 \times 8 \times \frac{8}{2}\right) - (5 \times 2) - (3 \times 8) = 0$$

$$R_B = 16.33\,KN$$

Substituting for $R_B = 16.33$ in (1)

$$R_A = 24 - 16.33$$

$$R_A = 7.67KN$$

Shear Force

@ A = 7.67KN

@ $D_1 = 7.67 - 4 = 3.67$ KN

@ $D_2 = 3.67 - 5 = -1.33$ KN

@ $B_1 = -1.33 - (2 \times 4) = -9.33$KN

@ $B_2 = -9.33 + 16.33 = 7$ KN

@ $C_1 = 7 - (2)2 = 3$KN

@ $C_2 = 3$KN

Bending Moment

@ A = 0

@ D = $(7.67 \times 6) - (2 \times 2 \times \frac{2}{2}) = 11.34$ KNm

@ B = $(7.67 \times 2) - (5 \times 4) - (2 \times \frac{6^2}{2}) = -9.98$ KNm

@ C = 0

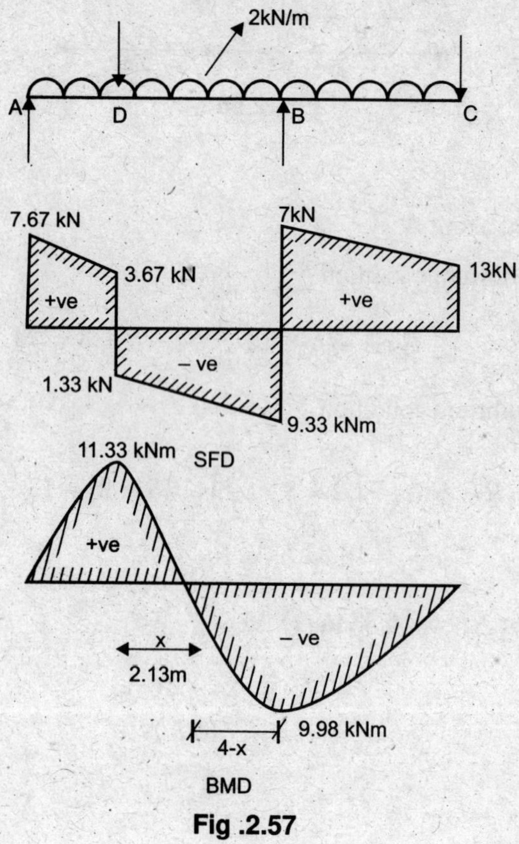

Fig .2.57

Determine of point of inflexion (or)contra flexure

$$\frac{x}{11.34} = \frac{4-x}{9.98}$$

$$0.88x = 4 - x \quad \Rightarrow \quad 0.88x + x = 4$$

$$x = \frac{4}{1.88}$$

$x = 2.13m$ $\therefore$ from point D

13. **A beam 12m long is supported at two points 2m from each end, so that there are two equal overhanging portions. It carriers concentrated load of 4KN, 3KN and 5KN at 1m, 8m and 12m respectively from left end. Draw SFD and BMD.**

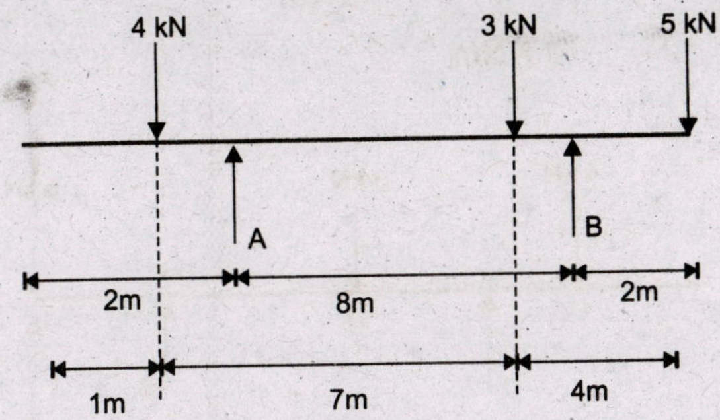

Fig .2.58

Solution

Applying or equilibrium condition $\sum V = 0$

$$R_A + R_B = 4 + 3 + 5 = 12KN \quad (1)$$

Applying or equilibrium condition $\sum M = 0$

$$-(4 \times 1) + (3 \times 6) - (R_B \times 8) + (5 \times 10) = 0$$

$$R_B = \frac{64}{8} = 8KN$$

Substituting for R_B in (1)

$$R_A = 12 - 8 = 4KN$$

Shear Force(SF)

Shear Force @ F = 0

@ C = – 4KN

@ A = – 4 + 4 = 0

@ D = 0 – 3 = – 3KN

@ B = – 3 + 8 = 5KN

@ E = 5KN

Bending Moment(BM)

@ F = 0

@ C = 0

@ A = $-4 \times 1 = -4$ KNm

@ D = $(-4 \times 7) + (4 \times 6) = -4$ KNm

@ B = $-5 \times 2 = -10$ KNm

@ E = 0

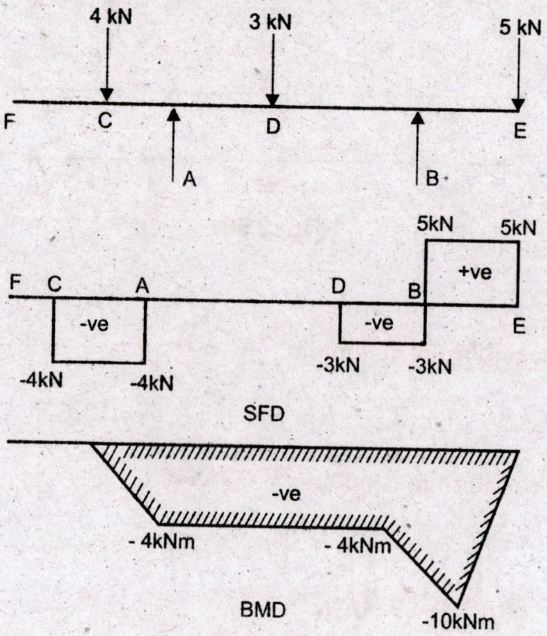

Fig .2.59

14. **Draw SFD and BMD of following over hanging beam.**

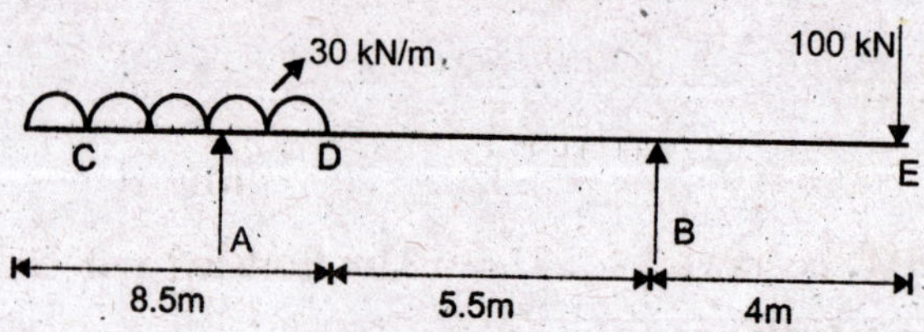

Fig .2.60

Solution

Applying or equilibrium condition $\sum V = 0$

$$R_A + R_B = 355KN \quad (1)$$

Applying or equilibrium condition $\sum M_A = 0$

$$R_B \times 11 = 100 \times 15 + \left(30 \times 5.5 \times \frac{5.5}{2}\right) - (\mathbf{30 \times 3 \times \frac{3}{2}})$$

$$R_B = 165.34KN$$

Substituting for R_B in (1)

$$R_A = 355 - 165.34 = 189.7KN$$

Shear Force

@ C = 0

@ A_1 = $(-30 \times 3) = -90$KN

@ A_2 = $-90 + 189.7 = 99.7$KN

@ D = $99.7(-30 \times 5.5) = -65.3$KN

@ $B_1 = -65.3$KN

@ $B_2 = -65.3 + 165.3 = 100$ KN

@ $E = 100$KN

Bending Moment

@ C = 0

@ A = $(-30 \times \frac{3^2}{2}) = -135$KNm

@ D = $(-30 \times \frac{8.5^2}{2}) + (189.7 \times 5.5) = -40.4$KNm

@ B = $-100 \times 4 = -400$KNm

@ E = 0

Max BM occurs at point where SF is zero

$$\frac{99.65}{x} = \frac{65.34}{5.5 - x}$$

$$x = 3.32m \qquad (\because \text{ from} A)$$

$\therefore$ The Max BM occurs @ 3 + 3.32 = 6.32m from left end 'c'

Now, maximum bending moment

$$= -30 \times \frac{(6.32)^2}{2} + R_A \times 3.332$$

Substituting for RA

$$BM_{(max)} = -599.14 + 629.8$$

$$\therefore\ max_m\ BM = 30.66KNm$$

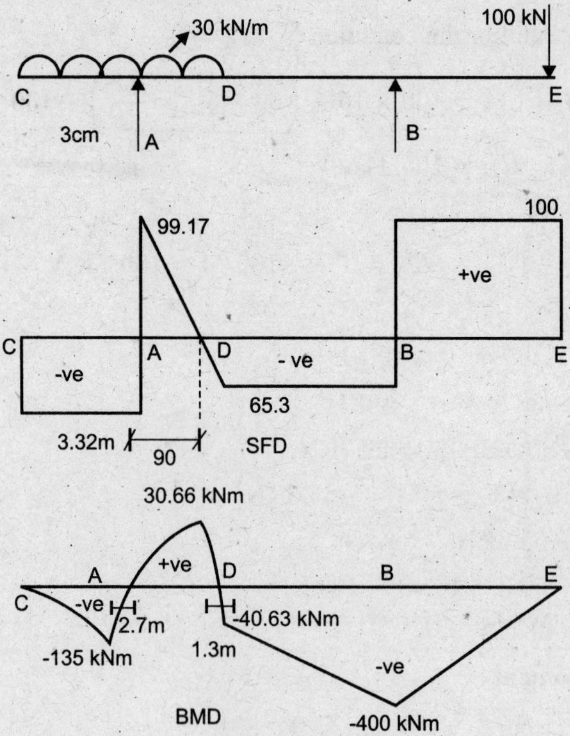

Fig .2.61

Point of contra flexure occurs at point where BM is Zero; here two point of contra flexure occurs.

1^{St} Point

$$\frac{135}{x_1} = \frac{30.66}{3.32 - x_1} \quad x_1 = 2.7m \quad \text{(from point } A)$$

2^{St} Point

$$\frac{40.63}{x_2} = \frac{30.66}{2.18 - x_2} \quad x_2 = 1.3m \quad \text{(from point } B)$$

15. **A Beam ABCD of length 8m is simply supported at B and C length AB = 1m, BC = 5m and CD = 2m. The beam carries a concentrated load of 10KN at A and another concentrated load of 20KN at E, 2m from end A. In addition the beam carries a UDL of 15KN/M from a point E to C. The beam also carries a concentrated load of 5KN at the free end D. Draw the Shear Force and Bending Moment diagrams.**

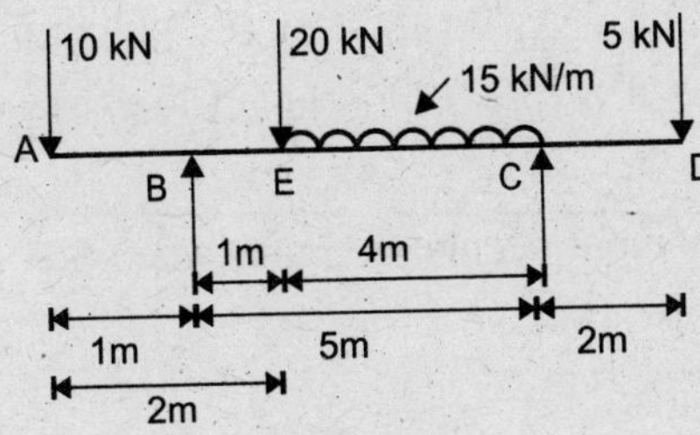

Fig .2.62

Solution

Applying or equilibrium condition $\sum v = 0$

$$R_B + R_C = TDL \text{ (Total Down ward Load)}$$
$$R_B + R_C = (15 \times 4) + 10 + 20 + 5$$
$$R_B + R_C = 95K_N \qquad (1)$$

Applying or equilibrium condition $\sum M = 0$

Taking Moment at B, $(20 \times 1) + (15 \times 4 \times 3) + (5 \times 7) - 10 \times 1R_C \times 5$

$$\therefore \quad R_C = 45K_N \quad R_B + R_C = 95$$
$$\therefore \quad R_B = 95 - 45$$
$$\boxed{R_B = 50K_N}$$

Shear Forces (SF)

$@\ D = 5K_N$
$@\ C = 5 - 45 = -40K_N$

Without Point Load	**With Point Load**
SF @ E = $5 - 45 + 15(4)$ =20kN	S.F @ E = $5 - 45 + 15(4) + 20 = 40$kN

$$SF@B = 5 - 45 + 15(4) + 20 - 50 = -10kN$$
$$SF@A = -10kN$$

Bending Moments (BM)

@ D = 0
@ C = $5 \times 2 = 10$KNM

@ E = $(5 \times 6) + (15 \times 4 \times \frac{4}{2}) - (45 \times 4) = -30\text{kNM}$
@ **B** = $(5 \times 7) + (15 \times 4 \times 3) + (20 \times 1) - (45 \times 5) = 10\text{kNM}$
@ A = 0.

Maximum Bending moment occurs @ point, when S.F. is zero. It occurs between E and C.

$$\frac{20}{x} = \frac{40}{4-x} = \frac{4-x}{x} = 2 \quad 4 - x = 2x \;\Rightarrow\; 3x = 4$$

$$\therefore \; \boxed{x = 1.33m} \quad \text{from 'E'.}$$

Maximum Bending moment @ 3.33m from 'A'

$$BM_{Max} = -\left[(-10 \times 3.33) + (50 \times 2.33) - (20 \times 1.33) - \left(15 \times \frac{1.33^2}{2}\right)\right]$$
$$= 33.3 - 116.5 + 26.6 + 13.27 = -43.33 \; KNM$$

Result

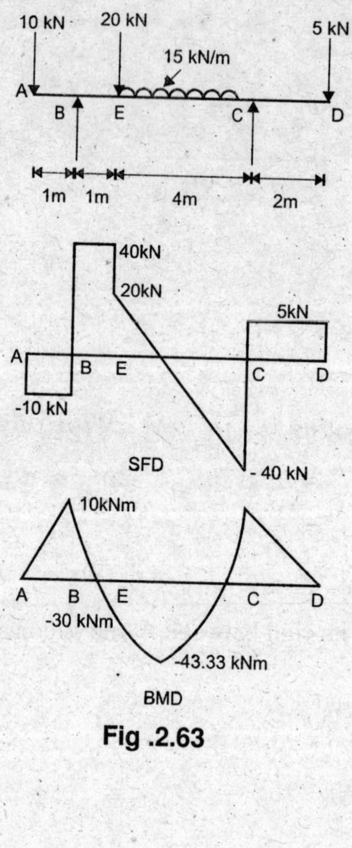

Fig .2.63

16. **An Overhanging beam is loaded as shown in fig below. Construct the SFD and BMD for the beam. Also calculate the maximum bending moment and locate the point of Contra flexure.**

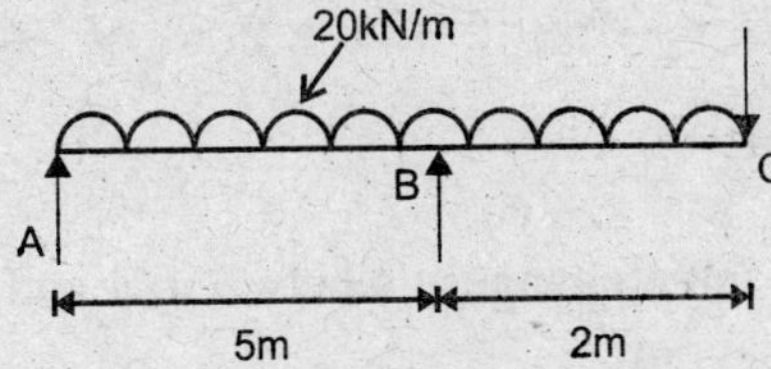

Fig .2.64

Solution

Applying or equilibrium condition $\sum V = 0$

$$R_A + R_B = (20 \times 7) + 10 = 150 \qquad (1)$$

Applying or equilibrium condition $\sum M = 0$

Taking Moment about A,

$$5R_B = (20 \times 7 \times \frac{7}{2}) + 10 \times 7$$

$$5R_B = 560$$

$$\boxed{R_B = 112KN}$$

Substituting for R_B in (1) $R_A + 112 = 150 \ \therefore \ \boxed{R_A = 38KN}$

Shear Forces (SF)

@ $C = 10K_N$

@ $B = 10 + 20 \times 2 - R_B = 10 + 40 - 112 = -62\ KN$ with reacts @ B

@ $B = 10 + (20 \times 2) = 50KN$ (Without reaction B)

@ $A = 10 + (20 \times 7) - R_B = 10 + 140 - 112 = 38KN$

Bending Moment BM at $A = 0$

In SFD, The SF changes its sign between A and B consider 'x' is the distance from the left support.

$$20.x - R_A = 0$$

$$\therefore \ 20x = R_A \quad x = \frac{38}{20} - 1.9m$$

Bending Moment

$$@ \quad C = 0$$

$$@ \quad B = -10 \times 2 - 20 \times 2 \times \frac{2}{2} = -60KNM$$

$$@ \quad A = 0$$

From SFD, The shear farce change it's between A and B consider 'x' is the distance from the left support.

SF at that point is

$$R_A - 20x = 0$$

$$38 = 20x$$

$$\boxed{x = 1.9m}$$

The maximum BM Occurs at 1.9m front left support. Taking Moment of all force about the point to the left sides.

$$\text{Maximum } BM = (R_A \times 1.9) - \left(20 \times 1.9 \times \frac{1.9}{2}\right) = (38 \times 1.9) - (36.1)$$

$$\text{Maximum} \quad BM = 36.1KNM$$

Point of Contra Flexure

The point where the bending moment changes its sign or zero is called the point of contra flexure.

The point of contra flexure occurs at $y \;\; m$ from the left end between A and B. Taking moment about that point and equating it to zero.

Taking moment of all force places left side of that point.

$$R_A.y - 20.y.\frac{y}{2} = 0$$

$$38y - 10y^2 = 0 \Rightarrow 38 = \frac{10y^2}{y}$$

$$\therefore \; y = \frac{38}{10} = 3.8m$$

$$\boxed{y = 3.8m}$$

The point of contra flexure occurs at a distance of 3.8m from left end.

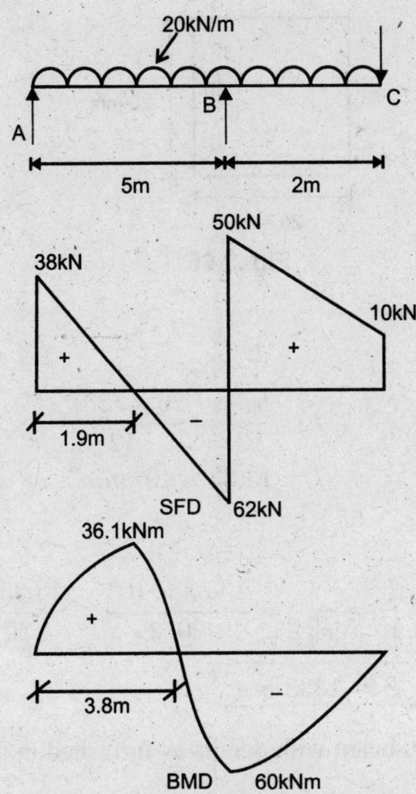

Fig .2.65

Problems the Bending Stresses

17. **A cast iron test beam $2cm \times 2cm$ in section and 1m long and supported at the ends fails when a central load of 64Kg is applied. What uniformly distributed load will break a cantilever of the same material 5cm wide, 10cm deep and 2m long?**

Given Data

Length of the cast iron Beam $(\ell) = 1\text{m} = 1000\text{mm}$

Breadth of cast iron beam $(b) = 2\text{cm} = 20\text{mm}$

Depth of cast iron beam $(d) = 2\text{cm} = 20\text{mm}$

Solution

We know that,

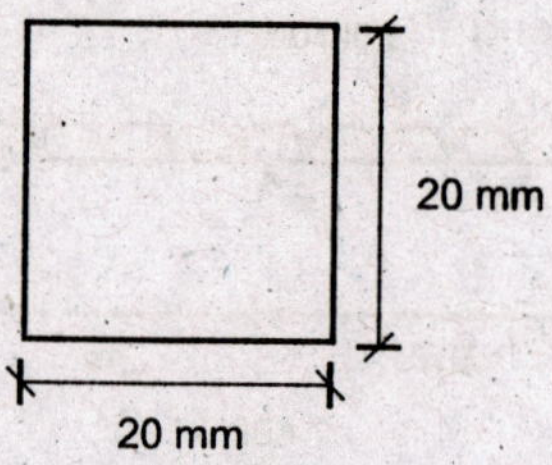

Fig .2.66

Moment of inertia:

$$I = \frac{bd^3}{12} = \frac{20 \times (20)^3}{12}$$
$$I = 13.33 \times 10^3\ mm^4$$

Section Modulus

$$Z = \frac{I}{Y} = \frac{I}{\left(\frac{d}{2}\right)} = \frac{13.33 \times 10^3}{20/2} = \frac{13.33 \times 10^3}{(20 \times 2)}$$
$$\boxed{\therefore Z = 1333 mm^3}$$

For Simply Supported beam with details as furnished in the problem is as below.

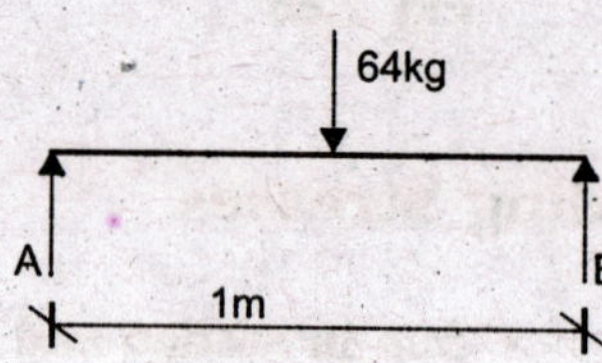

Fig .2.67

$$W = 64Kg = 64 \times 9.81 = 627.84N$$
$$\ell = 1m = 1000mm.$$

For a simply supported beam with a point load acting @ the centre, the Bending moment

$$BM = \frac{Wl}{4} = \frac{627.84 \times 1000}{4}$$
$$\boxed{M = 156.96 \times 10^3 Nmm}$$

From the theory of simple bending equation,

$$\frac{M}{I} = \frac{\sigma_b}{y}$$

We can write the Eqn. for bending stress as

$$(\sigma_b) = \frac{M.y}{I}$$

$$\Rightarrow \frac{\frac{M}{I}}{y} = \frac{M}{Z}$$

Where M = Bending Moment (BM) and 'z' is the section modulus.

Substitute the values for M and 'z' we get

$$= \frac{M}{Z} = \frac{156.96 \times 10^3}{1333}$$

$$\sigma_b = 117.74 N/mm^2$$

$$\boxed{\text{Maximum bending stress, } (\sigma_{\max}) = 117.74 N/mm^2}$$

For cantilever beam, as detailed in the problem, it can be drawn as below.

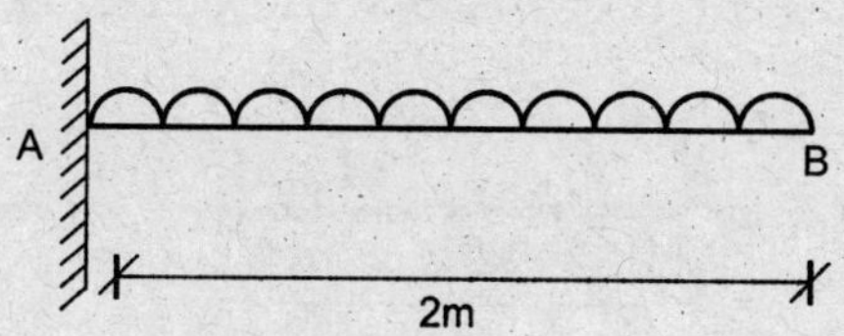

Fig .2.68

Given details for cantilever beam

Breadth (b) = 5cm = 50mm

Depth (d) = 10cm = 100mm

Length (ℓ) = 2m = 2000mm

For cantilever beam Bending moment at fixed end is

$$M = \frac{Wl^2}{2} = \frac{W \times (2000)^2}{2}$$

$$M = 2 \times 10^6 W Nmm$$

We know that, Moment of Inertia for rectangular section

$$(I) = \frac{bd^3}{12} = \frac{50 \times (100)^3}{12}$$

$$\boxed{I = 4.166 \times 10^6 mm^4}$$

section in modulus $(Z) = \frac{I}{Y} = \frac{I}{\frac{d}{2}} = \frac{4.166 \times 10^6}{50}$ $(\because\ d = 100mm)$

$$\boxed{Z = 83.33 \times 10^3 mm^3}$$

We know that Bending equation,

$$\frac{M}{I} = \frac{\sigma_b}{y}$$

$$\therefore\quad M = \frac{I}{Y} \times \sigma_b$$

Now, substituting for values (M in terms of 'w' and other values) $[\because\ Z = \frac{I}{y}]$

$$2 \times 10^6 W = Z \times \sigma_b$$

$$2 \times 10^6 W = 83.33 \times 10^3 \times 117.74$$

$$W = \frac{83.33 \times 10^3 \times 117.74}{2 \times 10^6}$$

$$\boxed{W = 4.9N/mm}$$

$W \rightarrow$ uniformly distributed load

18. **A rectangular beam 300mm deep is simply supported over the span of 4m. Determine the uniformly distributed load per meter which the beam may carry if the bending stress should not exceed 120N/mm^2. Take $I = 8 \times 10^6 mm^4$**

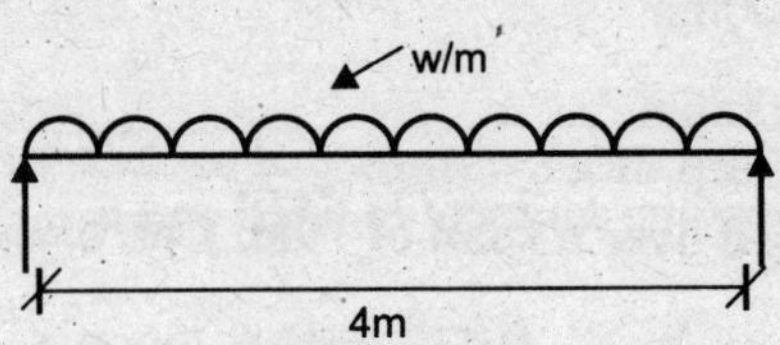

Fig .2.69

Given Data

Depth of beam (d) = 300mm

Length of beam (ℓ) = 4m = 4000mm

Bending stress in beam (σ) = $120N/mm^2$

Moment of Inertia $(I) = 8 \times 10^6 mm^4$

To Find:

Uniformly Distributed Load (UDL)

We know that, Bending equation

Solution

$$\frac{M}{I} = \frac{\sigma_b}{y}$$
$$\therefore \quad M = \frac{\sigma_b \times I}{y}$$

Now substituting for values

$$= \frac{120 \times 8 \times 10}{\left(\frac{300}{2}\right)} \left(\therefore \quad y = \frac{d}{2}\right)$$
$$\boxed{M = 6.4 \times 10^6 N.mm}$$

We know that, for a simply supported beam BM with uniformly disturbed load, The moment is

$$\frac{W\ell^2}{8}$$

Now equating BM arrived to $\dfrac{w\ell^2}{8}$

$$\frac{W \times 4000^2}{8} = 6.4 \times 10^6$$
$$\therefore \quad W = 3.2N/mm$$

19. **A beam 500mm deep of a symmetrical section has (I = 1 $\times$ 10^8 mm^4 and is simply supported over a span of 10m. Calculate,**

    i) The UDL it carry if the maximum bending stress is not to exceed $150N/mm^2$

    ii) The maximum bending stress, if the beam carries a central point load of 25kN.

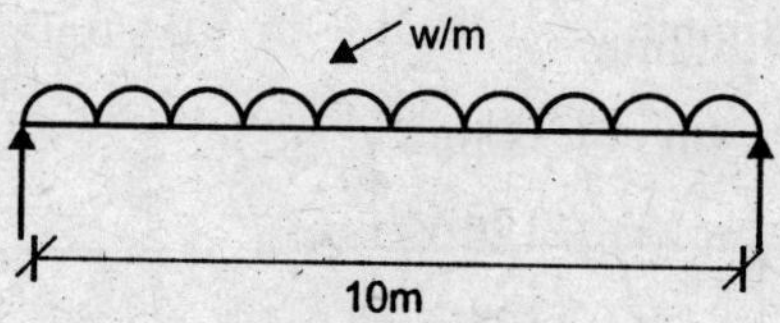

Fig .2.70

Given Data

Case - 1

Dept of the beam, (d) = 500mm

Width of the beam, (b) = 500mm[Symmetrical Section]

Moment of Inertia (I) = $1 \times 10^8 mm^4$

Solution

i) To find the UDL when bending stress is 150N/mm^2 using simple bending theory.

$$\frac{M}{I} = \frac{\sigma}{y}$$

Substituting for

$$\sigma = 150N/mm^2$$
$$y = \frac{d}{2} = \frac{500}{2} = 250mm$$
$$I = 1 \times 10^8 mm^4$$

$\therefore$ Maximum bending moment $(M) = \dfrac{\sigma \times I}{y} = \dfrac{150}{250} \times 1 \times 10^8$

$$\therefore \boxed{M = 60 \times 10^6 Nmm}$$

We know that,

For simply supported beam with UDL,

Maximum Bending moment = $\dfrac{w\ell^2}{8}$

Now, substituting for bending Moment 'M' and Length (L)

$$M = \frac{w\ell^2}{8}$$
$$60 \times 10^6 = \frac{W \times 10000^2}{8}$$

Maximum UDL (W) = 4.8N/m

ii) To find maximum bending stress for central point load of 25KN.

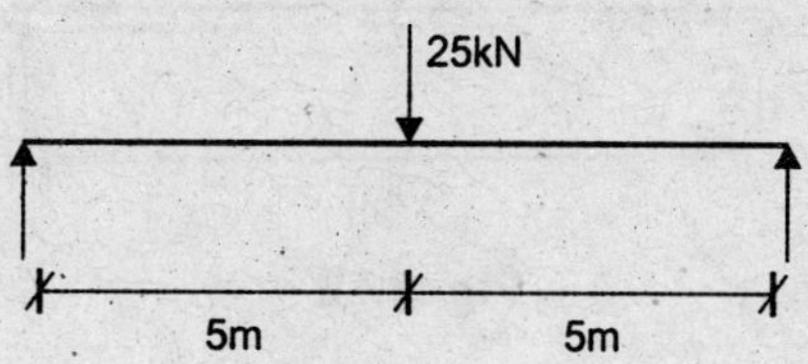

Fig .2.71

We know that, Bending Moment due to central point load on a beam is $\frac{w\ell}{4}$

$$M = \frac{W\ell}{4} = \frac{25 \times 10}{4} = 62.5\ KNm$$

Maximum bending moment, $M = 62.5 \times 10^6 Nmm$

Using bending theory equation,

$$\frac{M}{I} = \frac{\sigma}{y}$$

$$\therefore \quad \sigma = \frac{M}{I}.y$$

On substituting the values,

$$= \frac{62.5 \times 10^6 \times 250}{1 \times 10^8}$$

$$\boxed{\sigma = 156.25 N/mm^2}$$

20. **A simply supported beam which is having rectangular section 60 × 35mm and 3m long carrying a load of 5KN at mid span. Determine the maximum bending stress induced in the beam.**

Given Data

Width of the beam (b) = 60mm

Depth of the beam (d) = 35mm

Length of the beam (L) = 3m

Solution

Load (W) = 5kN at mid span of simply supported beam

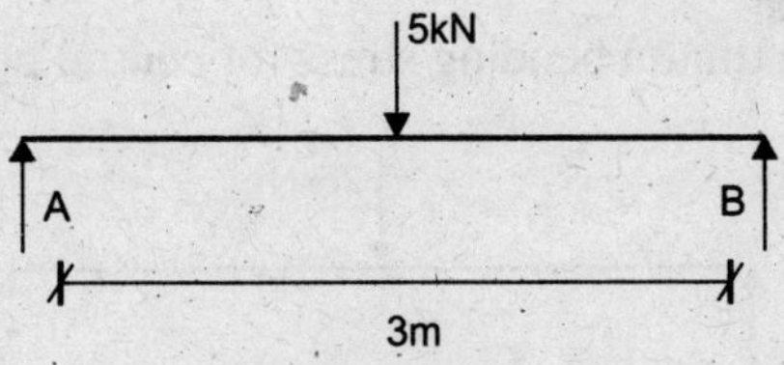

Fig .2.72

$$\text{Moment of Inertia } I = \frac{bd^3}{12} = \frac{60 \times 35^3}{12}$$
$$= 214.38 \times 10^3 mm^4$$

For point Load @ mid span, maximum bending moment

$$(M) = \frac{W\ell}{4} = \frac{5 \times 10^3 \times 3000}{4}$$
$$= 3.75 \times 10^6 Nmm$$

We know that Bending equation,

$$\frac{M}{I} = \frac{\sigma_b}{y}$$

$$\therefore \quad \sigma_b = \frac{M.y}{I} = \frac{3.75 \times 10^6 \times \left(\frac{35}{2}\right)}{214.38 \times 10^3} \qquad (\because \quad y = d/2)$$

$$\therefore \quad \boxed{\sigma_b = 306.12 N/mm^2}$$

21. **A timber of rectangular section is to support a total load of 20KN uniformly distributed over a span of 3.6m when the beam is simply supported if the depth is twice the width of the section and the stress in timber is not to exceed 3.5N/mm^2 find dimensions of the cross section?**

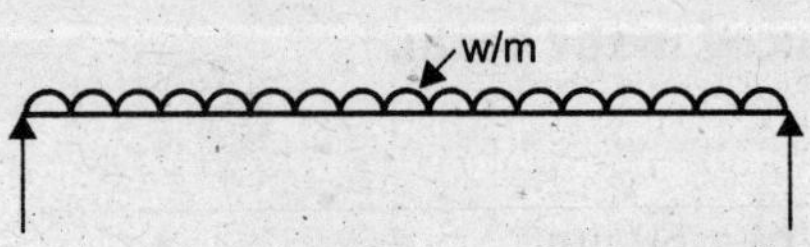

Fig .2.73

Given Data

Length of Beam (simply supported) = 3.6m

UDL over entire span = 20 kN/m

Bending Stress (σ_b) = 3.5 N/mm^2

Breath (b) = 2 d (depth of Beam)

Solution

We know that

$$\frac{M}{I} = \frac{\sigma_b}{y} = \frac{8.91 \times 10^6}{\frac{bd^2}{12}} = \frac{3.5}{\frac{d}{2}}$$

$$\frac{8.91 \times 10^6 \times 12}{bd^2} = 3.5 \times 2$$

$$bd^2 = 15.27 \times 10^6$$

$$\text{given} \quad d = 2b \quad \text{Substituting}$$

$$b(2b)^2 = 15.27 \times 10^6$$

$$4b^3 = 15.27 \times 10^6$$

$$b^3 = 3.82 \times 10^6$$

$$\boxed{\begin{aligned} \therefore b &= 156\ mm \\ d &= 312\ mm \end{aligned}}$$

22. **A cast iron water pipe 12 meters long, of 500mm inside diameter and 25mm wall thickness runs fall of water and is supported at its ends. Calculated the maximum stress in the metal if density of cast iron is 7200 kg/m^3 and that of water is 1000kg/m^3**

Given Data

Inside dia of cast iron pipe (d) = 500mm = 0.5m

Thickness of cast iron pipe (t) = 25m

Outside dia of cast iron pipe (D) = d + 2t

$=500 + (2 \times 25) = 550mm = 0.55m$

Solution

Length of cast iron pipe (L) = $12m = 12 \times 10^3 mm$

$$\text{Density of cast iron} = 7200\ kg/m^3$$
$$= 7200 \times 9.81 N/m^3$$
$$= 70632 N/m^3$$
$$= \frac{70632}{(10^3)^3} N/mm^3$$
$$\text{Density of cast iron} = 70.6 \times 10^{-6} N/mm^3$$
$$\text{Density of Water} = 1000 Kg/m^3$$
$$= 1000 \times 9.81 N/m^3$$
$$= 9810 \times 10^{-9} N/mm^3$$

We know that

C/S area of cast iron pipe

$$\frac{\pi}{4}\left[D^2 - d^2\right] = \frac{\pi}{4}\left[550^2 - 500^2\right]$$
$$= 41.23 \times 10^3 mm^2$$
$$= 0.412 m^2$$
$$\text{Weight} = \text{Area} \times \text{density}$$
$$(W_1) = \left(70.6 \times 10^{-6}\right) \times \left(41.23 \times 10^3\right) = 2.91 N/mm$$

C/S area of water

$$\frac{\pi}{4} \times \left(d^2\right) = \frac{\pi}{4}(500)^2$$
$$= 196.35 \times 10^3 mm^2$$
$$\text{Weight, } (W_2) = \left(9.8 \times 10^{-6}\right)\left(196.35 \times 10^3\right)$$
$$= 1.92 N/mm$$

Total weight of cast iron and water section

$$W = W_1 + W_2 = 2.91 + 1.92 = 4.83 N/mm$$

We also know the max BM at center of the beam subjected to a UDL

$$M = \frac{W\ell^2}{8} = \frac{4.83 \times \left(12 \times 10^3\right)}{8}$$
$$M = 86.94 \times 10^6 N.mm$$

Section Modulus of hollow circular section

$$Z = \frac{\pi}{32D} \times \left[D^2 - d^4\right]$$
$$= \frac{\pi}{32 \times 550}\left[550^4 - 500^4\right]$$
$$= 5.2 \times 10^6\ mm^3$$

∴ Max bending stress

$$\sigma_{b(\max)} = \frac{M}{Z} = \frac{86.94 \times 10^6}{5.2 \times 10^6} = 16.72\ N/mm^2$$

23. **A timber beam 150mm × 250mm in cross-section is simply supported at its ends and has a span of 3.5 m. The maximum safe allowable stress in bending is 7500 kN/m^2. Find the maximum safe UDL which the beam carry. What is the maximum shear in the beam for the UDL calculated?**

Given Data

Width of beam (b) = 150mm

Depth of beam (d) = 250mm

Length of Beam Bending stress (L) = 3.5m

$(\sigma_b) = 7500kN/m^2 = 7500 \times 10^{-3}N/mm^2 = 7.5N/mm^2$

To Find

1. Max safe load (UDL)
2. Max Shear

Solution

To Find Safe Max UDL 'W/m'

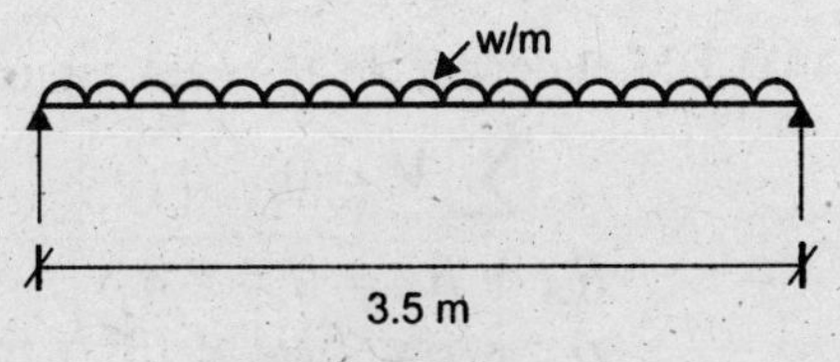

Fig .2.74

$$\frac{M}{I} = \frac{\sigma}{y}$$

$$M = \frac{wl^2}{8}$$

$$M = \frac{w \times 3500^2}{8}$$

$$= 1.53 \times 10^6 W Nmm$$

$$I = \frac{bd^3}{12} = \frac{150 \times 250^3}{12} = 195.31 \times 10^6 mm^4$$

$$Z = \frac{I}{y} = \frac{195.31 \times 10^6}{\frac{d}{2}}$$

$$= \frac{195.31 \times 10^6}{\frac{250}{2}} = 1562.48 \times 10^3 mm^3$$

$$\frac{M}{I} = \frac{\sigma}{y}$$

$$M = \sigma . \frac{I}{y}$$

$$= \sigma (Z)$$

$$1.53 \times 10^6 W = 7.5 \times 1562.48 \times 10^3$$

$$W = 7.6 N/mm$$

$$\boxed{W = 7.6 KN/m}$$

To Find Shear Stress Applying for equilibrium condition

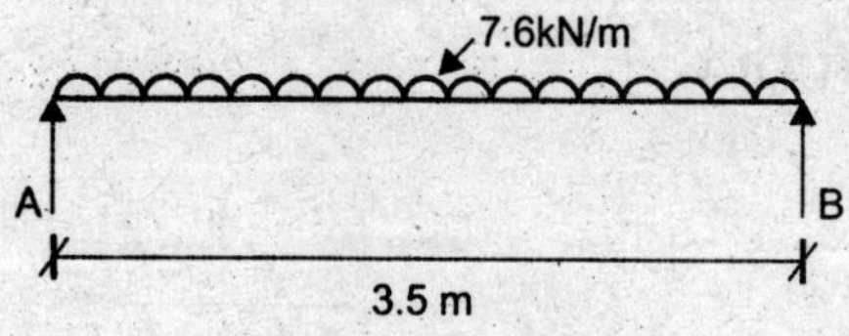

Fig .2.75

$$\sum V = 0$$

$$R_A + R_B = 7.6 \times 3.5$$

$$R_A + R_B = 26.6 \qquad (1)$$

Applying for equilibrium condition

$$\sum M_A = 0$$

$$= 7.6 \times \frac{3.5^2}{2} - 3.5R_B = 0$$

$$3.5R_B = 46.55$$

$$R_B = \frac{46.55}{3.5}$$

$$\boxed{R_B = 13.3kN}$$

$$\text{Sub } R_B \text{ in}(1)$$

$$\therefore \; R_A + 13.33 = 26.6$$

$$R_A = 26.6 - 13.3$$

$$\boxed{R_A = 13.3kN}$$

$$\text{Max shear force} = 13.3KN$$

$$\text{Shear Stress} = \frac{\text{Shear force}}{\text{Shear Area}}$$

$$= \frac{13.3 \times 10^3}{150 \times 250} = \frac{13.3 \times 10^3}{37500}$$

$$\boxed{\text{Shear Stress} = 0.35N/mm^2}$$

24. **Determine the dimensions of timber for span 8m to carry a brick wall 200mm thick and 5m high. If the brick work is 1850kg/m^3 and maximum permissible stress is limited to 7.5N/m^2, depth of joist is twice the width.**

Given Data

Maximum permissible stress

$$(\sigma) = 7.5MN/m^2 = \frac{7.5 \times 10^6 N}{10^6 mm^2} = 7.5N/mm^2$$

Length of beam ℓ = 8m = 8000mm

Brick wall thickness(t) = 200mm

Height of wall = 5m = 5000mm

Density of brick = 1850kg/m^3

$$= 1850 \times 9.81N/m^3 = 18.14 \times 10^3 N/m^3$$

$$= 18.14kN/m^3$$

Solution

Given that, depth of beam = twice the breadth of beam $d = 2b$

We know that, from theory of Bending

$$\frac{M}{I} = \frac{\sigma}{y} \quad \text{(self weight } (w) = \ell \times b \times t \times \text{Density of brick)}$$

$$W = 8 \times 0.20 \times 5 \times 18.14$$

$$\therefore \text{ Load Per meter} = \frac{145.12}{8} = 18.14KN/m$$

$$y = \frac{d}{2}$$

Maximum BM

$$M = \frac{W\ell^2}{8}$$

$$M = \frac{18.14 \times 8^2}{8}$$

$$M = 145.12KNm$$

$$M = 145.12 \times 10^6 N.mm$$

Now substituting the value

$$\frac{M}{I} = \frac{\sigma_b}{y} \Rightarrow \frac{145.12 \times 10^6}{\frac{bd^3}{12}} = \frac{7.5}{\frac{d}{2}}$$

$$= \frac{145.12 \times 10^6}{\frac{b \times (2b)^3}{12}} = \frac{7.5}{\frac{2b}{2}}$$

$$= \frac{145.12 \times 10^6}{0.666\, b^4} = \frac{7.5}{b}$$

$$\boxed{b = 307mm}$$

$$d = 2b$$

$$d = 2 \times 307mm$$

$$\boxed{d = 614mm}$$

25. A section of a simply supported beam gas the following dimensions as shown below.

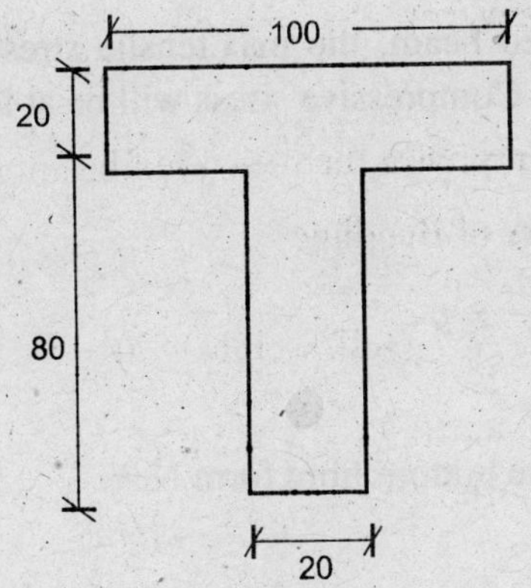

Fig .2.76

(a) Determine the max.stress in the beam when a bending moment at 12000Nm is acting on the section.

Solution

Let $\bar{y}$ be the distance of C.G. of the section from the bottom

$$\bar{y} = \frac{A_1 y_1 + A_2 y_2}{A_1 + A_2}$$

$$= \frac{(100 \times 20)\left(80 + \frac{20}{2}\right) + (80 \times 20)\left(\frac{80}{2}\right)}{(100 \times 20) + (80 \times 20)}$$

$$\bar{y} = 67.78mm$$

$\therefore$ Neutral axis lies at a distance of 67.78mm from the bottom face (or)$100 - 67.78 = 32.22$mm from Top.

Now moment of Inertia of the section about N.A

$$I = I_1 + I_2$$

$$= \frac{b_1 {d_1}^3}{12} + a_1 {h_1}^2 + \frac{b_2 {d_2}^3}{12} + a_2 {h_2}^2$$

$$= \frac{100 \times 20^3}{12} + 100 \times 20 \times \left(32.23 - \frac{20}{2}\right)^2$$

$$+ \frac{20 \times 80^3}{12} + (20 \times 80)\left(67.78 - \frac{80}{2}\right)^2$$

$$= 3142222.4\,mm^4 = 3.14 \times 10^6 mm^4$$

For a simply supported beam, the max.tensile stress will be at the extreme bottom fibre and max. Compressive stress will be at the extreme top fibre.

We know that, $\frac{M}{I} = \frac{\sigma}{y}$

Bending stress, $\sigma = \frac{M}{I} \times y$

For max. tensile stress,

y = Distance of extreme bottom fibre form N.A.

y = 67.78 mm

$$\therefore \ \sigma = \frac{1200/10^3}{3142222.4}$$
$$= \frac{12000 \times 10^3 \times 67.78}{3.14 \times 10^6}$$
$$\boxed{\sigma = 259N/mm^2}$$

For max. compressive stress,

y = Distance of extreme top fibre from N.A = 32.22mm

[∵ 100 - 67.78 = 32.22]

$$\therefore \ \sigma = \frac{M}{I} \times y$$
$$= \frac{1200 \times 10^3 \times 32.22}{3.14 \times 10^6}$$
$$\boxed{\sigma = 123.13N/mm^2}$$

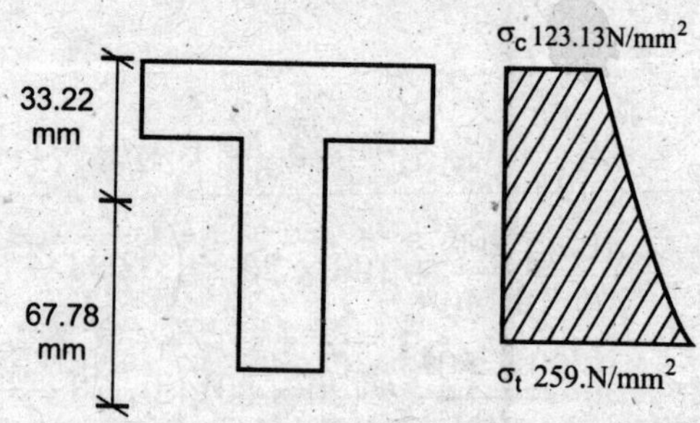

Fig .2.77

26. **An I- section with Rectangular ends has dimensions as shown in fig.**

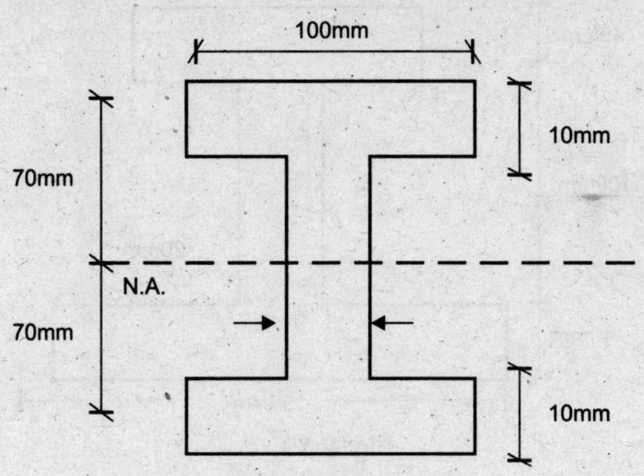

Fig .2.78

If this section is subjected to a bending moment of 6 kNm and a shearing force of 5kN. Find the max. Tensile and shear stress on it.

Solution

Moment of Inertia about N.A.

$$I = \frac{100 \times 140^3}{12} - \frac{90 \times 120^3}{12} = 9.9 \times 10^6 mm^4$$

Max. tensile stress be σ_t

$$\frac{M}{I} = \frac{\sigma_t}{y} \quad \sigma_t = \frac{M}{I} \times y = \frac{6 \times 10^6 \times 70}{9.9 \times 10^6} = 42.4 N/mm^2$$

$$\sigma_t = 42.4 \times 10^3 kN/m^2$$

shear stress is maximum in the web at N.A.

$$\tau_{\max} = \frac{SA\bar{y}}{IB}$$

$$A\bar{y} = (100 \times 10) \times \left(60 + \frac{10}{2}\right) + (10 \times 60)\left(\frac{60}{2}\right)$$

$$= 83000 mm^3$$

$$b = 10mm$$

$$\therefore \ \tau_{\max} = \frac{5 \times 10^3 \times 83000}{9.9 \times 10^6 \times 10}$$

$$\boxed{\tau_{\max} = 4.19 N/mm^2}$$

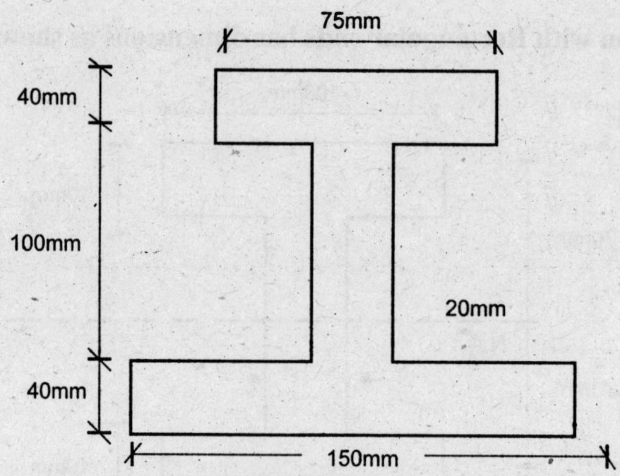

Fig .2.79

27. **A beam has an I-section with dimensions as shown in fig.**

If tensile stress is not to exceed 30N/mm^2 and compressive stress 80N/mm^2. What is the max. UDL, the beam can carry over a simply supported span of 6m.

Solution

$$\bar{y} = \frac{a_1y_1 + a_2y_2 + a_3y_3}{a_1 + a_2 + a_3}$$

$$a_1 = 75 \times 40 = 3000mm^2$$

$$a_2 = 20 \times 100 = 2000mm^2$$

$$a_3 = 150 \times 40 = 6000mm^2$$

$$y_1 = 40 + 100 + \frac{40}{2} = 160mm$$

$$y_2 = 40 + \frac{100}{2} = 90mm$$

$$y_3 = \frac{40}{2} = 20mm$$

$$\bar{y} = \frac{(3000 \times 160) + (2000 \times 90) + (6000 \times 20)}{3000 + 2000 + 6000}$$

$$\boxed{\bar{y} = 70.9mm}$$

$$I = \frac{b_1 {d_1}^3}{12} + a_1 {h_1}^2 + \frac{b_2 {d_2}^3}{12} + a_2 {h_2}^2 + \frac{b_3 {d_3}^3}{12} + a_3 {h_3}^2$$

$$\therefore \ h_1 = \bar{y} \sim y_1 = 70.9 \sim 160 = 89.1mm$$

$$h_2 = \bar{y} \sim y_2 = 70.9 \sim 90 = 19.1mm$$

$$h_3 = \bar{y} \sim y_3 = 70.9 \sim 20 = 50.9mm$$

$$\therefore I = \frac{75 \times 40^3}{12} + (3000 \times 89.1^2) + \frac{20 \times 100^3}{12} + (2000 \times 19.1^2) + \frac{150 \times 40^3}{12} + (6000 \times 50.9^2)$$

$$I = 42.95 \times 10^6 mm^4$$

In simply supported beam, tension occurs@ bottom and compression @ top.

$$y_t = \bar{y} = 70.9mm$$

$$y_c = 180 - 70.9 = 109.1mm$$

$$\frac{M}{I} = \frac{\sigma}{y}$$

On considering tensile strength, moment

$$M = \sigma_t \times \frac{I}{y_t} = 30 \times \frac{42.95 \times 10^6}{70.9}$$

$$= 18.17 \times 10^6 \Rightarrow 18.17kNm$$

On considering compressive stress

$$M = \sigma_c \times \frac{I}{y_c}$$

$$= 80 \times \frac{42.95 \times 10^6}{109.1} = 31.49 \times 10^6 Nmm$$

$$= 31.49kNm$$

Actual moment carrying capacity is the smaller value.

Max.BM in a simply supported beam of span 6m due to UDL of W kN/m is,

$$\frac{w \times 6^2}{8} = 18.17$$

$$W = \frac{18.17 \times 8}{6^2}$$

$$\boxed{w = 4.04kN/m}$$

Problems on Flitched Beams

28. **A flitched beam consist of a wooden joist 100mm wide and 240mm deep and it is strengthened by steel plates 10mm thick and 240mm deep on either side of the joist. Determine the moment of resistance. The max.stress in steel is $\sigma_s = 140N/mm^2$ and the max stress in wood, σ = 7N/mm^2**

Solution

Moment of resistance of the wooden joist.

$$M_w = \sigma_w z_w = \sigma \times \frac{bd^2}{6}$$
$$= 7 \times \frac{100 \times 240^2}{6}$$
$$= 6.72 \times 10^6 N\ mm = 6.72kNm$$

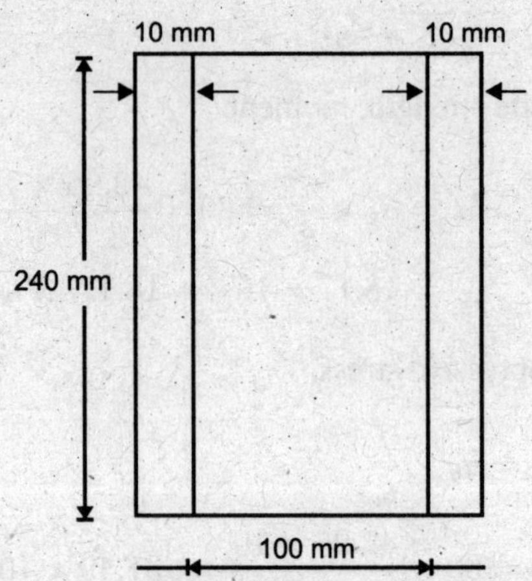

Fig .2.80

Moment of resistance of steel plate

$$M_s = \sigma_s z_s = \sigma_s \times \frac{2td^2}{6} \quad (z_s = \frac{td^2}{6}\text{section modulus})$$
$$= 140 \times \frac{2 \times 10 \times 240^2}{6}$$
$$= 26.89 \times 10^6 Nmm = 26.89kNm$$

Moment of resistance of flitched section = Moment of Resistance offered by wood + Moment of resistance offered by steel

$$M_F = M_w + M_s$$
$$= 6.72 + 26.89$$
$$\boxed{= 33.61kNm}$$

29. **A flitched beam consist of a wooden joist 150mm width and 300mm depth and strengthened by a steel plate 12mm thick and 300mm deep one on either side of the joist. If the maximum stress in the wooden joist is 7N/mm^2, find the corresponding maximum stress attained in steel. Find also the moment of resistance of the section take if E_s = 20 Eω.**

Given Data

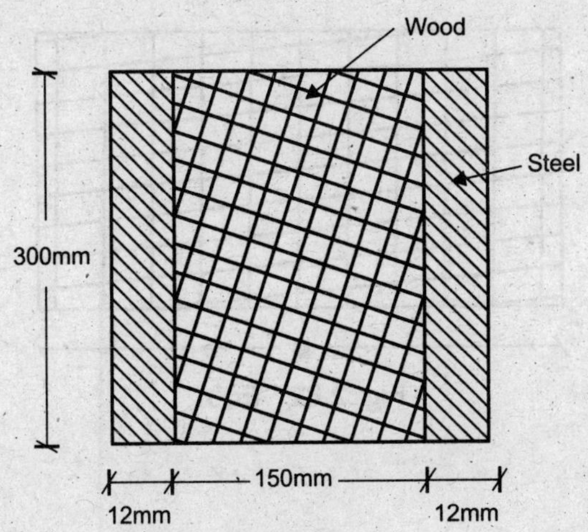

Fig .2.81

Solution

Maximum stress in wood ($\sigma\omega$) = $7N/mm^2$

Maximum stress in steel$(\sigma_s) = 20 \times 7 = 140N/mm^2 \quad (\because \quad E_s = 20E_w)$

Moment of resistance in wooden joint $= \dfrac{\sigma_w bd^2}{6}$

$$= \frac{7 \times 150 \times 300^2}{6} = 157.75 \times 10^6 Nmm = 15.75KNm$$

Moment of resistance on steel joist

$$= 2.\frac{\sigma_s bd^2}{6} = \frac{2 \times 140 \times 12 \times 300^2}{6} = 50.4 \times 10^6 N/mm = 50.4kNm$$

$\therefore$ Total Moment of resistance $= 15.75 + 50.4 = 66.15kNm$

Now, Width of the equivalent wooden section (B)

$$= b + m(2t)$$
$$= 150 + 20(2 \times 12) = 630mm$$

($\because$ Moment of resistance of equivalent wooden section has to be equal to the sum of the moment of resistance of the wooden section and steel section.)

$\therefore$ Moment of Resistance of equivalent section $= \frac{\sigma_w}{6} Bd^2$

Where B is illustrated in the following fig.which is without steel strip fixed to it both the ends.

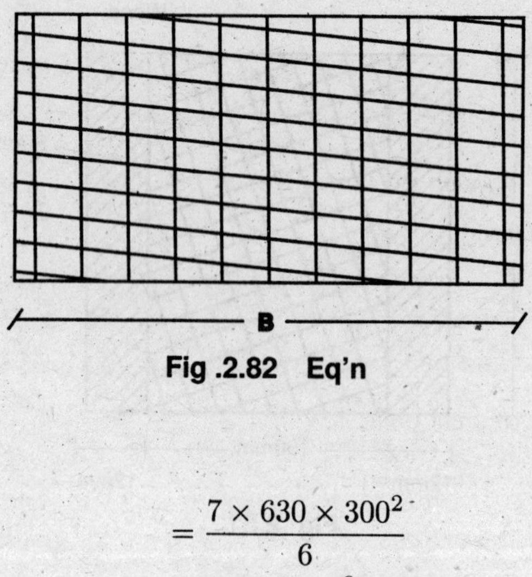

Fig .2.82 Eq'n

$$= \frac{7 \times 630 \times 300^2}{6}$$
$$= 66.15 \times 10^6 N.mm$$
$$= 66.15kN$$

30. **A flitched beam consists of two timber joist 100mm width and 240mm depth with a steel plate 180mm deep and 10mm thick placed symmetrically between the timber joist and well clamped. Determine if the modular ratio is 18.**

    (a) The maximum fibre stress in steel when the fibre stress in wood is 80 kg/cm^2

    (b) The combined moment of resistance.

Given Data

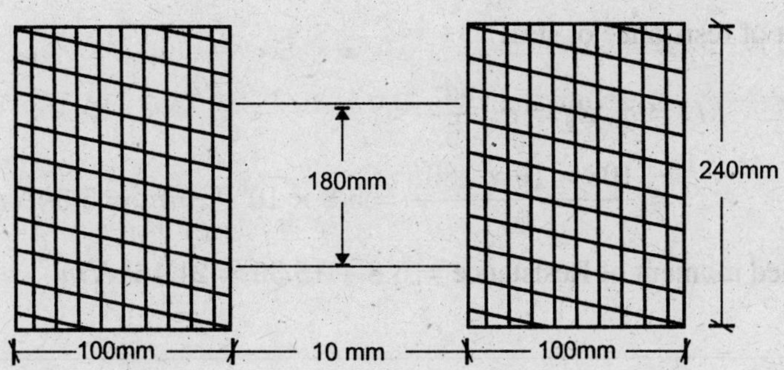

Fig .2.83

The problem can be understood well if a diagram is drawn based on the data given as below.

Solution

Maximum fibre stress (σ_w) in wood = 80 kg/cm^2 = 800 N/cm^2 = 8 N/mm^2

If the Depth of wooden joist and steel strip are not equal, then maximum stress in steel can be given by the following equation. $(\sigma_s) = m.\sigma_w.\frac{d}{D}$

Where,

'm' is the modular ratio

'd' is the depth of steel plate

'D' is the depth of the wooden joist

On substituting the values.

$$\therefore \ \sigma_s = 18 \times 8 \times \frac{180}{240} = 108N/mm^2$$

Moment of resistance of wooden joist

$$M_w = \sigma_w.Z_w$$

$$= \sigma_w \times 2 \times \frac{bd^2}{6}$$

$$= \frac{8 \times 2 \times 100 \times 240^2}{6}$$

$$= 15.36 \times 10^6 N.mm \simeq 15.36kNm$$

Moment of resistance of steel

$$M_s = \sigma_s.Z_s = \sigma_s.\frac{bd^2}{6}$$

$$= \frac{108 \times 10 \times 180^2}{6} = 5.8 \times 10^6 N.mm = 5.8kNm$$

Combined moment of Resistance $= 5.8 + 15.36 = 21.16kNm$

Review Questions

PART A

1. What is the differential relation between bending moment, shear force and the applied load?
2. Sketch the shear stress variation for symmetrical I section
3. What do you meant by point of contra flexure?
4. Write any four assumptions in the theory of simple bending
5. Differentiate between hogging and sagging bending moment.
6. A cantilever beam of span 4m is subjected to auld of 2 kN/m over its entire length.Sketch the bending moment diagram for the beam.
7. How would you find the bending stress in unsymmetrical sections?
8. How do you locate the point of maximum bending moment?
9. What do you understand by neutral axis & moment of resistance? How do you locate Neutral axis?
10. A beam subjected to a bending stress of 5N/mm^2 and the section modulus is 3530 cm^3.What is the moment of resistance of the beam?

PART B

1. A simply supported beam of length 10m carries the uniformly distributed load and two point loads as shown in Fig. Draw the S.F and B.M diagram for the beam and also calculate the maximum bending moment.
2. A rectangular beam 300mm deep is simply supported over the span of 4m. Determine the uniformly distributed load per metre which the beam may carry, if the bending stress should not exceed 120N/mm^2. Take $I = 8 \times 10^6 mm^4$.
3. A simply supported beam of span 4m carries audl of 6kN/m over the entire span. If the maximum allowable stress due to bending is restricted to 150 N/mm^2, determine the cross sectional dimensions if the section is;

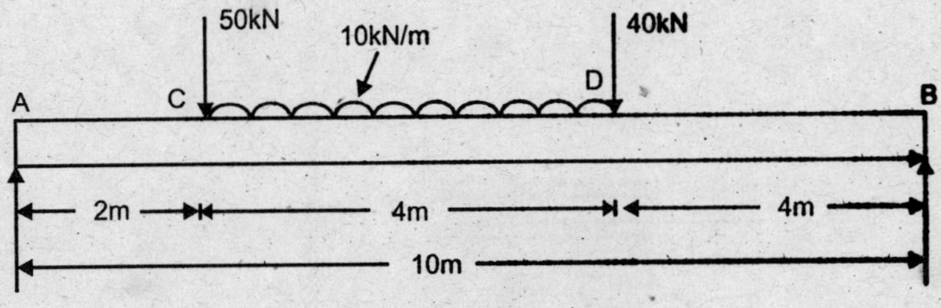

Fig .2.84

(a) Rectangular with depth twice the breadth
(b) Solid circular section
(c) Hollow circular section having a diameter ratio of 0.6

4. Draw shear force and **bending moment diagram for the beam shown in Fig.**

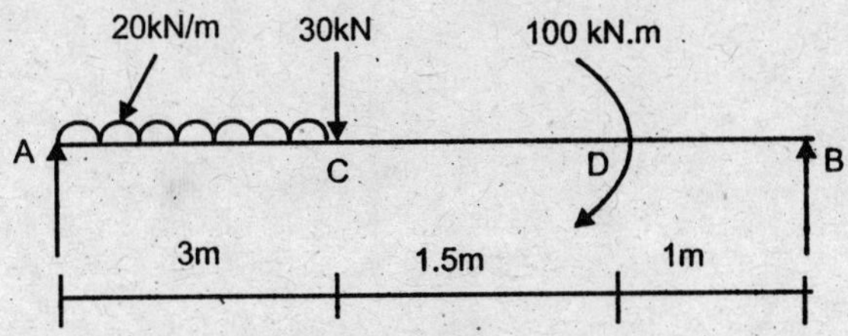

Fig .2.85

5. A flitched beam consists of two timber joist 100mm wide and 240mm deep with a steel plate 180mm deep and 10mm thick placed **symmetrically** be-tween the timber joists and well clamped. Determine

   (a) The maximum fibre stress when the maximum fibre **stress in wood** is 80 kg/cm^2.
   (b) The combined moment of resistance if the modular **ratio is** 18.

6. The cross section of T beam is as follows: Flange thickness = **10mm;** width of the flange= 100mm; thickness of the web = 10mm; depth of the web = 120mm; If a shear force of 2kN is acting at a particular section of the beam draw the shear stress distribution across the section.

7. An overhanging beam ABC is simply supported at A & B **over a span** of 6m and B Cover hangs by 3m. If the supported span AB **carries a central** concentrated load of 8kN and over hang span BC carries **2kN/m draw the** shear force and bending moment diagram.

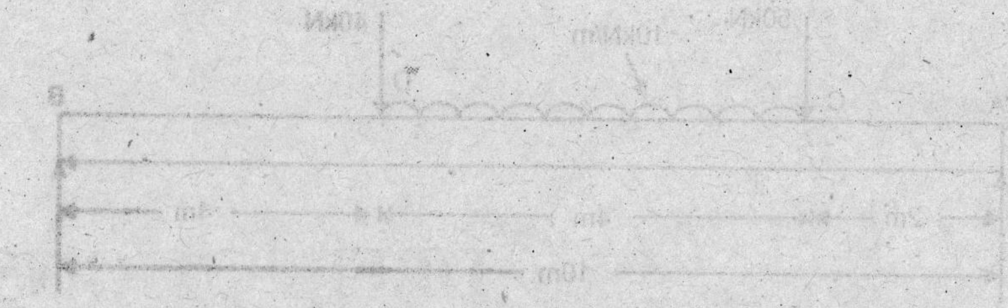

Chapter 3

DEFLECTIONS

Double integration method- Macaulay's methods - Area moment method - conjugate beam method for computation of slopes and deflections of determinant beams.

3.1 Introduction

External loads upon a beam make the beam to deflect. Unlike the shear force and bending moment, though the deflection can be measured directly by fixing dial gauge, it is necessary for a designer to compute it theoretically.

There are two important criteria available as of design of a beam is concerned. They are

(i) Strength criterion (ii) Stiffness criterion

As per the strength criterion, the beam cross section should be sufficient enough so as to resist the bending and shear stresses. Whereas the stiffness criterion warrants that the deflection caused in a beam because of external load should be within the permissible limit for safety purpose. This expects the beam to be stiff enough. Deflection is normally mentioned in terms of millimeter. Another important co-existing aspect along with deflection parameter is slope, which refers to the angle of rotation at the end support.

3.2 Relationship between Slope, Deflection, Bending Moment, Shear force and rate of loading and radius of curvature of a beam (Governing differential equations)

Under the action of beam, the beam will deflect and the deflection profile of a beam is known as its elastic curve. If the beam is subjected to pure bending, it will bend as a circular arc and the radius of bending or the radius of curvature is given by $\frac{M}{I} = \frac{E}{R}$.

Therefore $R = \frac{EI}{M}$.

If the beam is not subjected to pure bending, the radius of curvature will not be constant at all the points.

Now, let us consider a segment of deflected (Curved) portion of beam 'PQ' of infinitesimal length 'ds' as shown in Figure 3.1. Let 'R' be the radius of curvature and '$d\theta'$' be the subtended angle of segment 'PQ'', then the length $ds = R.d\theta$

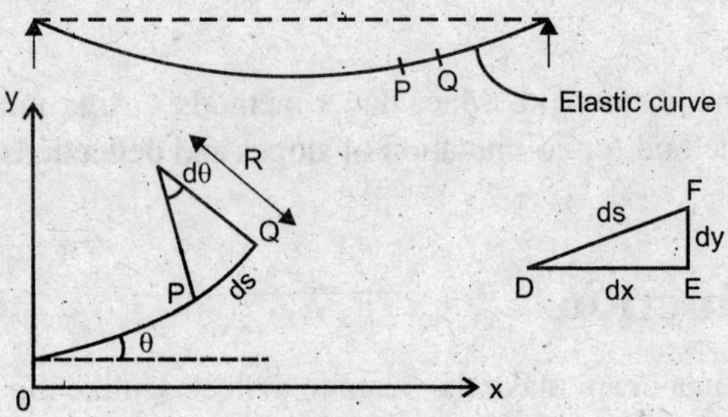

Fig .3.1

As ds is an infinitesimal length, it can be assumed to be hypotenuse of a right angle DEF as shown in the figure. The slope of the curve at the point 'P' with coordinates (x, y) is given by $\tan\theta = \frac{dy}{dx} \rightarrow (1)$.

Where dx and dy represent the projected lengths of the segment ds along X and Y- axes respectively. Differentiating (1) with respect to x.

$$\sec^2\theta.\frac{d\theta}{dx} = \frac{d^2y}{dx^2}$$

(or)
$$\sec^2\theta.\frac{ds}{R}.\frac{1}{dx} = \frac{d^2y}{dx^2}$$

(or)
$$\frac{\sec^2\theta}{R} = \frac{d^2y}{dx^2} \left(\because \frac{ds}{dx} = \sec\theta\right)$$

(or)
$$\frac{d^2y}{dx^2} = \frac{(1+\tan^2\theta)^{\frac{3}{2}}}{R}$$

Normally the radius of curvature 'R' is very large, when compared to span of the beam and therefore $\tan\theta$ or dy/dx, the slope of the tangent to the curve at any point is very very small. Further the square of such a small value be further small and hence it can be neglected.

$$\therefore \quad \boxed{\frac{d^2y}{dx^2} = \frac{1}{R} = \frac{M}{EI}}$$

The above equation is the governing differential equation of the beam and takes into account the effect of bending moment only. The effect of shear on the deflection is so small and so it is neglected.

It can be inferred from the above governing differential equation, the bending the moment borne by the beam element is equal to EI (flexural rigidity). More the EI, more is moment.

On differentiating the above governing equation along the length of beam 'x' we get shear force.

$$\frac{d_M}{d_x} = EI\,\frac{d^3y}{dx^3}$$

Now, Recalling from the theory of simple bending, $\dfrac{d_M}{dx} = F$

Thus, $$F = EI\,\frac{d^3y}{dx^3}$$

On differentiating the above shear force expression, we get the rate of loading

$$i.e \quad = \frac{d_F}{d_x} = W$$

$$\therefore \quad W = EI\,\frac{d^4y}{dx^4}$$

Therefore if 'y' is the deflection of the beam which is loaded, following relationships between, slope, bending moment, shear force and load intensity can be arrived.

$$\boxed{\text{Slope} = \frac{dy}{dx}}$$

$$\boxed{\text{B.M} = EI\,\frac{d^2y}{dx^2}}$$

$$\boxed{\text{Shear force (SF)} = EI\,\frac{d^3y}{dx^3}}$$

$$\boxed{\text{Load distribution (W)} = EI\,\frac{d^4y}{dx^4}}$$

3.3 Various methods for finding the slope and deflection of beam

The slope and deflection of the loaded beam can be obtained by various methods. They are as below.

1. Double integration method
2. Macaulay's method
3. Moment area method
4. Conjugate beam method

3.4 Double integration method

In this method, the governing differential equation of deflection is integrated two times. At the end of the first integration, slope is obtained. On further integration deflection value is derived.

$$M = EI\frac{d^2y}{dx^2} \; or \; \frac{M}{EI} = \frac{d^2y}{dx^2}$$

On integrating we get, slope equation

$$i.e., \frac{dy}{dx} = \int \frac{M}{EI}dx + A$$ this equation gives the slope of the Loaded beam

Integrating once again, we get the deflection.

$$i.e., \quad y = \iint \frac{M}{EI}\, dx + Ax + B$$

Where A and B are constants of integration to be evaluated from the known conditions of slope and deflections for the particular value of x.

Double Integration method for different beams under different loading conditions

1. **Cantilever Beam Subjected to Point Load at Free End**

   Consider a cantilever beam of span 'ℓ' subjected to concentrated load 'W' at the free end B as shown in Figure 3.2

   Consider a section $x - x$ on the beam at a distance x from end A

   B.M at section $x - x = -W\,(\ell - x)$

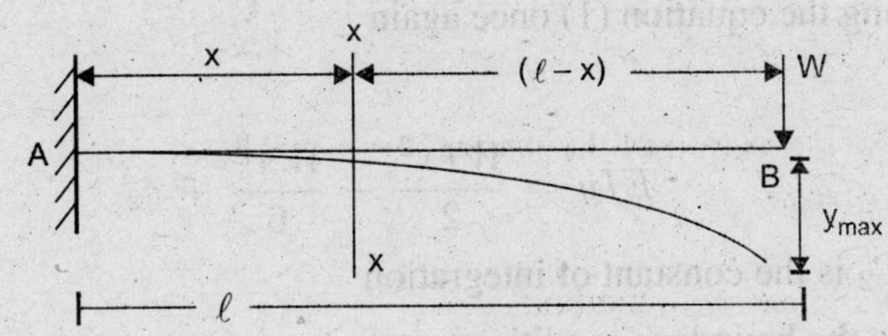

Fig .3.2

Substituting B.M value in the differential equation

$$EI\,\frac{d^2y}{dx^2} = W\,(\ell - x) = W\ell - Wx$$

[$-ve$ sign in the fundamental equation is retained as we have located the section $x - x$ from B, By measuring $[\ell - x)]$

Integrating the above equation once,

$$EI \cdot \frac{dy}{dx} = Wlx - \frac{Wx^2}{2} + c_1 \qquad (1)$$

Where 'C' is the constant of Integration.

We know that, in case of cantilevers, the slope at the fixed end is zero, and the boundary conditions are as below.

$$\therefore At \;\; x = 0\,; \; \frac{dy}{dx} = 0$$

Substituting in equation (1)

$$0 = 0 - 0 + c_1 \; \therefore c_1 = 0$$

Substituting for $C_1 = 0$, We get,

$$\therefore \quad EI\,\frac{dy}{dx} = W\ell x - \frac{Wx^2}{2}$$

$$\text{At}\, x = ;\, \ell, \quad \frac{dy}{dx} = \theta_B \text{ (Free end slope)}$$

Substituting in eqn.(1), we get

$$\therefore EI\theta_B = W\ell^2 - \frac{W\ell^2}{2}$$

$$\therefore\, \theta_B = \frac{W\ell^2}{2EI}$$

Integrating the equation (1) once again

We get

$$EIy = \frac{W\ell x^2}{2} - \frac{Wx^3}{6} + c_2$$

Where C_2 is the constant of integration

We know the boundary conditions

$$At \quad x = 0,\ y = 0$$

Substituting in equation (2), we get

$$\therefore \quad 0 = 0 + 0 + C_2 : C_2 = 0$$

We know the general equation for deflection is

$$EIy = \frac{W\ell x^2}{2} - \frac{Wx^3}{6}$$

We know the boundary conditions

$$At\, x = \ell,\ y = y_{\max}$$

$$\therefore EIy_{\max} = \frac{W\ell^3}{2} = \frac{W\ell^3}{6} = \frac{W\ell^3}{3}$$

$$\boxed{\therefore y_{\max} = \frac{W\ell^3}{3EI}}$$

2. **Cantilever Beam Subjected to UDL Over Entire Span**

   Let us consider a cantilever beam of span ℓ subjected to UDL of intensity 'W' kN/m as shown in Figure 3.3.

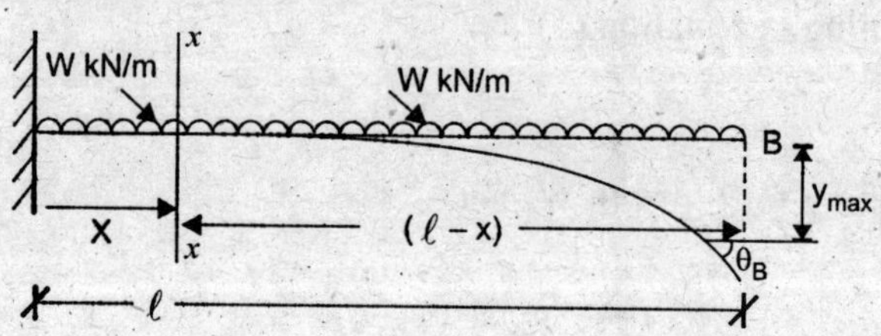

Fig .3.3

BM at section x - x

$$= \frac{W(\ell - x)(\ell - x)}{2}$$

$$BM = \frac{W(\ell - x)^2}{2}$$

Substituting 'BM' equation in the differential equation

$$EI\frac{d^2y}{dx^2} = \frac{W(\ell - x)^2}{2} = \frac{W\ell^2}{2} - w\ell x + \frac{Wx^2}{2} \rightarrow \quad (1)$$

Integrating the equation above

$$EI \cdot \frac{dy}{dx} = \frac{W\ell^2 x}{2} - \frac{W\ell x^2}{2} + \frac{Wx^3}{6} + C_1$$

Where 'C', is the constant of Integration

$$\text{At} \quad x = 0, \frac{dy}{dx} = 0\, C_1 = 0$$

General equation for slope is

$$EI \cdot \frac{dy}{dx} = \frac{W\ell^2 x}{2} - \frac{W\ell x^2}{2} + \frac{Wx^3}{6} \rightarrow \quad (2)$$

Considering the boundary condition,

$$\text{At}\, x = \ell; \frac{dy}{dx} = \theta_3$$

Substituting in equation (2)

$$\therefore EI\theta_B = \frac{W\ell x^2\theta_B}{2} - \frac{W\ell^3}{2} + \frac{Wx^3}{6}$$

$$\Rightarrow\therefore EI\theta_B = \frac{W\ell^3}{2} - \frac{W\ell^3}{2} + \frac{W\ell^3}{6}$$

$$\therefore \quad \theta_B = \frac{W\ell^3}{6EI}$$

Integrating equation(1) once again

$$EIy = \frac{W\ell^2 x^2}{4} - \frac{W\ell^2 x^3}{6} + \frac{Wx^4}{24} + C_2$$

Where 'cz' is the considering the boundary condition,

$$\text{At} \quad x = 0, y = 0 \quad \therefore \quad C_2 = 0$$

The general equation for deflection is

$$EIy = \frac{W\ell^2 x^2}{4} - \frac{W\ell x^3}{6} + \frac{Wx^4}{24} + C_2$$

Where 'C_2' is the constant of Integration

Considering the boundary condition,

$$\text{At } x = \ell,\ y = y_{\max}$$

$$\therefore EIy_{\max} = \frac{W\ell^2x^2}{4} - \frac{W\ell x^3}{6} + \frac{Wx^4}{24}$$

$$\text{At } x = \ell;\ y = y_{\max}$$

$$\therefore EIy_{\max} = \frac{W\ell^4}{4} - \frac{W\ell^4}{6} + \frac{W\ell^4}{24}$$

$$= \frac{6W\ell^4 - 4W\ell^4 + W\ell^4}{24} = \frac{3W\ell^4}{24}$$

$$\boxed{\therefore\ y_{\max} = \frac{W\ell^4}{8EI}}$$

3. **Simply supported beam loaded with the central load.**

Let us consider a simply supported beam loaded with a concentrated load 'W' as shown in Figure 3.4

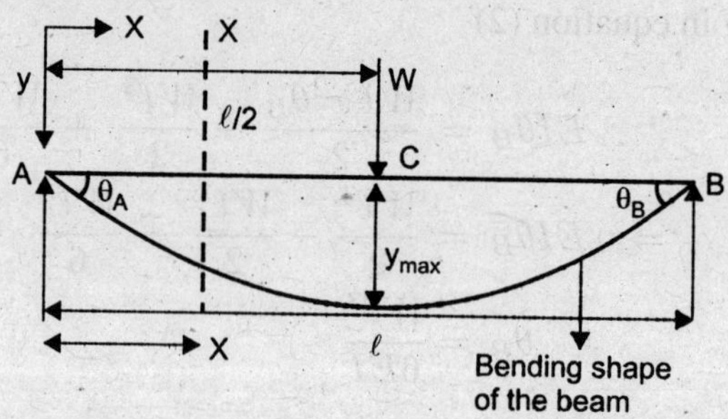

Fig .3.4

Now, consider a section x - x on the beam at distance 'x' measured in positive x direction.

Reaction of $A = \frac{W}{2}$ (Because of symmetry)

$\therefore$ BM at section x - x $= \frac{W}{2} \cdot x$

Now if we consider differential equation of elastic curve

$$EI.\frac{d^2y}{dx^2} = -M = -\frac{W}{2}x \qquad (1)$$

Integrating equation the above, we get

$$EI\,.\frac{dy}{dx} = \frac{W/2}{2} + C_1 = \frac{Wx^2}{4} + C_1 \qquad (1)$$

Where C_1 is constant of integration. Its value can be obtained by using boundary condition. By virtue of symmetry of loading, the deflection at the center of the beam was maximum. Then, a tangent drawn on the curve at this point will be horizontal.

$$i.e., \frac{dx}{dy} = 0; \text{thus at } x = \ell/2, \frac{dy}{dx} = 0$$

Considering the boundary condition for the above equation, we get

$$0 = \frac{Wl^2}{16} + C_1$$

$$\therefore C_1 = \frac{Wl^2}{16}$$

Therefore the general equation of slope is

$$EI \cdot \frac{dy}{dx} = -\frac{Wx^2}{4} + \frac{W\ell^2}{16}$$

Now,in order to obtain slope at A, put $x = 0$.

$$\therefore At\, x = 0; \frac{dy}{dx} = \theta_A$$

$$EI\theta_A = \frac{W\ell^2}{16}$$

$$\therefore \quad \theta_A = \frac{W\ell^2}{16EI}$$

$$\text{By Symmetry } \theta_B = \frac{W\ell^2}{16EI}$$

Integrating equation (1) yet again,we get

$$EIy = -\frac{Wx^2}{12} + \frac{W\ell^3}{16}x + C_2 \qquad (2)$$

Again 'C_2' is constant of integration, to be determined using boundary condition At $x = 0$ [at support point A], there can be no deflection,

$$\text{i.e,} \quad y = 0 = 0 + 0 + C_3$$

$$\therefore \quad C_2 = 0$$

Thus,the general equation of deflection is

$$EI \cdot y = \frac{-Wx^3}{12} + \frac{W\ell^2}{16} + C_2$$

$$\text{At } x = \frac{\ell}{2} \text{ (Center of Beam) y} = \text{y}_{\text{max}}$$

$$\therefore \; EI\, y_{\max} = \frac{W\ell^3}{96} + \frac{W\ell^3}{32} = \frac{2W\ell^3}{96} = \frac{W\ell^3}{48}$$

$$\therefore \; y_{\max} = \frac{W\ell^3}{48EI}$$

4. **Simply supported beam loaded with UDL over entire span**

Let us consider a simply supported beam of span 'ℓ' loaded with udl over the entire span as shown in Figure 3.5

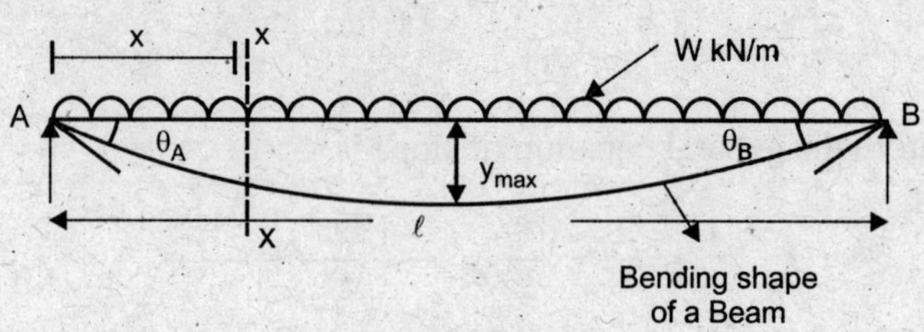

Fig .3.5

Now consider a section x - x on the beam at a distance x measured in positive 'X' direction, Reaction @ $A = \frac{WL}{2}$

$$BM \text{ @ section x} - \text{x} = M = \frac{W\ell}{2} - x - \frac{Wx^2}{2}$$

Substituting value of 'M' in differential equation.

$$EI\,.\frac{d^2y}{dx^2} = \frac{W\ell x}{2} + \frac{Wx^2}{2} \tag{1}$$

Integrating equation (1) once, we get

$$EI\,.\frac{dy}{dx} = \frac{W\ell x^2}{4} + \frac{Wx^3}{6} + C_1$$

Where C_1 is constant of integration

[As the deflection at center happens to be maximum]

$$\therefore \theta = -\frac{W\ell^3}{16} + \frac{W\ell^3}{48} + C_1$$

$$\therefore C_1 = \frac{W\ell^3}{24}$$

The general equation for slope is:

$$EI.\frac{dy}{dx} = \frac{W\ell x^2}{4} + \frac{Wx^3}{6} + \frac{W\ell^3}{24} \qquad (3)$$

$$@x = 0, \frac{dy}{dx} = \theta_A$$

$$\therefore EI\theta_A = \frac{W\ell^3}{24}$$

$$\boxed{\theta_A = \frac{W\ell^3}{24EI}}$$

$$@\, x = \frac{\ell}{2}; \frac{dy}{dx} = 0$$

Similarly, $\theta_B = \dfrac{W\ell^3}{24EI}$

Integrating equation (2) again

$$EIy = -\frac{W\ell x^3}{12} + \frac{Wx^4}{24} + \frac{W\ell^3 x}{24} + C_2$$

Where C_2 is the constant of integration

'C_2' can be determined by applying boundary conditions

$$@\, x = 0,\ y = 0 \quad \therefore\ ,0 = 0 + 0 + 0 + C_2 \quad \therefore C_2 = 0$$

The general equation for deflection now becomes

$$EIy_{Max} = -\frac{W\ell x^3}{12} + \frac{Wx^4}{24} + \frac{W\ell^3}{24}x$$

At $x = \dfrac{\ell}{2}; y = y_{\max}$

$$\therefore EIy_{max} = -\frac{W\ell^4}{96} + \frac{W\ell^4}{384} + \frac{W\ell^4}{48}$$

$$= \frac{-4W\ell^4 + W\ell^4 + 8W\ell^4}{384} = \frac{5W\ell^4}{384}$$

$$\therefore \boxed{y_{Max} = \frac{5Wl^4}{384EI}}$$

5. **Simply supported beam loaded with end moments**

Let us consider a simply supported beam loaded with equal and opposite moments M as shown in Figure 3.6.

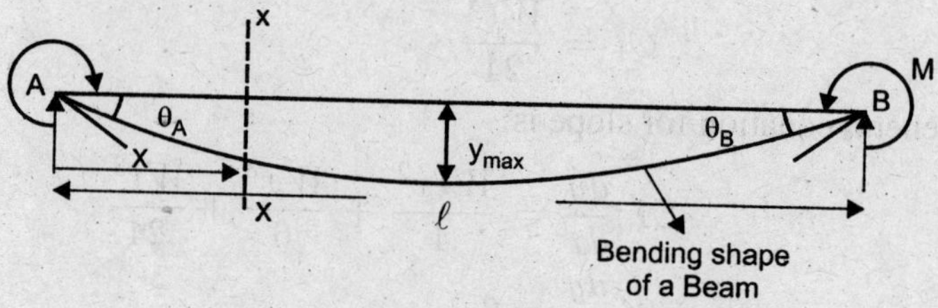

Fig .3.6

Consider a section x - x at a distance x measured from 'A' towards right.

$$R_A = R_B = 0 \text{ [As the Moments cancel away]}$$

$$\therefore BM \text{ at section } x - x = M$$

Substituting in differential equations we get;

$$EI.\frac{d^2y}{dx^2} = M \qquad (1)$$

Integrating equation (1) once;

$$EI.\frac{dy}{dx} = -Mx + C_1$$

Considering the Boundary conditions

$$\text{At } x = \frac{\ell}{2};\ \frac{dy}{dx} = 0$$

$$\therefore 0 = \frac{M\ell}{2} + C_1$$

$$\therefore C_1 = \frac{M\ell}{2}$$

The general equation for slope is

$$EI.\frac{dy}{dx} = -Mx + \frac{M\ell}{2} \qquad (2)$$

$$\text{At } \quad x = 0;\ \frac{dy}{dx} = \theta_A$$

$$\therefore EI\theta_A = \frac{M\ell}{2}$$

$$\therefore \theta_A = \frac{M\ell}{2EI}$$

$$\text{Similarly } @x = \alpha,\ \theta_B = \frac{M\ell}{2EI}$$

$$\text{Similarly at } x = \ell,\ \theta_B = \frac{M\ell}{2EI}$$

Integrating equation(2) once again

$$EI.y = \frac{Mx^2}{2} + \frac{M\ell x}{2} + C_2$$

Where 'C_2' is the constant of Integration

$$@x = 0;\ y = 0\ \therefore C_2 = 0$$

General equation for deflection is

$$EI.y = \frac{Mx^2}{2} + \frac{M\ell x}{2}$$

Considering the boundary conditions

$$At\, x = \frac{\ell}{2} : y = y_{\max} \text{ (owing to symmetry)}$$

$$\therefore EI\, y_{\max} = \frac{-M\ell^2}{8} + \frac{M\ell^2}{4} = \frac{M\ell^2}{8}$$

$$\therefore y_{\max} = \frac{M\ell^2}{8EI}$$

3.5 Macaulay's Method

If the deflections have to be computed for beams with many loads by double integration method, lengthy equation have to be solved involving many integration constants. For such cases, Macaulay's method is useful one, in which a single equation of moment is formed for all loads on the beam in such a way that the constants of integration are valid for all section beams.

3.5.1 Simply supported beam with a point load at the centre of the beam

Now, let us consider a simply supported beam with a central point load as shown below.

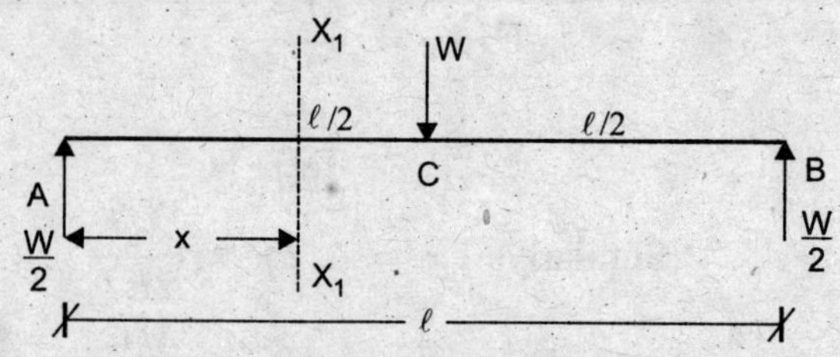

Fig .3.7

Let us cut a section $X_1 - X_1$ @ a distance x from A

$$M_X = +\frac{W}{2}x - W\left(x - \frac{\ell}{2}\right)$$

$$EI\,\frac{d^2y}{dx^2} = -M_X$$

$$= -\frac{W}{2}x + W\left(x - \frac{\ell}{2}\right)$$

Integrating equation (1) we get

$$EI\,\frac{dy}{dx} = \frac{-W}{2}\,\frac{x^2}{2} + C_1 + \frac{W}{3 \times 2}\left(x - \frac{\ell}{2}\right)^2 \tag{2}$$

Integrating equation (2) we get

$$EIy = \frac{-Wx^3}{12} + C_1x + C_2 + \frac{W}{3 \times 2}\left(x - \frac{\ell}{2}\right)^3 \tag{3}$$

Considering boundary conditions

$$@x = 0;\ y = 0 \qquad \text{Substituting in}(2)$$

$$\Rightarrow 0 = 0 + C_1\,(0) + C_2$$

$$@x = 1;\ y = 0$$

$$\boxed{0 = C_2}$$

$@x = \ell;\ y = 0$

$$O = \frac{-W\,(\ell^3)}{12} + C_1\,(\ell) + 0 + W\left(\frac{\ell}{2}\right)^3$$

$$C_1 = \frac{-W\ell^3}{8} + \frac{W\ell}{12}$$

$$EI\tfrac{dy}{dx} = \tfrac{-Wx^2}{4} + \tfrac{W\ell^2}{16}\Big| \quad + \frac{W}{2}\left(x - \frac{\ell}{2}\right)^2$$

$@x = 0$

$$EI\left(\frac{dy}{dx}\right)_A = \frac{W\ell^2}{16}$$

$$\boxed{\left(\frac{dy}{dx}\right)_A = \frac{W\ell^2}{16EI}}$$

Intergrading

$$EI.y = \frac{-Wx^3}{12} + \frac{W\ell^2 x}{16} + 0 + \frac{W}{6}\left(x - \frac{\ell}{2}\right)^3$$

$$@x = \frac{\ell}{2}$$

$$EI\, y_{\max} = \frac{-W\ell^3}{96} + \frac{W\ell^3}{32} + 0 = \frac{2W\ell^3}{96} = \frac{W\ell^3}{48}$$

$$\therefore\ y_{max} = \frac{W\ell^3}{48EI}$$

3.5.2 Simply Supported beam with an eccentric load

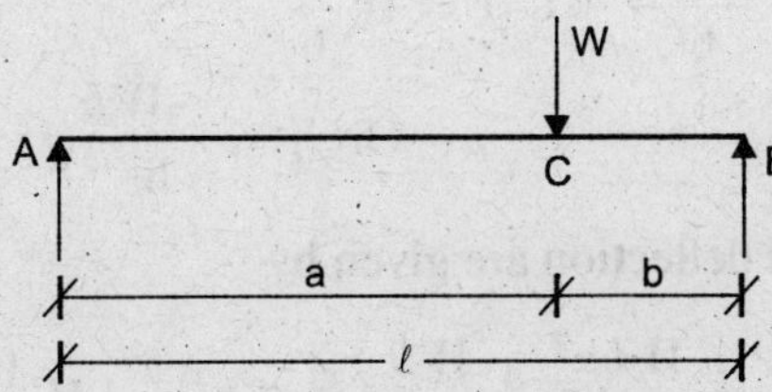

Taking moments about B,

$$R_a \cdot \ell = Wb$$

$$R_a = \frac{Wb}{\ell}$$

$$III^{ly}, R_b = \frac{Wa}{\ell}$$

Bending moment for any segment

$$= \frac{Wbx}{\ell}\Bigg| - W(x-a) \text{ (or) } EIy'' = \frac{Wbx}{\ell}\Bigg| - W(x-a)$$

The portion to the left of this line is valid for the portion AC whereas the whole expression is valid for the portion CB.

Integrating,

$$EIy' = \frac{Wbx^2}{2\ell} + C_1 \Big| - W\frac{(x-a)^2}{2}$$

Integrating again,

$$EIy' = \frac{Wbx^2}{6\ell} + C_1x + C_2 \Big| - W\frac{(x-a)^3}{6}$$

At A, $x = 0, y = 0;$

The equation for the elastic curve for the portion AC is

$$EIy = \frac{Wbx^3}{6\ell} + C_1x + C_2$$

$$\therefore \frac{Wbx^3}{6\ell} + C_1x + C_2 = 0$$

$$\text{or} \quad C_2 = 0$$

At B, $x = \ell, y = 0;$

The whole equation is valid for the portion CB.

$$\therefore 0 = \frac{Wb\ell^2}{6\ell} + C_1\ell = W\frac{b^3}{6}$$

$$\text{Or} C_1 = \frac{-Wb}{6\ell}(\ell^2 - b^2)$$

Thus, the slope and deflection are given by

$$EIy' = \frac{Wbx^2}{2\ell} - \frac{Wb}{6\ell}(\ell^2 - b^2) \Big| - W\frac{(x-a)^3}{2}$$

$$\text{and} \quad EIy = \frac{Wbx^3}{6\ell} - \frac{Wb}{6\ell}(\ell^2 - b^2)x \Big| - W\frac{(x-a)^3}{6}$$

Deflection under the load,

$$EIy = EIy' = \frac{Wba^2}{6\ell} - \frac{Wb}{6\ell}(\ell^2 - b^2)\,a$$

$$= -\frac{Wab}{6\ell}(\ell^2 - b^2 - a^2)$$

$$= \frac{-Wab}{6\ell}\left[(a+b)^2 - b^2 - a^2\right]$$

$$= \frac{-Wab}{6\ell}\left[a^2 + b^2 + 2ab - b^2 - a^2\right]$$

3.6 Moment - Area Methods - Mohr's Theorem

This method is based on the Mohr's first and second moment area theorem.

Mohr's I Theorem of moment area method (For finding slope)

Mohr's I theorem of moment area method states that the difference (Change) of slope between any two points on an elastic curve of a beam is equal to the net area of the bending moment diagram between these two points divided by flexural rigidity (EI).

Mohr's II Theorem of moment area method (For finding deflection)

Mohr's II theorem of moment area method states that the total deflection between any two points on an elastic curve of a beam is equal to the net moment taken about that line of the area of the bending moment diagram between the two points divided by flexural rigidity (EI).

This method is useful in problems of the following cases.

1. Cantilever beams (Zero Slope at fixed ends)
2. Symmetrically loaded simply supported beams (Zero slope at the centre)
3. Fixed beams (Built beams) (Zero slope at each ends)

3.7 Conjugate Beam Method

This method is useful in the case of beams where the flexural rigidities are different along the length of the beam.

Conjugate Beam is an imaginary beam of length equal to that of original beam but the load upon that imaginary beam is equal to the bending moment at the point (M) divided by flexural rigidity (EI) i.e. load is M/EI.

The slope and deflection can then be computed by the following two theorems which are also known as Mohr's theorem for conjugate beams.

Mohr's I Theorem of conjugate beam (For Computing slope)

Mohr's I theorem of conjugate beam states that the slope at any section for the given beam is equal to the shear force at the corresponding section of the conjugate Beam.

Mohr's II Theorem of conjugate beam (For Computing deflection)

Mohr's II theorem of conjugate beam states that the deflection at any section for the given beam is equal to the bending moment at the corresponding section of the conjugate Beam.

Part - A

Two Marks Questions and Answers

1. **State the Significance of deflection**

   If deflections in beams are beyond the allowable limits, they will cause cracks in beams. Further presence of deflection gives an indication to the occupants of the building regarding the safety of such structural members.

2. **Define Strength and stiffness criterion**

   Designing a beam for bending and shear stress consideration is called as strength criterion; whereas designing a beam for deflection aspect is known as stiffness criterion.

3. **Define elastic curve**

   Deflection profile of a beam is known as its elastic curve.

4. **Write the Governing differential equation of a beam under the action of bending moment.**

   $$EI.\frac{d^2y}{dx^2} = M$$

5. **What are the different methods for finding out the slope and deflection of determinant beams?**

   i) Double integration method
   ii) Macaulay's method
   iii) Moment area method
   iv) Conjugate beam method

6. **State Double Integration Method**

   In double Integration method, the equation of the elastic curve is integrated twice to obtain deflection of the beam at any cross section. The Integration constants are arrived by applying the boundary (end) conditions. The equation is arrived in such a way that the integration constants are applicable throughout length of the beam. This method is convenient method for simple cases (Single loading).

7. **Write the significance of Macaulay's Method**

   In Macaulay's method, a Single bending moment equation is arrived for all the segment of beams. The equation is arrived in such a way that the integration constants are applicable throughout the length of beam. This is useful method for complex cases (Many loadings).

8. **What are the units of slope and deflection?**

Unit of slope is radians and the unit of deflection is mm.

9. **State Flexural rigidity**

The Product EI is an index of bending (flexural) strength of an element and is called as the flexural rigidity.

10. **What are the assumptions of double integration method?**

i) The equation is based on the bending moment
ii) The effect of shear force is very small and thus neglected.
iii) Beams are uniform
iv) Inertia is uniform
v) Material is homogenous.

11. **Name the method which employs bending moment diagram for the calculation of slope and deflection.**

Moment area method.

12. **State Mohr Theorems of moment area method.**

Mohr's Theorem of moment area method (For computing slope)

The change of slope between any two points is equal to the net area of the Bending moment diagram between those points divided by flexural rigidity "EI".

Mohr's II Theorem of moment area method (For computing deflection)

The total deflection between any two points is equal to the moment of bending moment area between those points divided by flexural rigidity "EI".

13. **What is meant by conjugate beam and how to determine the slope and deflection of real beam from conjugate beam? Or State Mohr's theorems of conjugate beam method.**

Conjugate beam is an imaginary secondary beam, which when loaded with $\frac{M}{EI}$ diagram of the real beam yield directly the slope and deflection of the real beam in the form of shear force and deflection of the real beam in the form of bending moment of the conjugate beam.

Mohr's I theorem of conjugate beam method (For finding slope)

The slope at any section for the given beam is equal to the shear force at the corresponding section of the conjugate beam.

Mohr's II theorem of Conjugate beam method (For finding deflection)

The deflection at any section for the given beam is equal to the bending moment at the corresponding section of conjugate beam.

14. **Write the sign conventions for deflections.**

Consider a simply supported beam and a condition beam as below.

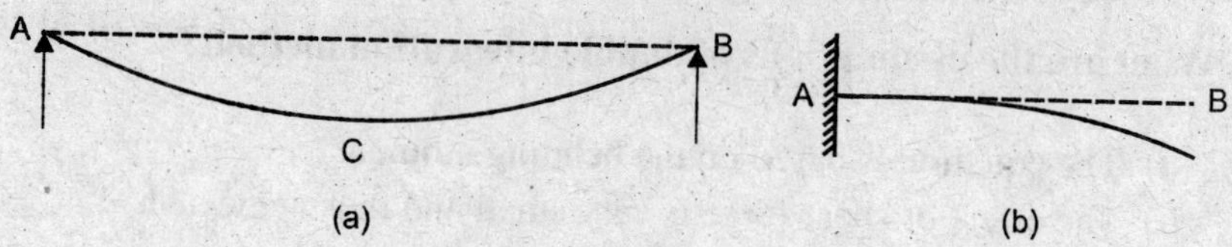

Fig .3.8

Origin @ A: If origin is taken at 'A', 'x' is considered as +ve towards right and 'y' is +ve towards up wards. Under such circumstances.

Deflection is $-ve$ in both the cases. Slope is $-ve$ between A & C and +ve between C & B for simply supported beam. For cantilever beam the slope is $-ve$ throughout.

Origin @ B: If origin is taken at B, 'x' is considered as $+ve$ towards left upwards and y is $+ve$ towards. Upwards then deflection is $-ve$ in both the cases.

The slope is $+ve$ between A & C and $-ve$ between C & B for the simply supported beam and $+ve$ for cantilever throughout.

15. **Calculate the slope and deflection at the free end of a cantilever beam of length ℓ subjected to a concentrated load (W) at the free end.**

Solution

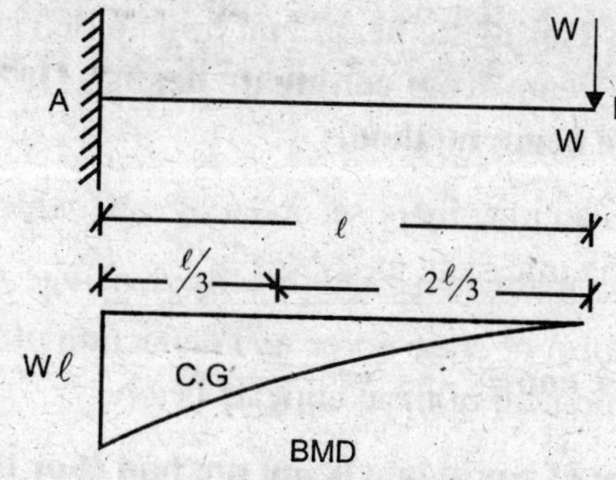

Fig .3.9

Area of the BMD $= \frac{1}{2}.\ell. \times W\ell = \frac{W\ell^2}{2}$

Slope at the free end $= -\frac{A}{EI} = \frac{-Wl^2}{2EI}$

Deflection at the free end $= \frac{A\bar{x}}{EI} = \frac{W\ell^2}{2EI}.\frac{2\ell}{3} = \frac{W\ell^3}{3EI}$

16. **Calculate the slope and deflection of a cantilever of length at the free end subjected to uniformly distributed load.**

Solution

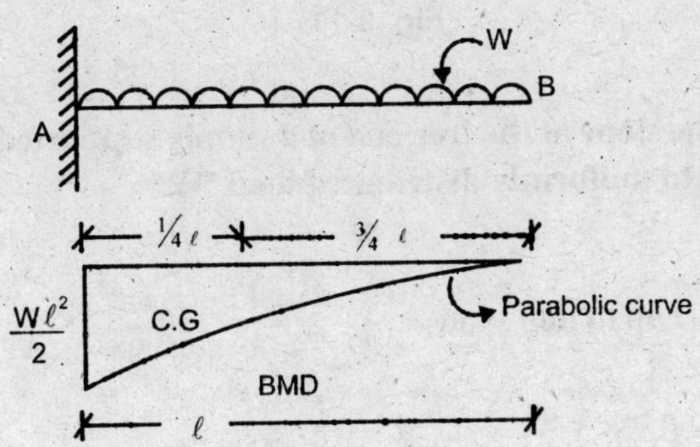

Fig .3.10

Area of the BMD $= \frac{1}{3}.\ell \times \frac{W\ell^2}{2} = \frac{W\ell^3}{6}$

Slope at the free end $-\frac{A}{EI} = \frac{-Wl^3}{6EI}$

Deflection at the free end $\frac{A\bar{x}}{EI} = \frac{W\ell^3}{6EI}.\frac{3\ell}{4} = \frac{W\ell^4}{8EI}$

17. **Calculate the slope the free end and deflection at the mid span of a simply supported beam of length ℓ subjected to a point load (W) at the mid span.**

Solution

Area of up to the mid span the BMD $= \frac{1}{2}.\frac{\ell}{2}.\frac{W\ell}{4} = \frac{W\ell^2}{16}$

Slope at the free end $= \frac{A}{EI} = \frac{-W\ell^2}{16EI}$

Let $\bar{x}$ is the distance of centroid of the BMD from $= \frac{2}{3}\frac{\ell}{2} = \frac{\ell}{3}$

$\therefore$ Deflection at mid span $= \frac{A\bar{x}}{EI} = \frac{wl^2}{16EI}.\frac{l}{3} = \frac{wl^3}{48EI}$

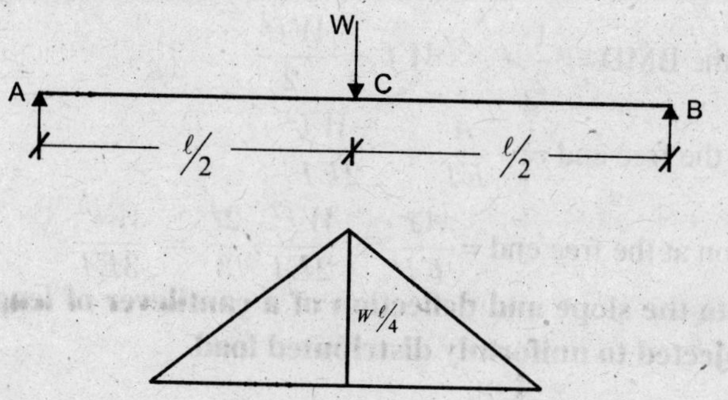

Fig .3.11

18. **Calculate the slope at the free end of a simply supported beam of length ℓ subjected to uniformly distributed load 'W'**

Solution

Area of BMD up to mid span $= \dfrac{2}{3} \cdot \dfrac{\ell}{2} \cdot \dfrac{w\ell^2}{8} = \dfrac{w\ell^3}{24}$

$\therefore$ Slope at the free end $= -\dfrac{A}{EI} = \dfrac{-W\ell^3}{24EI}$

19. **Calculate the deflection of simply supported beam of length ℓ at the mid span subjected to UDL.**

Solution

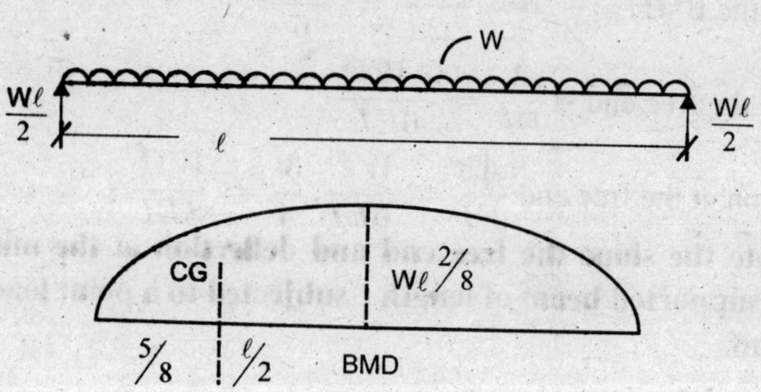

Fig .3.12

Area a BMD @ up to mid span = $A, = \dfrac{2}{3} \cdot \dfrac{\ell}{2} \cdot \dfrac{W\ell^2}{8} = \dfrac{W\ell^3}{24}$

Let $\overline{x}$ = distance of centroid of the BMD from 'A' $= \dfrac{5}{8}\dfrac{\ell}{2} = \dfrac{5\ell}{16}$

$\therefore$ Deflection at mid span $= \dfrac{A\overline{x}}{EI} = \dfrac{W\ell^3}{24EI} \cdot \dfrac{5\ell}{16} = \dfrac{5w\ell^4}{384EI}$

20. **A cantilever of length has a point load 'W' at the free end. Calculate the slope at the free end**

Solution

The Cantilever beam is as below.

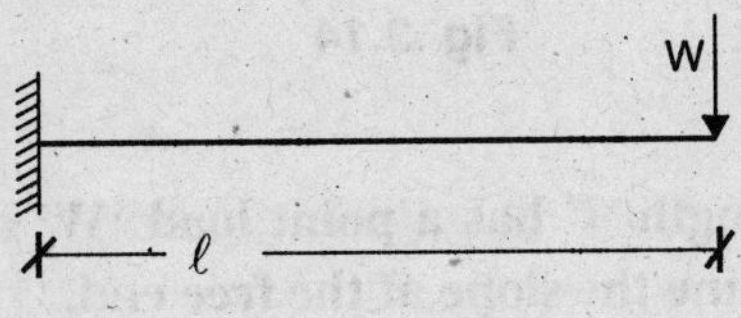

The conjugate beam is as below

Conjugate beam

Fig .3.13

Here the fixed end is transformed as free end and the free end as fixed end.

$$\therefore \text{ Slope at the free end} = \frac{W\ell}{EI} \cdot \frac{\ell}{2} = \frac{W\ell^2}{2EI}$$

21. **A cantilever of length ℓ has a point load 'W' at the free end. Calculate the deflection at the free end.**

Solution

The cantilever beam is as below.

The conjugate beam is as below.

Here the fixed end is transformed as free end and vice versa.

Deflection at the free end = Bending moment at 'B' for conjugate beam

$$= \frac{W\ell^2}{2EI} \cdot \frac{2}{3} . \ell = \frac{W\ell^3}{3EI}$$

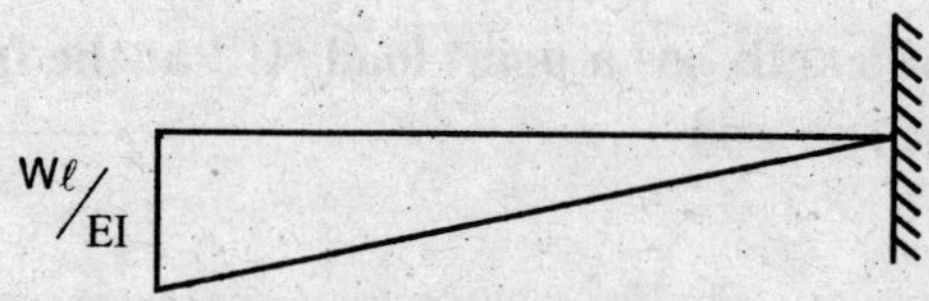

Conjugate beam

Fig .3.14

22. **A cantilever of length 'ℓ' has a point load 'W' at a distance from the fixed end. Determine the slope at the free end.**

Solution

The cantilever beam and conjugate beam are as below for this problem

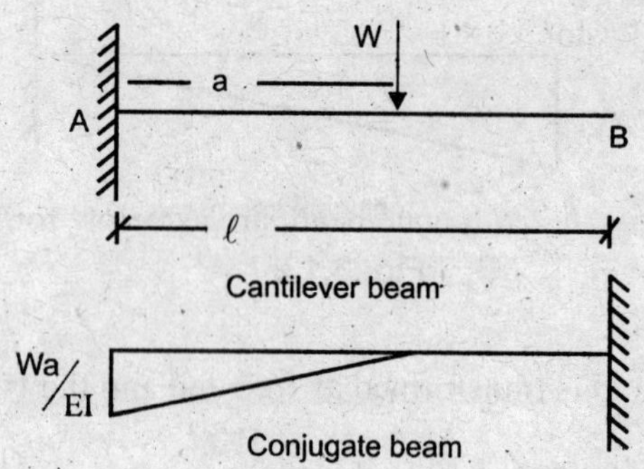

Fig .3.15

Slope at the free end shear free at 'B' = Shear force at 'B' for conjugate beam.

$$= \frac{Wa}{2} \cdot a \cdot \frac{1}{EI} = \frac{Wa^3}{2EI}$$

23. **A cantilever of length ℓ has a point load 'W' at a distance from the fixed end. Calculate the deflection at free end.**

Solution

The cantilever and conjugate beam are as below for this problem

Deflection at the free end = Bending moment at 'B' for conjugate beam

$$= \frac{Wa^2}{2EI}\left(\ell - \frac{a}{3}\right)$$

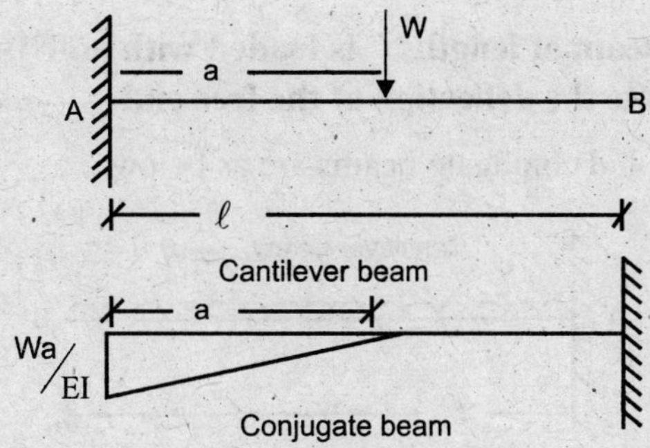

Fig .3.16

24. **A cantilever of length 'ℓ' is loaded with an UDL of 'W per unit length. Calculate the slope at the free end.**

Solution

The cantilever and conjugate beam are as below for the problem

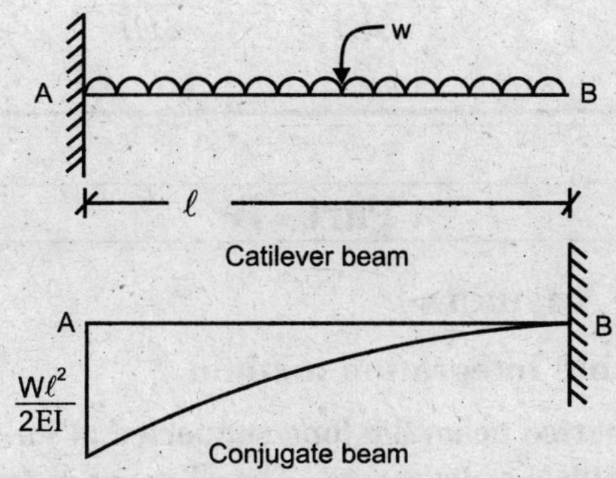

Fig .3.17

Slope at the free end = Shear force at B for conjugate beam

$$= \frac{1}{3}.\ell.\frac{W\ell^2}{2}.\frac{1}{EI} = \frac{W\ell^3}{6EI}$$

25. **A cantilever beam of length 'ℓ' is loaded with an UDL of 'W' per limit length, calculate the deflection at the free end**

The cantilever and conjugate beams are as below.

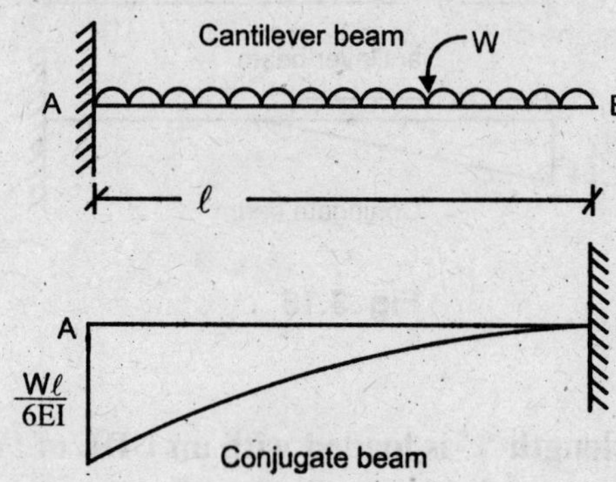

Fig .3.18

Deflection at the free end = BM at B for conjugate beam

$$= \frac{W\ell^3}{6EI} \cdot \frac{3\ell}{4} = \frac{W\ell^4}{8EI}$$

Part - B

Questions and Answers

Problems on Double Integration method

1. **A simply supported beam 8m long supported at its ends is carrying a point load of 50KN at its centre. The Moment of Inertia of the beam is equal to 78 $\times$ 10^6 mm^4. If E for the material of the beam is 2.1 $\times$ 10^5 N/mm^2. Find deflection using Double integration method.**

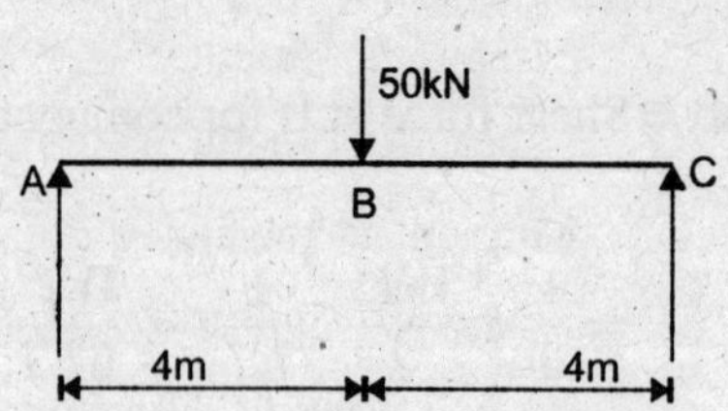

Fig .3.19

Given Data

$$\text{Length of Beam} \quad (\ell) = 8m = 8000mm$$
$$\text{Point Load} \quad (W) = 50KN = 50 \times 10^3\,N$$
$$\text{Moment of Inertia} \quad (I) = 78 \times 10^6\,mm^4$$
$$\text{Young's modulus} \quad (E) = 2.1 \times 10^5\,N/mm^2$$

To Find Deflection at the mid span.

Solution

By double Integration Method we know that the formula for freely deflection at the mid span for a $y = \frac{W\ell^3}{48EI}$.

Now, substituting the values,

Supply supported beaming with a point to and.

$$= \frac{50 \times 10^3 \times (8000)^3}{48 \times 2.1 \times 10^5 \times 78 \times 10^6}$$
$$y = 32.56mm$$

2. **A simply supported beam of length 10m is loaded as shown. Calculate slope and deflection by using double integration method. Assume E = 2.1 × 10^5 N/mm^2, I = 78 × 10^6 mm^4**

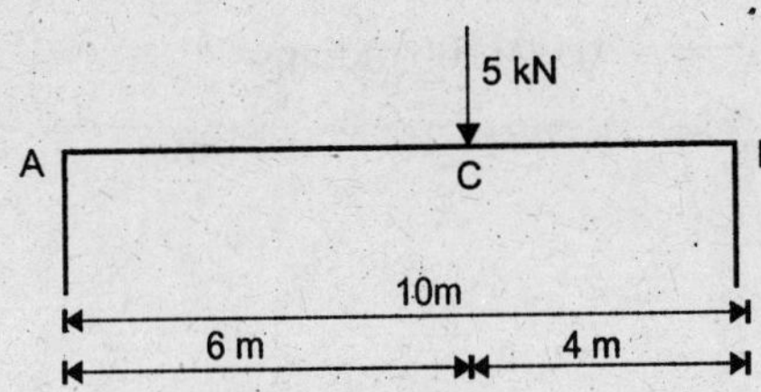

Fig .3.20

Given Data

$$\text{Young's modulus}(E) = 2.1 \times 10^5\,N/mm^2$$
$$\text{Length of beam} \quad (\ell) = 10m = 10000mm$$
$$\text{Moment of Inertia}(I) = 78 \times 10^6\,mm^4$$
$$\text{Distance of Load from left support}(a) = 6m = 6000mm$$
$$\text{Point Load}W = 5KN = 5000N$$
$$\text{Distance of Load from right support}(b) = 4m = 4000mm$$

To Find

Slope and Deflection by double integration Method

Solution

$$\text{Slope at B } (\theta_B) = -\frac{Wab\,(b+2a)}{6EI\ell}$$

$$= -\frac{5000 \times 6000 \times 4000 \times (4000 + (2 \times 6000))}{6 \times 2.1 \times 10^5 \times 78 \times 10^6 \times 10000}$$

$$= -\frac{1.92 \times 10^{15}}{9.828 \times 10^{17}}$$

$$-\ 1.953 \times 10^{-3} \text{ radians}$$

$$\theta_B = -1.953 \times 10^{-3} \text{ Radians}$$

$$\text{Slope at A, } (\theta_A) = -\frac{Wab\,(a+2b)}{6EI\ell}$$

$$= -\frac{5000 \times 6000 \times 4000\,(6000 + 2 \times 4000)}{6 \times 2.1 \times 10^5 \times 78 \times 10^6 \times 10000}$$

$$= -\frac{1.68 \times 10^{15}}{9.828 \times 10^{17}}$$

$$\theta_A = -0.001709 \text{ radians}$$

$$= -1.709 \times 10^{-3} \text{ radians}$$

Slope at C, $(\theta_c) = 0$

Deflection at A, B = 0

$$\text{Deflection at C, } (y_c) = \frac{Wa^2 b^2}{3EI\ell}$$

$$= \frac{5000 \times (6000)^2 \times (4000)^2}{3 \times 2.1 \times 10^5 \times 78 \times 10^6 \times 10000}$$

$$= \frac{2.88 \times 10^{18}}{4.914 \times 10^{17}}$$

$$y_c = 5.86mm$$

Result

$$\text{Slope at support 'A'} \quad (\theta_A) = 1.709 \times 10^{-3}, \text{radians}$$
$$\text{Slope at support 'B'} \quad (\theta_B) = 1.953 \times 10^{-3} \text{radians}$$
$$\text{Slope at support 'C'} \quad (\theta_C) = 0$$
$$\text{Deflection @ 'A'} \quad (y_A) = 0$$
$$\text{Deflection @ 'B'} \quad (y_B) = 0$$
$$\text{Deflection @ 'C'} \quad (y_C) = 5.86\, mm$$

3. **A horizontal kN/m beam is freely supported at its ends 10m apart and carries a UDL of 15KN/M over the entire Span Find the maximum deflection using Double integration method. Take E = 2 × 10^5 N/mm^2 and I = 2 × 10^9 mm^4**

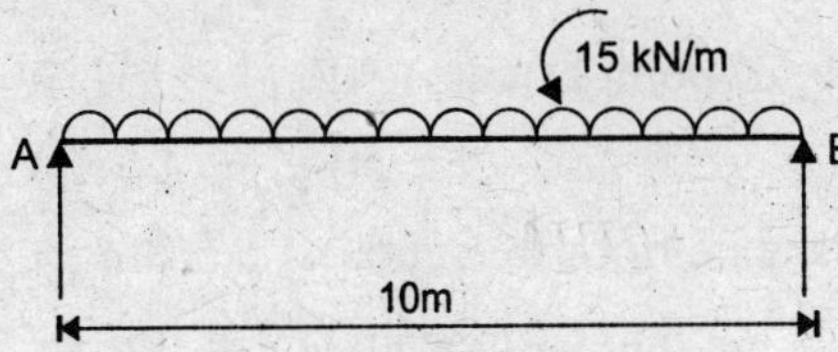

Fig .3.21

Given Data

$$\text{Length of beam} \quad (\ell) = 10000 mm$$
$$\text{udl} \quad (W) = 15 KN/m = 15 N/mm$$
$$\text{Young's modulus} \quad (E) = 2 \times 10^5\, N/mm^2$$
$$\text{Moment of Inertia} \quad (I) = 2 \times 10^9\, mm^4$$

To Find Maximum Deflection

By double integration method, the maximum deflection at the mid span

Solution

$$(y_c) = \frac{5W\ell^4}{384EI} = \frac{5 \times 15 \times (10000)^4}{384 \times 2 \times 10^5 \times 2 \times 10^9}$$
$$= \frac{7.5 \times 10^{17}}{1.536 \times 10^{17}}$$
$$y_c = 4.88 mm$$

Problems on Macaulay's Method

4. **A Steel cantilever 6m long carries two point loads, 15KN at the free end 25KN at a distance of 3m from the free end. Find Slope at the free end and deflection at the free by Macaulay's method**

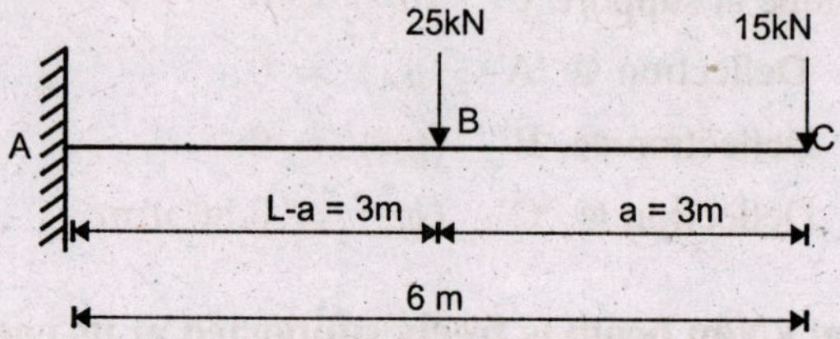

Fig .3.22

Given Data

$$\text{Intensity of point Loads} \quad (W_1) = 15KN = 15000N$$
$$(W_2) = 25KN = 25000N$$
$$\text{Distance from free end } (a) = 3m = 3000mm$$
$$\text{Length of Beam } (\ell) = 6m = 6000mm$$

To Find

Deflection @ free end using Maculay's method

Solution

Bending Moment, $M_X = -W_1 x - W_2 (x - a)$

Integrating above equation,

$[\because \ Mx = EI\dfrac{d^2y}{dx^2}]$

$$\text{Slope,} = EI\frac{dy}{dx} = -W_1\frac{x^2}{2} + C_1 - W_2\frac{(x-a)^2}{2} \tag{1}$$

Integrating once again,

$$EI.y = -W_1\frac{x^3}{6} + C_1 x + C_2 - W_2\frac{(x-a)^3}{6} \tag{2}$$

Applying the following boundary conditions for cantilever

i) When $x = 6m$, Slope $\dfrac{dy}{dx} = 0$

ii) When $x = 6m$, deflection $y = 0$

Applying first Boundary condition to slope equation

$$0 = \frac{-15000 \times (6000)^2}{2} + C_1 \frac{-25000\,(6000 - 3000)^2}{2}$$

$$= -2.7 \times 10^{11} + C_1 - 1.25 \times 10^{11}$$

$$= -3.825 \times 10^{11}\, C_1$$

$$C_1 = 3.825 \times 10^{11}$$

Applying second Boundary condition to the **deflection equation,**

$$0 = \frac{-15000 \times 6000^{\,3}}{6} + 3.825 \times 10^{11} \times 6000$$

$$+ C_2 \frac{-25000\,(3000)^3}{6}$$

$$= -5.4 \times 10^{14} + 2.295 \times 10^{15} + C_2 - 1.25 \times 10^{14}$$

$$= -6.525 \times 10^{14} + 2.295 \times 10^{15}$$

$$= 1.64 \times 10^{15}\, C_2$$

$$C_2 = -1.64 \times 10^{15}$$

Substituting the value of x,C_1 , C_2 in equation (I), and (II) we can get the slope and deflection.

At the free end,$x = 0$ and first terms only valid for that section.

∴ Slope,

$$\left(\frac{dy}{dx}\right) = \frac{C_1}{EI}$$

$$= \frac{3.825 \times 10^{11}}{2 \times 10^5 \times 1.3 \times 10^8}$$

$$= 0.0147 \text{ rad}$$

Deflection, (y)

$$= \frac{C_2}{EI}$$

$$= \frac{1.64 \times 10^{15}}{2 \times 10^5 \times 1.3 \times 10^8}$$

$$= -63.076mm$$

[$-v_e$ sign due to downward deflection]

5. **A cantilever of length 5m carries a UDL of 8KN/m length over the entire length. If the section is rectangular of 150 mm × 260 mm find the deflection and slope at the free end by Macaulay's Method. Take E = 2.1 × 10^5 N/mm^2**

Given Data

Length : (L) = 5m = 5000mm.

UDL, (W) = 8kN/m = 8N/mm.

Cross section of beam (Rectangle)

$$= 150mm \times 260mm$$

$$\text{Moment of Inertia}(I) = \frac{150 \times 260^3}{12} \qquad \left[\because I = \frac{bd}{12}\right]$$

$$= 2.197 \times 10^8 mm^4$$

$$E = 2.1 \times 10^5 N/mm^2$$

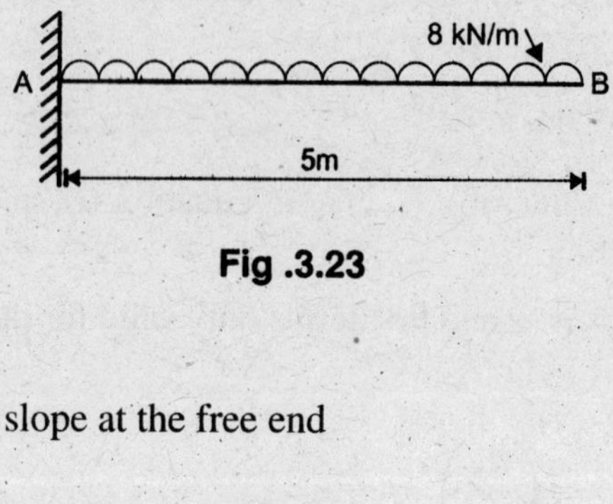

Fig .3.23

To Find

Deflection and slope at the free end

Solution

Bending Moment, $(M_X) = \frac{-Wx^2}{2}$

Integrating above equation,

Slope equation

$$EI\,\frac{dy}{dx} = \frac{-Wx^3}{6} + C_1 \qquad (1)$$

Integrating again, we get

Deflection equation

$$EI.y = \frac{-Wx^4}{24} + C_1.x + C_2$$

Applying the following boundary conditions for cantilever

(i) When $x = 5m$, slope $\dfrac{dy}{dx} = 0$

(ii) When $x = 5m$ Deflection $(y) = 0$

Applying the first Boundary Condition to the slope equation

$$0 = \frac{-8 \times 5000^3}{6} + C_1$$
$$C_1 = 1.66 \times 10^{11}$$

Applying the second Boundary Condition to the deflection **equation,**

$$\begin{aligned} 0 &= \frac{-8 \times 5000^4}{24} + 1.66 \times 10^{11} \times 5000 + C_2 \\ &= -2.083 \times 10^{14} + 8.3 \times 10^{14} + C_2 \\ 0 &= 6.217 \times 10^{14} + C_2 \\ C_2 &= -6.217 \times 10^{14} \end{aligned}$$

Substituting the value of x, C_1 and C_2 in equation (i) and (ii) we get the slope and deflection.

At the free end, $X = 0$.

(i) Slope

$$\begin{aligned} (\theta_B) &= \frac{dy}{dx} = \frac{C_1}{EI} \\ &= \frac{1.66 \times 10^{11}}{2 \times 10^5 \times 2.197 \times 10^8} \\ &= 0.00359 \text{ rad} \end{aligned}$$

(ii) Deflection

$$\begin{aligned} (y) &= \frac{C_2}{EI} \\ &= \frac{-6.217 \times 10^{14}}{2.1 \times 10^{15} \times 2.197 \times 10^8} \\ &= -13.47mm \end{aligned}$$

[$-v_e$ sign due to downward deflection]

6. **A cantilever projecting 4m from a wall carries an UDL of 12KN/m for a length of 2m from the fixed end and a point load 1.5KN at the free end. Find the deflection at the free end by Macaulay's method.**

Take E = 2.1 × 10^5 N/mm^2 and $I = 1 \times 10^8 mm^4$

Given Data

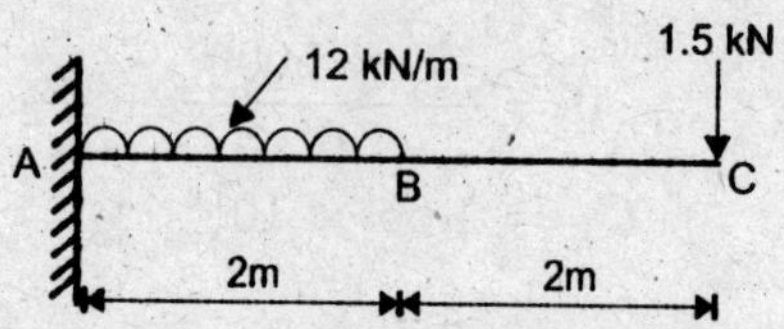

Fig .3.24

UDL for a length of $2m$

$$\begin{aligned}
\text{From the fixed end} \quad (W) &= 12KN/m, = 12N/mm \\
\text{Point Load @ free end} \quad (W) &= 1.5KN = 1500N \\
\text{Length of UDL from fixed end} \quad (a) &= 2m = 2000mm \\
\text{Total length of Cantilever} \quad (\ell) &= 4m = 4000mm \\
\text{Young's modulus} \quad (E) &= 2 \times 10^5 \, N/mm^2 \\
\text{Moment of Inertia} \quad (I) &= 1 \times 10^8 \, mm^4
\end{aligned}$$

To Find

Deflection at the free end by Macaulay's method

Solution

Bending Moment, $M_x = -W_x - \frac{W(x-a)^2}{2}$

(@ section X - X, which is 'x' distance from the support)

Integrating the above equation

We get equation

$$\text{Slope} = EI\frac{dy}{dx} = \frac{-Wx^2}{2} + C_1 - \frac{W(x-a)^3}{6}$$

Integrating again, We get

$$\text{Deflection equation, } EI.y = \frac{-Wx^3}{6} + C_1.x + C_2 - \frac{W(x-a)^4}{24}$$

The boundary conditions for this cantilever beam.

(i) When $x = 4m,$ Slope, $\frac{dy}{dx} = 0$

(ii) When $x = 4m,$ Deflection $y = 0$

Applying first B.C to the Slope equation:

$$0 = \frac{-1500 \times (4000)^2}{2} + C_1 \frac{-12 \times (2000)^3}{6}$$

$$0 = -1.2 \times 10^{10} + C_1 - 1.6 \times 10^{10}$$

$$0 = -2.8 \times 10^{10} + C_1$$

$$C_1 = 2.8 \times 10^{10}$$

Applying Second B.C to the deflection equation,(2)

$$0 = \frac{-1500 \times (4000)^3}{16} + (2.8 \times 10^{10})(4000)$$

$$+ C_2 - \frac{12 \times (2000)^4}{24}$$

$$= -6 \times 10^{12} + 1.12 \times 10^{14} - 8 \times 10^{12} + C_2$$

$$= -1.4 \times 10^{13} + 1.12 \times 10^{14} + C_2$$

$$0 = 9.8 \times 10^{13} + C_2$$

$$C_2 = -9.8 \times 10^{13}$$

Substituting the value of x, C_1 and C_2 in equation (i) and (ii) we get the slope and deflection.

We can get slope and deflection,

At the free end, $x = 0$ and first term is only valid for that section

$$\therefore \text{ Slope } = \theta_C = \frac{dy}{dx} = \frac{C_1}{EI}$$

$$= \frac{2.8 \times 10^{10}}{2 \times 10^5 \times 1 \times 10^8}$$

$$= 0.0014\text{rad}$$

$$\therefore \text{ Deflection} = y_c = \frac{C_2}{EI}$$

$$= \frac{-9.8 \times 10^{13}}{2.1 \times 10^{15} \times 1 \times 10^8}$$

$$= -4.99mm$$

[$-V_e$ sign due to downward deflection]

7. **A cantilever beam of length 8m carries a UDL of 18KN/m over length of 4m from the free end along with a point load 2KN at 4m from free end. Determine the deflection at the free end of the beam by Macaulay's method E = 2.1 × 10^5 N/mm^2 and I = 1.2 × 10^{8}mm^4**

Given Data

$$(L) = 8m = 8000mm$$

Distance of Point Load from the free end $(a) = 4m = 4000mm$

$$UDL, (W) = 18KN/m = 18N/mm$$

Point Load $(W) = 2KN = 2000N$

Moment of Inertia $(I) = 1.2 \times 10^8 mm^4$

Young's modulus $(E) = 2.1 \times 10^5 N/mm^2$

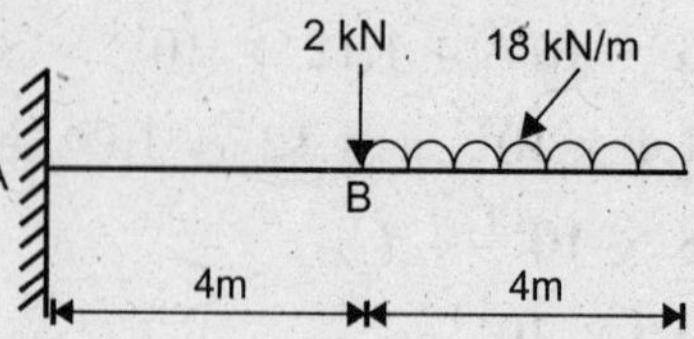

Fig .3.25

To Find

Deflection at the free end by Maculay's method

Solution

Bending moment, @ a section X - X, which is at a distance 'x' from fixed end.

$$M_X = -W(x-a) - \frac{Wx^2}{2} + \frac{W(x-a)^2}{2} \qquad (1)$$

Note: For Calculating BM, We are Considering the UDL acting up to the right end and negative upward load acting upto the point B from left end.

On Integrating equation (1) once, we get equation for

$$\text{Slope, } EI\frac{dy}{dx} = -\frac{W(x-a)^2}{2} + C_1 - \frac{Wx^3}{6} + \frac{W(x-a)^3}{6}$$

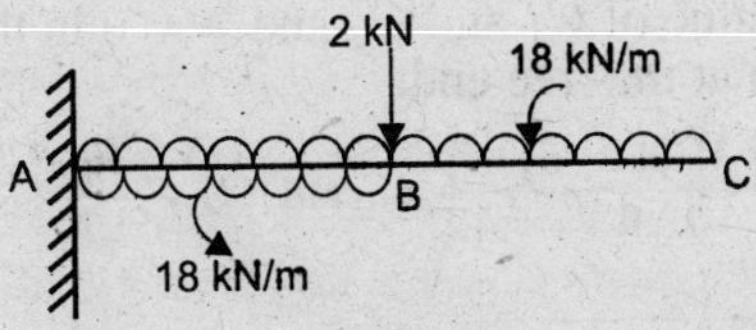

Fig .3.26

Integrating once again, we get the equation for deflection

$$\text{i.e., } EI.y = -\frac{W(x-a)^3}{6} + C_1 x + C_2 - \frac{Wx^4}{24} + \frac{W(x-a)^4}{24}$$

Boundary conditions (BC) for the cantilever beam

(i) At $x = 8000mm$, Slope $\frac{dy}{dx} = 0$

(ii) At $x = 8000mm$, deflection $y = 0$

Applying the first Boundary conditions (B.C) to the slope equation

$$0 = \frac{-2000 \times (8000 - 4000)^2}{2} + C_1 - \frac{18 \times (8000)^3}{6} + \frac{18 \times (8000 - 4000)^3}{6}$$

$$= -1.6 \times 10^{10} + C_1 - 1.536 \times 10^{12} + 1.92 \times 10^{11}$$

$$= -1.36 \times 10^{12} + C_1$$

$$\therefore \quad C_1 = 1.36 \times 10^{12}$$

Applying the Second B.C to the deflection equation

$$0 = \frac{-2000 \times (8000 - 4000)^2}{2} + 1.36 \times 10^{12} \times 8000 + C_2 - \frac{18 \times (8000)^4}{24} + \frac{18 \times (8000 - 4000)^4}{24}$$

$$= -1.6 \times 10^{10} + 1.088 \times 10^{16} + C_2 - 3.072 \times 10^{15} + 1.92 \times 10^{14}$$

$$\therefore \; C_2 = -7.97 \times 10^{15}$$

Substituting the value of C_1 and C_2 and $x = 0$ in the deflection equation to find the deflection at the free end.

$$y_c = \frac{1}{EI}\left(\frac{-W(x-a)^3}{6} + C_1 x + C_2 - \frac{Wx^4}{24} + \frac{W(x-a)^4}{24}\right)$$

$$= \frac{1}{1.2 \times 10^8 \times 2.1 \times 10^5}$$

$$\left[\frac{-2000 \times (0-4000)^3}{6} + 1.36 \times 10^{12} \times 0 - 7.97 \times 10^{15} - \frac{18 \times (0)^4}{24} + \frac{18 \times (0-4000)^4}{24}\right]$$

$$= 3.968 \times 10^{-14}\left[-(2.13 \times 10^{13}) - 7.97 \times 10^{15} + 1.92 \times 10^{14}\right]$$

$$y_c = -307.78mm$$

8. **A horizontal girder having uniform cross-section is 14m long and is simply supported @ its end. It carries two concentrated load as shown the following fig. Find deflection @ C & D.**

Take E = 200 Gpa; I = 160 $\times$ 10^{6}mm^4

Given Data

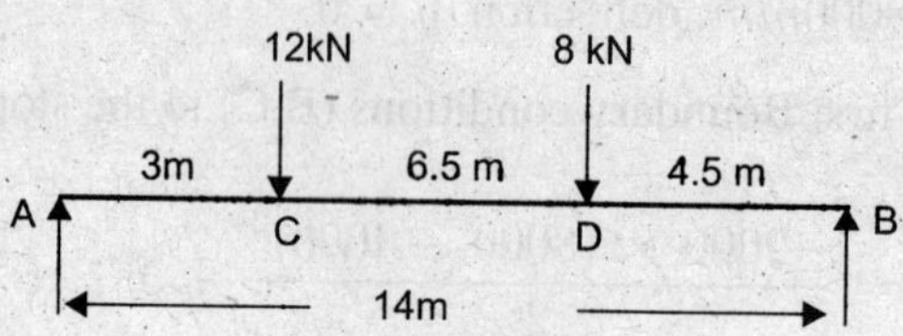

Fig .3.27

Cross - section of Girder Uniform

Length (L) = 14m

Solution

Considering equilibrium conclusion

$$\sum V = 0$$

$$R_A + R_B = 12 + 8 = 20 \qquad (1)$$

Considering the equilibrium condition

$$\sum M_A = 0$$

$$-R_B \times 14 + 8 \times 9.5 + 12 \times 3 = 0$$

$$R_B\,(14) = 76 + 36$$

$$R_B = \frac{112}{14}$$

$$\therefore \boxed{R_B = 8KN} \text{ on substituting} R_B \text{in}$$

We get $\boxed{R_A = 12KN}$

Now bending moment at a section X - X, which is at a distance of 'x'. from the support A.

Integrating(2)

$$\therefore \quad M_X = 12x| - 12\,(x - 3)\,| - 8\,(x - 9.5)$$

$$EI\frac{d^2y}{dx^2} = -12x| + 12\,(x - 3)\,|$$

$$+\,8\,(x - 9.5) \qquad \because M = EI\frac{d^2y}{dx^2} \tag{2}$$

Once, We get slope equation

$$EI\,\frac{dy}{dx} = \frac{-12x^2}{2} + C_1| + \frac{12}{2}\,(x - 3)^2 + \frac{8}{2}\,(x - 9.5)^2 \tag{3}$$

Integrating (3)

Once, we get equation for deflection

$$\therefore \quad EI.y = \frac{-12x^3}{6} + C_1\,x + C_2\,|$$

$$+\,\frac{12}{2 \times 3}(x - 3)^3\,| + \frac{8}{6}\,(x - 9.5)^3 \tag{4}$$

Boundary Conditions for this Beam

$$@\,x = 0,\; y = 0$$

$$@x = 14;\; y = 0$$

Substituting these BCs in (4) we get

$$EI\,(0) = 0 + 0 + C_2$$

$$\therefore \boxed{C_2 = 0}$$

$$EI(0) = \frac{-12(14)^3}{6} + c_1\,(14) + 0\Big| + 2\,(11)^3 + \frac{4}{3}\,(4.5)^3$$

$$0 = -54888 + 14c_1 + 2662 + 121.5 - 144 = 2704.5$$

$$\therefore \boxed{C_1 = +193.17}$$

Now substituting c_1 and c_2 in(4)

$$EIy = \frac{-12x^3}{6} + 193.17\,x\,| + 2\,(x - 3)^3| + \frac{4}{3}\,(x - 9.5)^3 \qquad (5)$$

Now, Deflections @ C & D

@ $x = 3$

substituting for $x = 3$ in (5)

$$EI\,y_C = \frac{-12\,(3)^3}{6} + 193.17\ (3)$$

$$E\,I\,y_C = \frac{-108 \times 3}{6} + 5.579.51$$

$$= -54 + 579.51$$

$$E\,I\,y_C = 525.51 \ \Rightarrow \ y_c = \frac{525.51}{EI}$$

$$\therefore \ y_C = \frac{525.51}{200 \times 10^3 \times 10^3 \times 160 \times 10^6 \times 10^{-12}}$$

$$= \frac{525.51}{200 \times 160}$$

$$y_C = 0.01642m$$

$$= 0.01642 \times 1000$$

$$\boxed{y_C = 16.42mm}$$

@$x = 9.5$

Substituting $x = 9.\ 5$ in equation (5), We get

$$EI\,y_D = \frac{-12x^3}{6} + 193.17x| + 2\,(x - 3)^3$$

$$= \frac{-12\,(9.5)^3}{6} + 193.17\,x\,9.5\,| + 2(6.5)^3$$

$$EI\,y_D = -1714.75 + 1835.12 + 549.25$$

$$EI\,y_D = 669.62 = \therefore \ \ y_D = \frac{669.62}{EI}$$

$$y_D = 200 \times 10^3 \times 10^3 \times 160 \times 10^6 \times 10^{-12} = \frac{669.62}{200 \times 160}$$

$$= 0.0209\,m$$

$$y_D = 0.0209 \times 1000$$

$$\boxed{y_D = 20.9mm}$$

9. **A simply supported beam of 8m length carries two point loads of 64 kN and 48 kN at 1m and 4m respectively from the left. Find the deflection under each load and the maximum deflection. E = 210 GPa and I = 180 × 10^6 mm^4.**

Solution

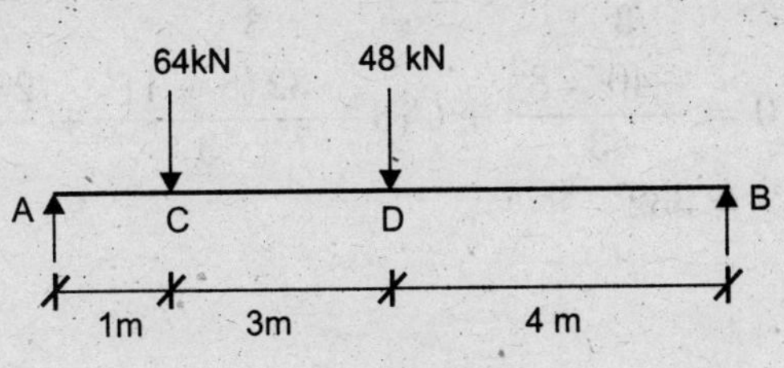

Fig .3.28

Taking moments about A,

$$R_b \times 8 = (64 \times 7) + (48 \times 4)$$
$$R_b = 80kN$$
$$\therefore \quad R_a = 64 + 48 - 80 = 32kN$$

At any section x from A,

$$EI\frac{d^2y}{dx^2} = -80x| + 64(x-1)| + 48(x-4)$$

Integrating,

$$EI\frac{dy}{dx} = \frac{-80x^2}{2} + C_1 \left| + \frac{64\,(x-1)^2}{2} \right| + \frac{48\,(x-4)^2}{2}$$
$$= -40x^2 + C_1 \left| +32\,(x-1)^2 \right| + 24\,(x-4)^2 \qquad (1)$$

Integrating again,

$$EIy = \frac{-40x^3}{3} + C_1x + C_2 \left| + \frac{32\,(x-1)^3}{3} \right| + \frac{24\,(x-4)^3}{3} \qquad (2)$$

At A, $x = 0, y = 0$;

The equation for the elastic curve for the portion AD is,

$$EIy = \frac{-40x^3}{3} + C_1x + C_2$$
$$\text{(or)} \quad C_2 = 0$$

At B ,$x = 8m, y = 0;$

The whole equation is valid for the portion DB.

$$EIy = \frac{-40x^3}{3} + C_1x + \frac{32\,(x-1)^3}{3} + \frac{24\,(x-4)^3}{3}$$
$$0 = \frac{-40 \times 8^3}{3} + C_18 + \frac{32\,(8-1)^3}{3} + \frac{24\,(8-4)^3}{3}$$
$$C_2 = 332$$

Thus, equation (2) becomes,

$$EIy = \frac{-40x^3}{3} + 332 \times x \left| + \frac{32\,(x-1)^3}{3} \right| + \frac{24\,(x-4)^3}{3}$$

Deflection under 64 kNs

$$EIy = \frac{-40 \times 1^3}{3} + 332 \times 1$$
$$= 318.7kN.m^3$$
$$= 318.7 \times 10^{12} N.mm^3$$
$$y = \frac{318.7 \times 10^{12}}{210000 \times 180 \times 10^6} = 8.43mm$$

Under 48 kN,

$$EIy = \frac{-40 \times 4^3}{3} + (332 \times 4) + \frac{32 \times 3^3}{3}$$
$$= 762.7kNm^3$$
$$= 762.7 \times 10^{12} N.mm^3$$
$$y = \frac{762.7 \times 10^{12}}{210000 \times 180 \times 10^6} = 20.18\,mm$$

For Maximum deflection, assuming it to be between DC and differentiating

(1),

$$0 = -40x^2 + 332 + 32\,(x-1)^2$$

$$0 = -40x^2 + 332 + 32x^2 + 32 - 64x$$

$$x^2 + 8x - 45.5 = 0$$

$$x = \frac{-8 \pm \sqrt{8^2 + (4 \times 45.5)}}{2}$$

$$\boxed{x = 3.84m}$$

$$EIy = \frac{-40 \times 3.84^3}{3} + (332 \times 3.84) + \frac{32 \times 2.84^3}{3}$$

$$= 764.2kNm^2 = 764.2 \times 10^{12} N.mm^2$$

$$y_{max} = \frac{764.2 \times 10^{12}}{210000 \times 180 \times 10^6} = 20.22mm$$

Problems on conjugate beam method

10. **A beam of simply supported 6m long supported at its end is carrying a point load of 50KN at its center. The moment of inertia of a beam is equal to 78 × 10^6 mm^4units. It E for the material of the beam is 2.1 × 10^5 N/mm^2 calculate the deflection at the center by conjugate beam method.**

Given Data

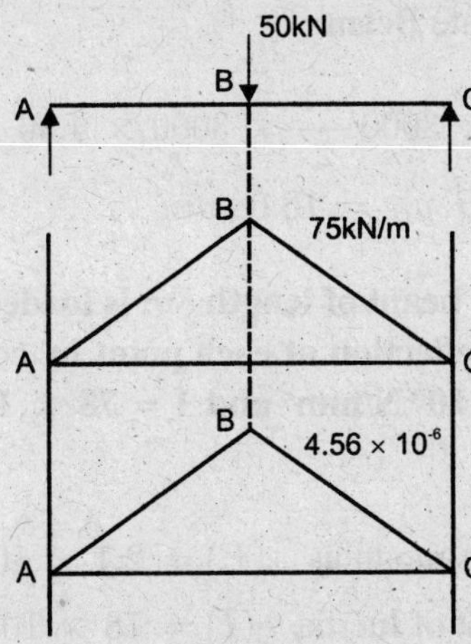

Fig .3.29

$$\text{Length of simply supported beam} \quad (\ell) = 6m = 6000mm$$
$$\text{Point Load @ center} \quad (W) = 50KN = 50000N$$
$$\text{Moment of Inertia} \quad (I) = 78 \times 10^6\, mm^4$$
$$\text{Young's modulus} \quad (E) = 2.1 \times 10^5\, N/mm^2$$

To Find

Deflection at the mid span of simply supported beam

Solution

Total load on the conjugate beam = Area of $\frac{M}{EI}$ diagram

$$\therefore \quad P = \frac{1}{2} \times 6000 \times 4.56 \times 10^{-6} = 0.01368\, kN$$

$\therefore$ Reactions, for this conjugate beam

$$R_A = R_B = \frac{P}{2} = \frac{0.01368}{2} = 6.84 \times 10^{-3}\, kN$$

As per the mohr's theorem(I) for conjugate Beam

$$\text{slope@}A(\theta_A) = R_A \text{ and } \text{slope@}B(\theta_B) = R_B$$
$$\therefore \quad \theta_A = \theta_C = 6.84 \times 10^{-3}\, rad$$

Deflection at $B(y_B) = BM$ at C for the conjugate beam as per the Mohr's theorem II of conjugate Beam.

$$= 6.84 \times 10^{-3} \times 3000 \frac{-1}{2} \times 3000 \times 4.56 \times 10^{-6} \times \frac{1}{3} \times 3000$$
$$\therefore \quad \text{Deflection@}B, y_B = 13.68mm$$

11. **A simply supported beam of length $8m$ is loaded as shown in fig calculate the slope and deflection at each point by conjugate beam method. Assume E = 2.1 × 10^5 N/mm^2 and I = 78 × 10^6 mm^4**

Given Data

$$\text{Young's modulus} \quad (E) = 2.1 \times 10^5\, N/mm^2$$
$$\text{Moment of Inertia} \quad (I) = 78 \times 10^6\, mm^4$$
$$\text{Point Load} \quad (W) = 50KN = 5000N$$

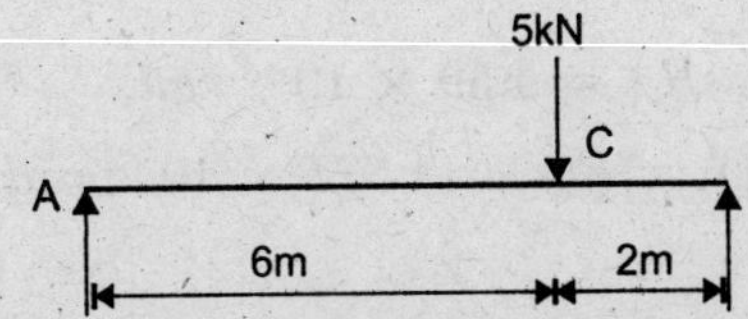

Fig .3.30

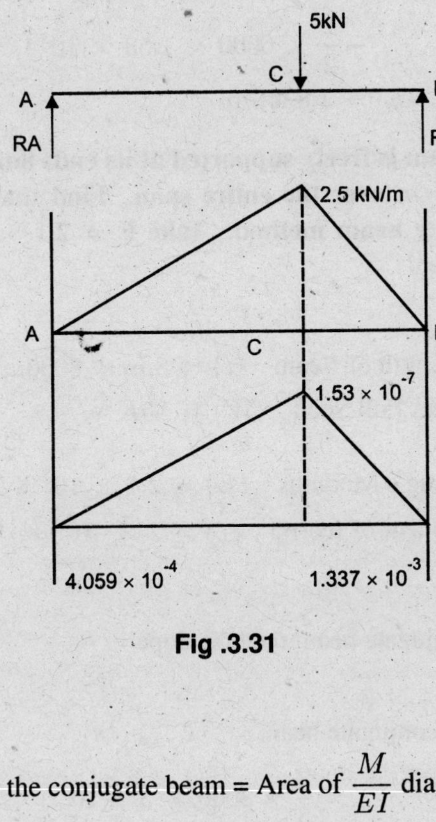

Fig .3.31

Solution

Total load on the conjugate beam = Area of $\dfrac{M}{EI}$ diagram

$$P = \frac{1}{2} \times 8000 \times 1.53 \times 10^{-7} = 6.12 \times 10^{-4} kN$$

$\therefore$ Reaction at the supports are,

$$R_A = p \times \frac{6000}{8000} = 6.12 \times 10^{-4} \times \frac{6000}{8000} = 4.59 \times 10^{-4}\, kN$$

$$R_B = 6.12 \times 10^{-4} - 4.59 \times 10^{-4} = 1.337 \times 10^{-3} kN$$

Slopes

As per morh's theorem I for conjugate beam

Slopes at A, $(\theta_A) = R_A = 4.59 \times 10^{-4}\,rad,$

The Slope at $B, (\theta_B) = R_B = 1.377 \times 10^{-3}\,rad$

Deflection

As per the mohr's theorem II for conjugate beam

$$\text{Deflection} \quad y_c = 4.59 \times 10^{-4} \times 6000 - \frac{1}{2} \times 6000 \times 1.53 \times 10^{-7} \times \frac{1}{3} \times 6000$$

$$y_c = 1.836mm$$

12. **A horizontal beam is freely supported at its ends 8m apart and carries a UDL of $15kN/m$ over the entire span. Find the maximum deflection by conjugate beam method. Take E = 2.1 × 10^5 N/mm^2 and I = 2 × 10^9 mm^4**

Given Data

$$\text{Length of Beam} \quad (\ell) = 8m = 8000mm$$
$$\text{UDL over Full Span} \quad (W) = 15KN/m = 15N/mm$$
$$\text{Young's Modulus} \quad (E) = 2.1 \times 10^5\,N/mm^2$$
$$\text{Moment of Inertia} \quad (I) = 2 \times 10^9\,mm^4$$

To Find

Deflection by conjugate beam method slope

Solution

Total load on the conjugate beam

$$= \frac{2}{3} \times 8000 \times 3 \times 10^{-7}$$
$$P = 1.6 \times 10^{-3}kN$$

Reactions at the supports are

$$R_A = R_B = \frac{P}{2} = \frac{1.6 \times 10^{-3}}{2} = 8 \times 10^{-4}kN$$

Deflection at $C, y_{max} = BM$ at C for the conjugate beam as per the mohr's Theorem -II for conjugate beam

$$= 8 \times 10^{-4} \times 4000 - 8 \times 10^{-4} \times \frac{3}{16} \times 8000 = 2mm$$

Maximum deflection $y_{max} = 2mm$

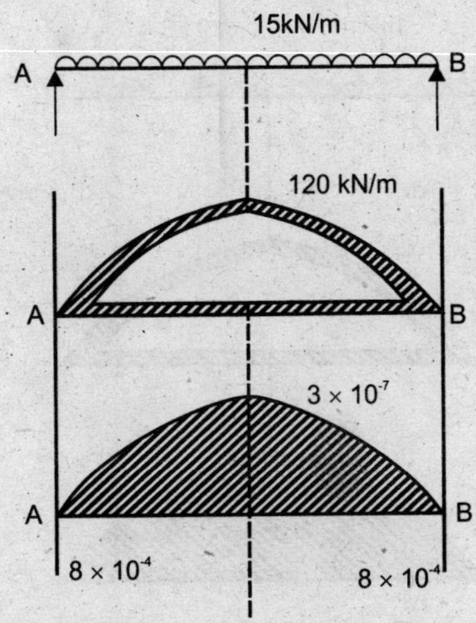

Fig .3.32

13. **A SSB of $8m$ span carries a point load of 10kN at its center. It is also subjected to an UDL of 1kN/m over its entire span. Find the maximum deflection of beam by conjugate beam method.**

Given Data

$$\text{Young's Modulus} \quad (E) = 200KN/mm^2$$
$$\text{Moment of Inertia} \quad (I) = 200 \times 10^6\, mm^4$$
$$\text{Point load @ center} \quad (W) = 10KN = 10000N$$
$$\text{UDL over entire span} \quad (W) = 1KN/m = 1N/mm$$
$$\text{Span of Beam} \quad (\ell) = 8m = 8000mm$$

To Find

$Maxml$ deflection of beam of conjugate beam.

Solution

Total load on the conjugate beam

$$P = \frac{2}{3} \times 8000 \times 7 \times 10^{-7}$$
$$P = 3.73 \times 10^{-3}$$

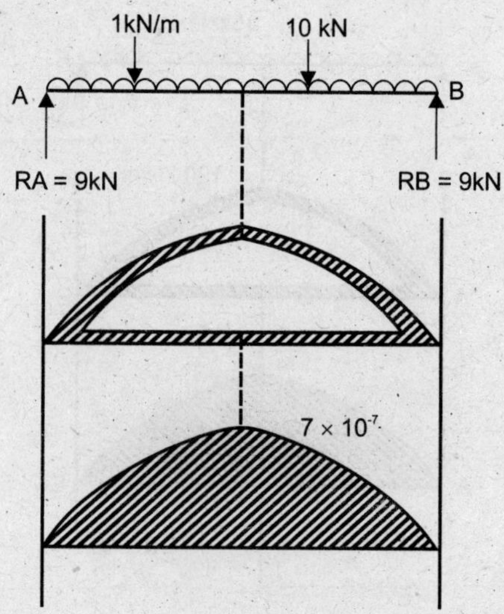

Fig .3.33

Slope at A = Slope at B

$$\theta_A = \theta_B = R_A$$

$$R_A = \frac{P}{2} = \frac{3.73 \times 10^{-3}}{2} = 0.001865\, rad$$

Deflection at C = BM of @ centre for the conjugate beam, as per mohr's theorem (II)

$$= 0.001865 \times 4000 - 0.001865 \times \frac{3}{6} \times 8000 = 4.66mm$$

Problems on moment area method

14. **A steel cantilever 6m long carriers two point loads, 15KN at the free end and 25KN at distance of 3m from the free end. Find slope at the free end and deflection at the free end by moment area method. Take 'E' as 2×10^5 N/mm^2 and 'I' as 1.3×10^8 mm^4.**

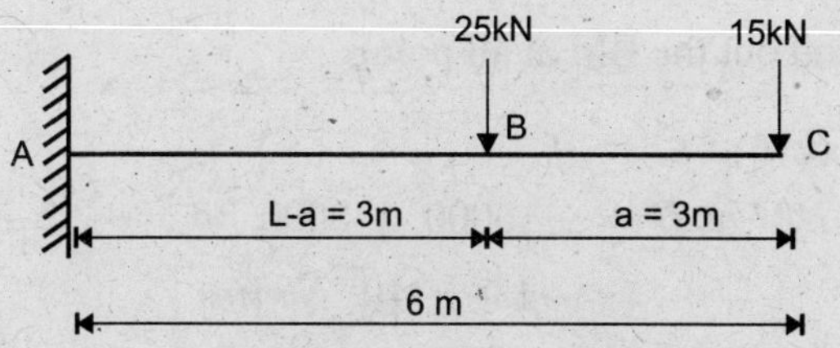

Fig .3.34

Given Data

$$\text{Point Load 1,} \quad (W_1) = 15KN = 15000N$$
$$\text{Point Load 2,} \quad (W_2) = 25KN = 25000N$$
$$\text{Distance of Load 2 from free end,} \quad (a) = 3m = 3000mm$$
$$\text{Span of Beam,} \quad (\ell) = 6m = 6000mm$$
$$\text{Youngs modulus}(E) = 2 \times 10^5 N/mm^2$$
$$\text{moment of inertia} \quad (I) = 1.3 \times 10^8 mm^4$$

To Find

Slope and deflection at the free end of cantilever beam by moment area method.

Solution

Moment area Method

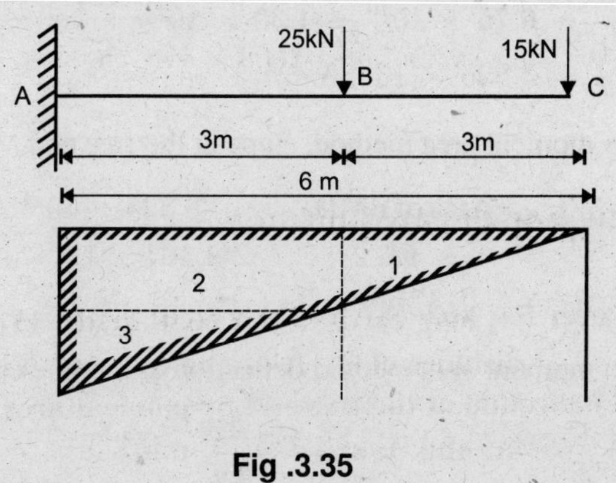

Fig .3.35

Let us first find out the BM at all points,

$$BM\ at\ C = 0$$

$$BM\ at\ B = -15000 \times 3000$$

$$= -4.5 \times 10^7\ Nmm$$

$$BM\ at\ A = -15000 \times 6000 - 25000 \times 3000$$

$$= -1.65 \times 10^8 Nmm$$

To find the area of BMD, divide the BMD into two triangles and one rectangle, as shown in Fig. 3.35.

Area of the First section,

$$A_1 = \frac{1}{2} \times 3000 \times 4.5 \times 10^7$$

$$A_1 = 6.75 \times 10^{10}\ Nmm^2$$

Area of the second section,

$$A_2 = 3000 \times 4.5 \times 10^7$$

$$A_2 = 1.35 \times 10^{11}\ Nmm^2$$

Area of the third Section

$$A_3 = \frac{1}{2} \times 3000 \times \left(1.16 \times 10^8 - 4.5 \times 10^7\right)$$

$$A_3 = 1.8 \times 10^{11}\ Nmm^2$$

Total area of BMD,

$$A = A_1 + A_2 + A_3$$

$$= 6.75 \times 10^{10} + 1.35 \times 10^{11} + 1.8 \times 10^{11}$$

$$A = 3.825 \times 10^{11}\ Nmm^2$$

According to moment area method, slope at the free end,

$$\theta_C = \frac{\text{Area of BMD}}{\text{EI}} = \frac{3.825 \times 10^{11}}{2 \times 10^5 \times 1.3 \times 10^8}$$

$$\theta_C = 0.0147\ \text{rad}$$

According to moment area method deflection at the free end 'C',

$$y_C = \frac{A\overline{x}}{EI}$$

The total moment of the BMD about B is given by

$$A\overline{x} = A_1\overline{x_1} + A_2\overline{x_2} + A_3\overline{x_3}$$
$$= 6.75 \times 10^{10} \times \frac{2}{3} \times 3000 + 1.35 \times 10^{11} \times \left(3000 + \frac{3000}{2}\right)$$
$$= 1.8 \times 10^{11} \times \left(3000 + \frac{2}{3} \times 3000\right)$$
$$= 1.35 \times 10^{14} + 6.075 \times 10^{14} + 9 \times 10^{14}$$
$$A\overline{x} = 1.64 \times 10^{15}$$
$$y_C = \frac{1.64 \times 10^{15}}{2 \times 10^5 \times 1.3 \times 10^8} = 63.17mm \quad (\because \frac{A\overline{x}}{Ei} = y_c)$$

15. **A cantilever projecting 4m from a wall carries an UDL of 12KN/m for length of 2m from the fixed end and a point load of 1.5KN at the free end. Find the deflection at the free end. Find the slope at the free end by moment area moment area method. Take $E = 2.1 \times 10^5\ N/mm^2$ and $I = 1 \times 10^8\ mm^4$.**

Given Data

$$\text{UDL for a Length of 2m, from fixed end } (W) = 12KN/m = 12N/mm$$
$$\text{Point load at the forced } (W) = 1.5KN = 1500N$$
$$\text{Length of UDL from fixed end } (a) = 2m = 2000mm$$
$$\text{Length of beam } (L) = 4m = 4000mm$$
$$\text{Young's modulus } (E) = 2 \times 10^5\ N/mm^2$$
$$\text{Moment of Inertia } (I) = 1 \times 10^8\ mm^4$$

To Find

Slope and Deflection of a cantilever beam by

Solution

Moment area Method

$$BM\ at\ C = 0$$
$$BM\ at\ B = -1500 \times 2000 = -3 \times 10^6\ N - mm$$
$$BM\ at\ A = -1500 \times 4000 - 12 \times \frac{2000}{2}$$
$$M_A = -3 \times 10^7\ N - mm$$

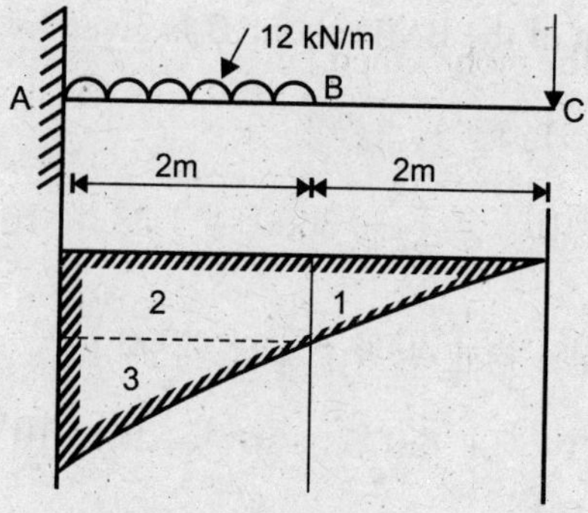

Fig .3.36

Area for the first section

$$A_1 = \frac{1}{2} \times 2000 \times 3 \times 10^6 = 3 \times 10^9\, N - mm^2$$

Area for second section,

$$A_2 = 2000 \times 3 \times 10^6 = 6 \times 10^9 Nmm^2$$

Area for the third section,

$$A_3 = \frac{1}{3} \times 2000 \times \left(3 \times 10^7 - 3 \times 10^6\right)$$
$$= 1.8 \times 10^{10}\, N - mm^2$$

Total area of BMD,

$$A = A_1 + A_2 + A_3$$
$$= 3 \times 10^9 + 6 \times 10^9 + 1.8 \times 10^{10}$$
$$A = 2.7 \times 10^{10}\, N - mm^2$$

Slope at the free end as per mohr's theorem (I) of moment area method $\frac{\text{Area of BMD}}{\text{EI}}$

$$\therefore \text{ Slope @ } (\theta_c) = \frac{\text{Area of BMD}}{\text{EI}}$$

$$\therefore \text{ Slope @ } (\theta_C) = \frac{\text{Area of BMD}}{\text{EI}}$$
$$= \frac{2.7 \times 10^{10}}{2 \times 10^5 \times 1 \times 10^8}$$
$$= 0.0013 \text{ rad}$$

Deflection as per the mohr's theorem (II) of moment area method, deflection $= \frac{A\overline{x}}{EI}$

$$\therefore \text{ @ c Deflection,} (y_c) = \frac{A\overline{x}}{EI}$$

$$\text{We know that, } A\overline{x} = A_1\overline{x_1} + A_2\,\overline{x_2} + A_3\,\overline{x_3}$$

$$= 3 \times 10^9 \times \frac{2}{3} \times 2000 + 6 \times 10^9$$

$$\times \left(2000 + \frac{2000}{2}\right) + 1.8 \times 10^{10}$$

$$\times \left(2000 + \frac{3}{4} \times 2000\right)$$

$$= 4 \times 10^{12} + 1.8 \times 10^{13} + 6.3 \times 10^{13}$$

$$A\overline{x} = 8.5 \times 10^{13}$$

$$y_C = \frac{8.5 \times 10^{13}}{2 \times 10^{15} \times 1 \times 10^8}$$

$$= 4.25mm$$

$$y_c = 4.25mm$$

16. **A cantilever beam of length $8m$ carries a UDL $18kN/m$ over length of $4m$ from the free end along with a point load $2KN$ at $4m$ from free end. Determine the deflection at the free end of the beam by moment area method. Take $E = 2.1 \times 10^5\,N/mm^2$ and $I = 1.2 \times 10^8\,mm^4$**

Given Data

$$\text{Length of Beam } (L) = 8m = 8000mm$$

$$\text{Length of UDL from } (a) = 4m = 4000mm$$

$$\text{Free end } UDL, (W) = 18KN/m = 18N/mm$$

$$\text{Point Load @ 4m from free end } (W) = 2KN, = 2000N$$

$$\text{Moment of Inertia } (I) = 1.2 \times 10^8\,mm^4$$

$$\text{Young's modulus } E = 2.1 \times 10^5\,N/mm^2$$

To Find

Deflection @ the free end by moment area method

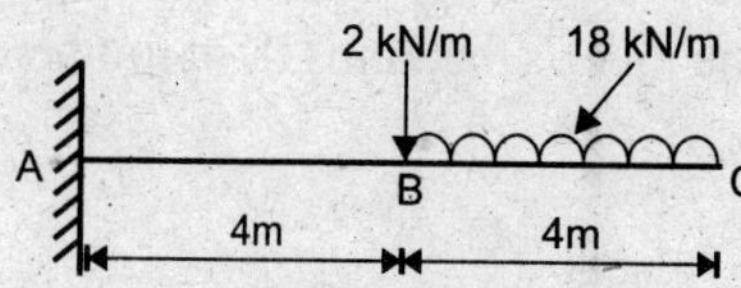

Fig .3.37

Solution

$$BM\, at\, C = 0$$

$$BM\, at\, B = -18 \times 4 \times \frac{4}{2} = -144 KNm$$

$$BM\, at\, A = -18 \times 4 \times \left(4 + \frac{4}{2}\right) - 2 \times 4$$

$$= 72 \times 6 - 8 = 440 KN - m$$

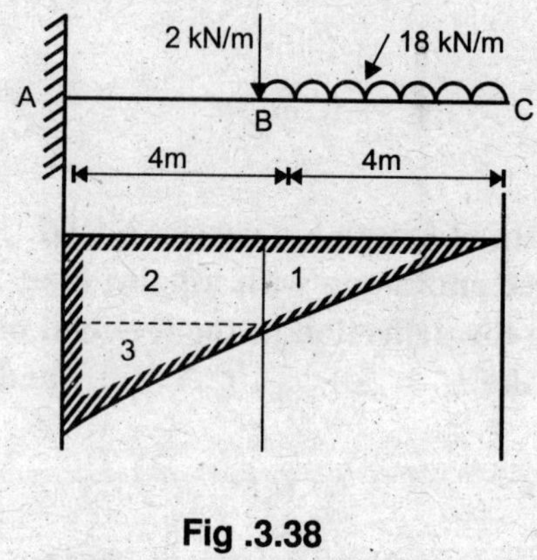

Fig .3.38

Area for the first section

$$A_1 = \frac{1}{3} \times 4000 \times \left(144 \times 10^6 - 0\right) = 1.92 \times 10^{11}\, Nmm^2$$

Area for second section,

$$A_2 = 4000 \times 144 \times 10^6 = 5.76 \times 10^{11} Nmm^2$$

Area for the third section,

$$A_3 = \frac{1}{2} \times 4000 \times (440 - 144) \times 10^6 = 5.92 \times 10^{11} Nmm^2$$

Total area of BMD,

$$
\begin{aligned}
A &= A_1 + A_2 + A_3 \\
&= 1.92 \times 10^{11} + 5.76 \times 10^{11} + 5.92 \times 10^{11} \\
A &= 1.36 \times 10^{12}\, N - mm^2 \\
A\overline{x} &= A_1\overline{x_1} + A_2\overline{x_2} + A_3\overline{x_3} \\
x_1 &= \frac{3}{4} \times 4000 = 3000mm \\
x_2 &= 4000 + \frac{4000}{2} = 6000mm \\
x_3 &= 4000 + \frac{2}{3} \times 4000 \\
x_3 &= 6666.67mm \\
A\overline{x} &= \left(1.92 \times 10^{11} \times 3000\right) \\
&\quad + \left(5.76 \times 10^{11} \times 6000\right) + \left(5.92 \times 10^{11} \times 6666.67\right) \\
&= 5.76 \times 10^{14} + 3.456 \times 10^{15} + 3.946 \times 10^{15} \\
A\overline{x} &= 7.978 \times 10^{15}
\end{aligned}
$$

As per mohr's theorem (2) of moment area method

Deflection at the free end,

$$(y_c) = \frac{A\overline{x}}{EI}$$

On substituting the appropriate values, Deflection at free end c (y_c)

$$= \frac{7.978 \times 10^{15}}{1.2 \times 10^{8} \times 2.1 \times 10^{5}} = 316.56mm$$

Problems on conjugate beam method

17. **A steel cantilever $6m$ long carries two point loads, $15kN$ at the free end $25KN$ at a distance of $3m$ from the free end find**

(i) Slope at the free end

(ii) Deflection at the free end

by conjugate beam method Take $I = 1.3 \times 10^{8}\, mm^4$ and $E = 2 \times 10^{5}\, N/mm^2$

Given Data

$$\text{Point Load 1} \quad (W_1) = 15KN = 15000N$$
$$\text{Point Load 2} \quad (W_2) = 25KN = 25000N$$
$$\text{Distance of 2nd Load} \quad (a) = 3m = 3000mm$$
$$\text{Length of Beam} \quad (\ell) = 6m = 6000mm$$

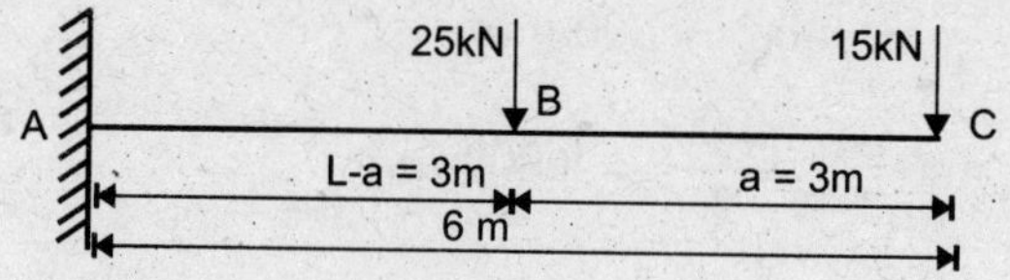

Fig .3.39

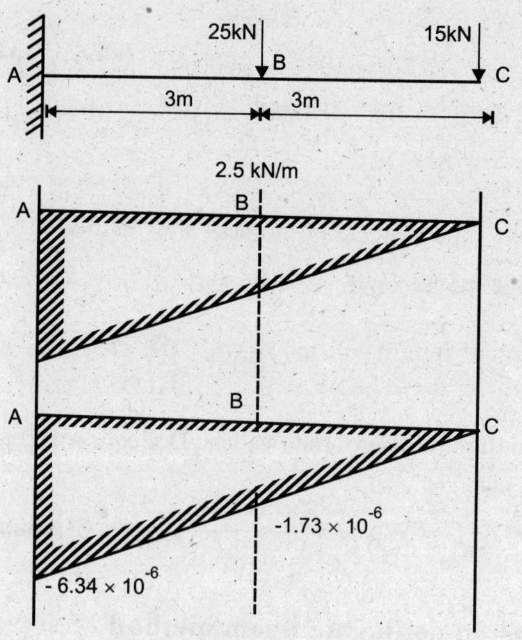

Fig .3.40

Solution

$$\frac{M}{EI} = \frac{-4.5 \times 10^7}{2 \times 10^5 \times 1.3 \times 10^8} = -1.73 \times 10^{-6}$$
$$\frac{M}{EI} = \frac{-1.65 \times 10^8}{2 \times 10^5 \times 1.3 \times 10^8} = -6.34 \times 10^{-6}$$

Total load on the conjugate beam

$$= \text{Area of the } \frac{M}{EI} \text{ diagram}$$

$$= \frac{1}{2} \times 1.73 \times 10^{-6} \times 3000 - 1.73 \times 10^{-6} \times 3000$$

$$- \frac{1}{2} \times 3000 \times \left(6.34 \times 10^{-6} - 1.73 \times 10^{-6}\right)$$

$$\therefore \quad P = -0.0147$$

$$\therefore \text{ Slope @ } \theta_C = -P = 0.0147$$

Deflection at c, (y_c) = BM at c, for the conjugate beam

$$BM \text{ at } c = \left[\frac{1}{2} \times 1.73 \times 10^{-6} \times 3000 \times \frac{2}{3} \times 3000\right]$$

$$+ \left[1.73 \times 10^{-6} \times 300 \times \left(3000 + \frac{3000}{2}\right)\right]$$

$$+ \left[\frac{1}{2} \times 3000 \times \left(6.34 \times 10^{-6} - 1 \times 10^{-6}\right),\right.$$

$$\left. \times \left(\frac{2}{3} \times 3000 + 3000\right)\right]$$

$$\text{Deflectionat } c \ \ (y_c) = 56.205mm$$

18. **A Cantilever of length $5m$ carries a UDL of $8KN/m$ length over the entire length. If the section is rectangular of $150mm \times 260mm$. Find the deflection and slope at the free end by conjugate method. Take $E = 2.1 \times 10^5 \, N/mm^2$**

Given Data

$$\text{Length, } L = 5m = 5000mm$$

$$\text{UDL, } W = 8KN/m = 8N/mm$$

$$\text{Cross section} = 150mm \times 260mm$$

$$\text{young's modulus } (E) = 2.1 \times 10^5 \, N/mm^2$$

$$\text{Moment and Section}(I) = \frac{bd^3}{12} = 2.197 \times 10^8 mm^4$$

To Find

Slope and deflection by conjugate beam method

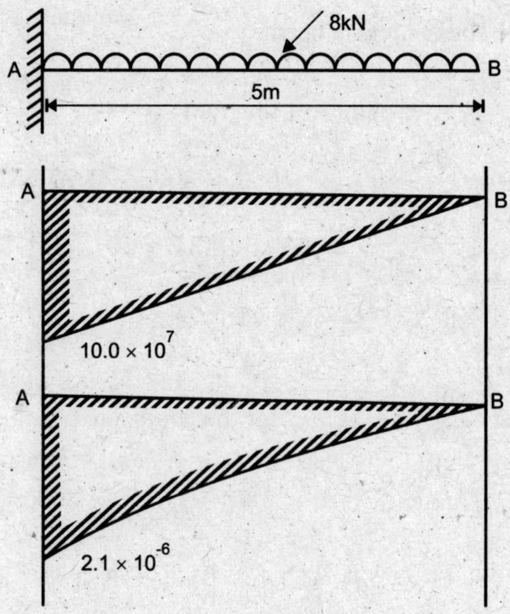

Fig .3.41

Solution

$$\frac{M}{EI} = \frac{10.0 \times 10^{-7}}{2 \times 10^5 \times 2.197 \times 10^8}$$
$$= 2.1 \times 10^6$$

Total load on the conjugate beam

$$= \text{Area of the } \frac{M}{EI} \text{ diagram}$$

$$= \frac{1}{3} \times 5000 \times \left(-2.1 \times 10^{-6}\right)$$
$$P = -0.0035$$
$$\theta_B = -P = 0.0035\,rad$$

Deflection at B, by Mohr's Theorem (2) for conjugate beam

$$\begin{aligned}(y_B) &= \text{BM at B for conjugate beam}\\ &= -P \times \frac{3}{4} \times 5000\\ &= -0.0035 \times \frac{3}{4} \times 5000\\ &= 13.125mm\end{aligned}$$

$$\text{Deflection } (y_B) = 13.125mm$$

19. **A cantilever projecting 4m from a wall caries a UDL of $12KN/m$ for a length of $2m$ from the fixed end and a point load of $1.5KN$ at the free end. Find the deflection at the free end by conjugate beam method. Take $E = 2.1 \times 10^5\ N/mm^2$ and $I = 1 \times 10^8\ mm^4$.**

Given Data

$$\begin{aligned}\text{UDL for a length of 2m from Fixed end } (W) &= 12KN/m = 12N/mm\\ \text{Point load @ free end } (W) &= 1.5KN = 1500N\\ \text{Distance of UDL } a &= 2m = 2000mm\\ \text{Span of beam } (\ell) &= 4m = 4000mm\\ \text{Young's modulus } (E) &= 2 \times 10^5\ N/mm^2\\ \text{Moment of Inertia } I &= 1 \times 10^8\ mm^4\end{aligned}$$

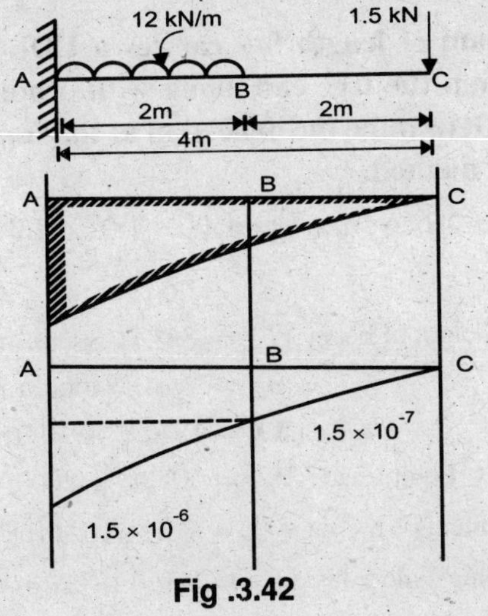

Fig .3.42

To Find

Slope and deflection of beam by conjugate beam method

Solution

$$\frac{M}{EI} = \frac{3 \times 10^6}{2 \times 10^5 \times 1 \times 10^8} = -1.5 \times 10^{-7}$$

$$\frac{M}{EI} = \frac{3 \times 10^7}{2 \times 10^5 \times 1 \times 10^8} = 1.5 \times 10^{-6}$$

Total load on the beam

$$= \text{Area of the } \frac{M}{EI} \text{ diagram}$$

$$P = \frac{1}{2} \times 2000 \times 1.5 \times 10^{-7} + 2000 \times 1.5 \times 10^{-7}$$

$$+ \frac{1}{3} \times 2000 \times \left(1.5 \times 10^{-6} - 1.5 \times 10^{-7}\right)$$

$$P = 0.00125\, rad$$

$$\text{Slope } \theta_C = -P = 0.00125\, rad$$

$$BM\, at\, C = \left[\frac{1}{2} \times 1.5 \times 10^{-7} \times 2000 \times \frac{2}{3} \times 2000\right]$$

$$+ \left[1.5 \times 10^{-7} \times 2000 \times \left(2000 + \frac{2000}{2}\right)\right]$$

$$+ \left[\frac{1}{3} \times 2000 \times (1.5 \times 10^{-6} - 1.5 \times 10^{-8}) \times \left(\frac{3}{4} \times 2000 + 2000\right)\right]$$

$$y_c = 4.565mm$$

20. **A cantilever beam of length $8m$ carries a UDL of $18KN/m$ over a length of 4m from the free end along with a point load $2KN$ at $4m$ from free end. Determine the deflection at the free end of the beam by conjugate beam method.**

Take $E = 2.1 \times 10^5\, N/mm^2$ and $I = 1.2 \times 10^8\, mm^4$.

Given Data

$$\text{Length of beam } (L) = 8m = 8000mm$$

$$(a) = 4m = 4000mm$$

$$UDL, (W) = 18KN/m = 18N/mm$$

$$\text{Point Load } (W) = 2KN = 2000N$$

$$\text{Moment of Inertia } (I) = 1.2 \times 10^8\, mm^4$$

$$\text{Young's modulus } (E) = 2.1 \times 10^5\, N/mm^2$$

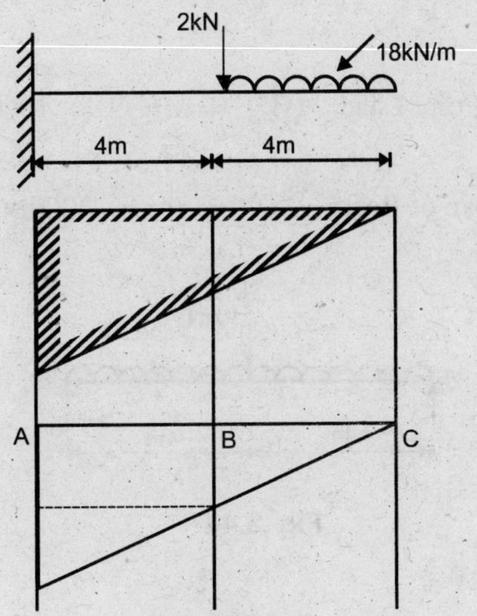

Fig .3.43

To Find

Deflection at the free end by conjugate beam method

Solution

$$\frac{M}{EI} = \frac{144}{2.1 \times 10^5 \times 1.2 \times 10^8} = 5.71 \times 10^{-6}$$

$$\frac{M}{EI} = \frac{332}{2.1 \times 10^5 \times 1.2 \times 10^8} = 1.31 \times 10^{-6}$$

Deflection at C, y, c = BM at C for the conjugate beam.

$$\text{BM at } C = \left[\frac{1}{2} \times 4000 \times 5.71 \times 10^{-6} \times \frac{3}{4} \times 4000\right]$$

$$+ \left[4000 \times 5.71 \times 10^{-6} \times \left(4000 + \frac{4000}{2}\right)\right]$$

$$+ \left[\frac{1}{2} \times 4000 \times (1.31 \times 10^{-6} - 5.71 \times 10^{-6}) \times \left(4000 + \frac{2}{3} \times 4000\right)\right]$$

$$y_C = 93.62 \times 10^{-3} mm$$

21. **A SSB of $8m$ span carries a point load of $10KN$ at its centre. It is also subjected to an UDL of $1KN/m$ over its entire span. Find the maximum deflection of beam by double integration method**

$E = 2.1 \times 10^5\ N/mm^2$ **and** $I = 1.2 \times 10^8\ mm^4$

Given Data

$$\text{Point Load} \quad (W) = 10KN = 1000N$$
$$\text{UDL} \quad (W) = 1KN/m = 1N/mm$$
$$\text{Span of Beam} \quad (L) = 8m = 8000mm$$

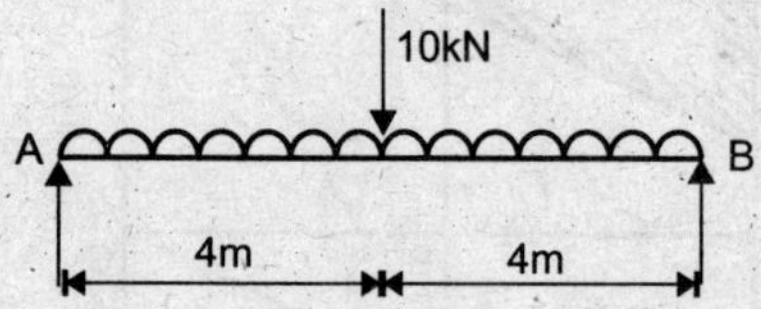

Fig .3.44

To Find

Maximum deflection by double integration method

Solution

Double Integration Method

$$y_{\max} = \text{Deflection due to point load} + \text{Deflection due UDL}$$

Now considering the formula

$$y_{max} = \frac{WL^3}{48EI} + \frac{5WL^4}{384EI}$$

On substituting the given values,we get

$$\text{max m deflection } y_{\max} = \frac{10000 \times 8000^3}{48 \times 200000 \times 200 \times 10^6} + \frac{5 \times 1 \times 8000^4}{384 \times 200000 \times 200 \times 10^6}$$
$$= 2.67 + 1.33$$
$$y_{\max} = 4mm$$

22. **A simply supported beam $8m$ long supported at its ends is carrying a point load of $50KN$ at its center. The moment of Inertia of the beam is equal to $78 \times 10^6\ mm^4$. If $E = 2.1 \times 10^6\ N/mm^2$. Calculate the deflection by using Macaulay's method**

Given Data

Span of Beam (L) = $8m$

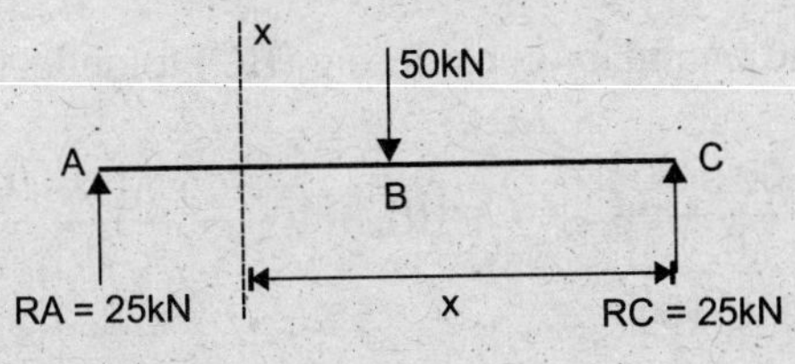

Fig .3.45

Point Load @ centre = $50kN$

Considering $\sum v = 0$

Solution

$$R_A + R_C = 50$$
$$R_A = R_C = 25KN = 25000N$$

BM at x,

$$EI\frac{d^2y}{dx^2} = M_X = |R_C x - W\left(x - \frac{L}{2}\right) \quad (1)$$

Integrating the above equation twice

Slope,

$$EI\frac{dy}{dx} = R_C\frac{x^2}{2} + C_1 = | - \frac{W\left(x - \frac{L}{2}\right)^2}{2} \quad (2)$$

Deflection Equation,

$$EIy = R_C\frac{x^3}{6} + C_1 x + C_2 | \frac{W\left(x - \frac{L}{2}\right)^3}{6} \quad (3)$$

The boundary conditions are,

(i) When $x = \frac{L}{2}m \qquad \frac{dy}{dx} = 0$

(ii) When $x = 0 \qquad y = 0$

Applying first boundary considering (BC) to slope equation,

$$c = \frac{25000 \times (4000)^2}{2} + C_1 - \frac{50 \times 10^3\,(4000 - 4000)^2}{2}$$
$$C_1 = -2 \times 10^{11}$$

Applying Second boundary considering (BC) to deflection equation,

$$O = \frac{25000 \times 0^3}{6} + C_1\,(0) + C_2 - \frac{50 \times 10^3\,(0 - 4000)^3}{6}$$

$$C_2 = 5.33 \times 10^4$$

Substituting the value of $C_1\,C_2$ and $x = \frac{L}{2} = 4000mm$ in deflection equation, we get

$$y_B = \frac{1}{78 \times 10^6 \times 2.1 \times 10^5}\left[\begin{array}{l} \frac{25000 \times (4000)^3}{6} - (2 \times 10^{11} \times 4000) \\ +5.33 \times 10^{14} - \frac{50000 \times (4000 - 4000)^3}{6} \end{array}\right]$$

$$y_B = 6.10 \times 10^{-14}\left[2.67 \times 10^{14} - 8 \times 10^{14} + 5.33 \times 10^{14} - 0\right]$$

$$\boxed{y_B = 6mm}$$

23. **A SSB of length $10m$ is loaded as shown calculate the slope and deflection at each point by Macaulay's method $E = 2.1 \times 10^5\,N/mm^2$ and $I = 78 \times 10^6\,mm^4$ ·**

Given Data

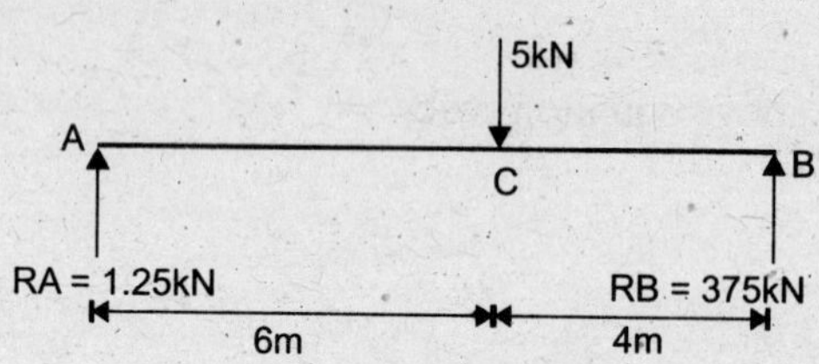

Fig .3.46

To Find

Slope and deflection by machlay's method

Solution

Considering equilibrium conditions

$$\sum M = 0 \;\; R_B \times 10 = 5 \times 6$$

$$R_B = 3KN$$

$$\sum v = 0 \;\; R_A + R_B = 5$$

$$R_A = 2KN$$

BM Equation

$$M_X = R_B - |W(x - b)$$

Integrating the above equation twice,

$$\text{Slope} EI \frac{dy}{dx} = R_B \frac{x^2}{2} + C_1 | \frac{W(x-b)^2}{2}$$

Deflection Equation

$$EI \frac{dy}{dx} = R_B \frac{x^3}{6} + C_1 x + C_2 - | \frac{W(x-b)^3}{6}$$

The Boundary conditions are

(i) When $x = 10m \quad y = 0$

(ii) When $x = 0 \quad y = 0$

Apply second BC to deflection equation,

$$0 = \frac{3 \times 10^3 \times 0^3}{6} + (C_1 \times 0)$$

$$+ C_2 - \frac{5000 \times (0 - 4000)^3}{6}$$

$$C_2 = 5.33 \times 10^{13}$$

Applying first BC to deflection equation,

$$0 = \frac{3 \times 10^3 (10000)^3}{6}$$

$$+ C_1 (1000) - 5.33 \times 10^{13} - \frac{5000\,(10000 - 4000)^3}{6}$$

$$C_1 = -2.667 \times 10^{10}$$

Deflection equation becomes

$$Y = \frac{1}{EI} \left[\frac{3 \times 10^3 x^3}{6} - 2.667 \times 10^{10} x - 5.33 \times 10^{14} - \frac{5000 \times (x - 4000)^3}{6} \right]$$

Deflection @ $A, B = 0$

Deflection @ C, When $x = 4000mm$

$$y_C = \frac{1}{2.1 \times 10^5 \times 78 \times 10^6} \begin{bmatrix} \frac{3000 \times (4000)^3}{6} - (2.667 \times 10^{10} \times 4000) \\ -5.33 \times 10^{13} - \frac{5000 \times (4000 - 4000)^3}{6} \end{bmatrix}$$

$$y_C = -7.80mm$$

Slope equation

$$\theta = \frac{dy}{dx} = \frac{1}{EI}\left[\frac{3000x^2}{2} - 2.667 \times 10^{10} - \frac{5000\,(x-4000)^2}{2}\right]$$

Slope @B, When $x = 0$,

$$\theta_B = \frac{1}{2.1 \times 10^5 \times 78 \times 10^6}\left[\frac{3000\,(0^2)}{2} - 2.667 \times 10^{10} - \frac{5000\,(0-4000)^2}{2}\right]$$

$$\theta_B = -0.00406 \text{ radians}$$

Slope @ A, when $x = 10000mm$,

$$\theta_A = \frac{1}{2.1 \times 10^5 \times 78 \times 10^6}\left[\frac{3000\,(10000^2)}{2} - 2.667 \times 10^{10} - \frac{5000\,(10000-4000)^2}{2}\right]$$

$$\theta_A = 0.0953 \text{ radians}$$

Slope @C, When $x = 4000mm$

$$\theta_c = \frac{1}{2.1 \times 10^5 \times 78 \times 10^6}\left[\frac{3000\,(4000^2)}{2} - 2.667 \times 10^{10} - \frac{5000\,(4000-4000)^2}{2}\right]$$

$$\theta_C = 0.000016 \text{ radians}$$

To find maximum deflection, put $\frac{dy}{dx} = 0$ in slope equation,

24. **A SSB of $8m$ span carries a point load of $10KN$ at its centre. It also subjects to a UDL of $1KN/m$ over its entire span. Find the maximum deflection of beam by Macaulay's method $E = 200\,KN/mm^2$, $I = 200 \times 10^6\,mm^4$.**

Given Data

$$\begin{aligned} \text{Span of Beam} \quad (L) &= 8m \\ \text{Point Load @ center} \quad (W) &= 10kN \\ \text{UDL one entire span} \quad (W_2) &= 1kN/m \\ \text{Young's modulus} \quad (E) &= 200kN/mm^2 \\ \text{Moment of Intertia} \quad (I) &= 200 \times 10^6 mm^4 \end{aligned}$$

Solution

Considering Equilibrium equation $\sum M = 0$ and $\sum v = 0$ Taking Moment

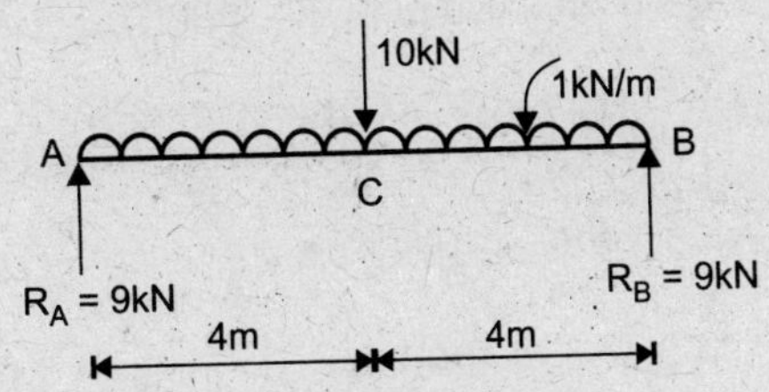

Fig .3.47

about A

$$\sum M_A = 0 \quad R_B \times 8 = (1 \times 8 \times 4) + (10 \times 4)$$

$$R_B = 9KN$$

$$\sum M_v = 0 \quad R_A + R_B = 10 + (1 \times 8)$$

$$R_A = 9KN$$

BM Equation

$$EI \frac{d^2y}{dx^2} = R_B x - \frac{Wx^2}{2} | - W(x - 4000) \qquad (1)$$

Integrating the above equation twice,

$$\text{Slope equal } EI \frac{dy}{dx} = \frac{R_B x^2}{2} - \frac{Wx^3}{6} + C_1 | - \frac{W(x - 4000)^2}{2} \qquad (2)$$

Deflection Equation

$$EIy = \frac{R_B x^3}{6} - \frac{Wx^4}{24} + C_1 x + C_2 - \frac{W(x - 4000)^3}{6} \qquad (3)$$

The boundary conditions, are

(i) When $x = 0 \quad y = 0$

(ii) When $x = 8000mm \quad y = 0$

Applying first BC to deflection equation, (3)

$$0 = \frac{R_B(0)^3}{6} - \frac{W(0)^4}{24} + C_1(0) + C_2 - \frac{W \times (0 - 4000)^3}{6}$$

$$\therefore C_2 = \frac{10000\,(0 - 4000)^3}{6} \quad C_2 = -1.067 \times 10^{14}$$

Applying the second Boundary condition to deflection equals

$$0 = \left[\frac{9000(8000)^3}{6} + C_1(8000) - 1.067 \times 10^{14} - \frac{1(8000)^4}{24} - \frac{10000\,(4000)^3}{6}\right]$$

$$C_1 = -4.7996 \times 10^{10}$$

Deflection at C, When $x = 4000mm$

$$y_{\max} = \frac{1}{200000 \times 200 \times 10^6}\left[\begin{array}{l}\frac{9000\,(4000)^3}{6}\;(4.7996 \times 10^{10} \times 4000) \\ -(1.067 \times 10^{14}) - \left(\frac{1(4000)^2}{24}\right) \\ = -\left(\frac{10000(4000-4000)^3}{6}\right)\end{array}\right]$$

$$= 2.5 \times 10^{-14}\left[9.6 \times 10^{13} - 1.919 \times 10^{14} - 1.067 \times 10^{14} - 1.067 \times 10^{13}\right]$$

$$= 2.5 \times 10^{-14}\left[-2.1324 \times 10^{14}\right]$$

$$y_{\max} = -5.33mm$$

Review Questions

Part - A

1. Why moment method is more useful when compared with double integration?
2. What is conjugate beam method?
3. State the two theorems in moment area method?
4. What are the boundary conditions for a simply supported end?
5. When Macaulay's method is preferred?
6. What is meant by double integration method?
7. When do you prefer Moment area method?
8. Write the relation between deflection of bending moment and flexural rigidity for abeam?
9. Explain the theorem of conjugate beam method
10. Write the basic equation of the elastic line of a deflected beam

11. Define method of Singularity functions?

12. What are the points to be worth for conjugate beam method?

Part - B

1. A beam of length 6 m is simply supported at its ends and carries two point loads of 48 Kn and 40 kN at a distance of 1 m and 3 m respectively from the left support. Find

   (a) Deflection under each load
   (b) Maximum deflection
   (c) The point at which the maximum deflection occurs.
   Take I = 85 × 10^6mm 4E = 2 × 10^5N/mm^2.

2. A steel joist, simply supported over a span of 6 m carries a point load of 50 kN at 1.2 m from the left hand support. Find the position and magnitude of the maximum deflection.Take EI = 14 × 10^{12} N/mm^2.

3. For the beam shown in fig show that the deflection at the free end is WL4/684EI. Use Macaulay's method.

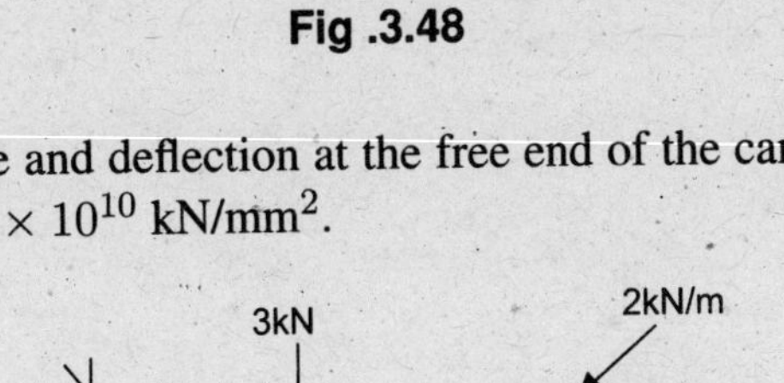

Fig .3.48

4. Find the slope and deflection at the free end of the cantilever shows in fig. Take $EI = 1 \times 10^{10}$ kN/mm^2.

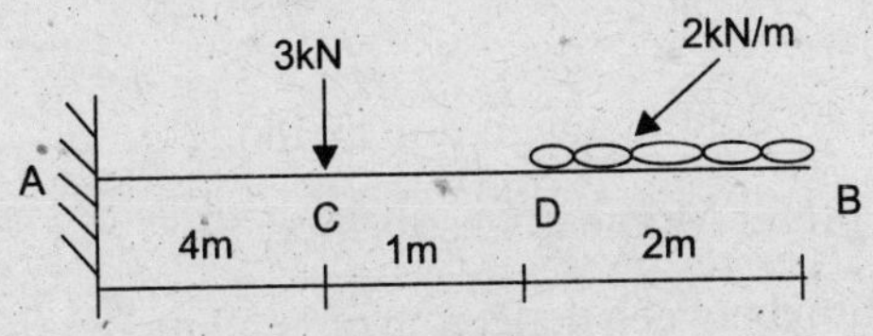

Fig .3.49

5. Using conjugate beam method, obtain the slope and deflections at A, B, C and D of the beam shown in fig. Take E = 200GPa and $I = 2 \times 10^{-2} m^4$.

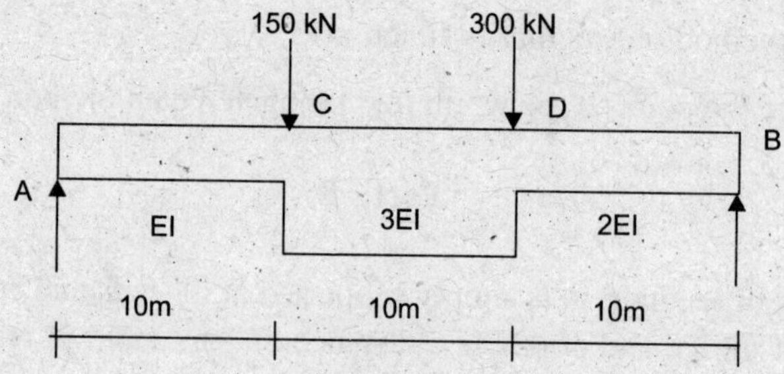

Fig .3.50

6. A steel cantilever of 2.5m effective length carries a load of 25kN at its free end. If the deflection at the free end is not exceed 40mm. What must be the I value of the section of the cantilever. Take $E = 210\ GN/m^2$ using moment area method.

Chapter 4

TORSION

Torsion of Circular and Hollow Shafts - Elastic Theory of Torsion - Stresses and Deflection in Circular Solid and Hollow Shafts - Combined bending moment and torsion of shafts - Strain energy due to torsion - Modulus of Rupture - Power transmitted to shaft - Shaft in series and parallel - Close and Open Coiled helical springs - Leaf Springs - Springs in series and parallel - Design of buffer springs.

4.1 Introduction

One of the classic examples for torsion is screw driver while tightening a screw. It involves the twisting rotation of screw rod or any prismatic bar about its longitudinal axis under the action of torsion to torque. Other examples are shafts transmitting power from the prime mover to the machine or drive shaft of an automobile. In all the above torsion examples, angular displacement of one end of cross - section with respect to the other end occurs besides shear stresses in the cross sections of the shaft perpendicular to its axis.

Torsion problems can be analysed by simplified analysis proposed by coulomb and exact analysis based on theory of elasticity. In this chapter we shall study the problems based on theory of elasticity. The assumptions made in torsion theory of elasticity are as here under.

1. The material is homogenous i.e., of uniform elastic properties exists throughout the material, isotropic and perfectly elastic.
2. The material obeys Hook's law and the stress remains within the limit of proportionality.
3. The twisting couples act in the transverse plane only.
4. The circular section remains circular and the parallel planes normal to the axis do not warp or distort because of torsion.
5. Cross section rotates as if rigid i.e., all diameters of a cross section rotate through the same angle.

4.2 Torsional Equation Derivation

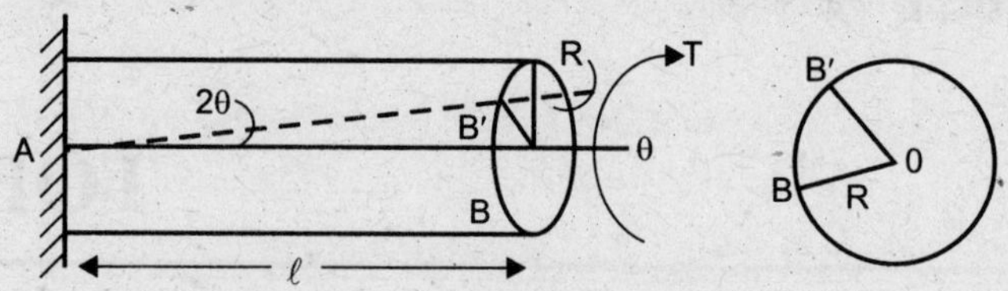

Fig .4.1 Twisting of a shaft

Let'0' be the centre of circular section and 'B' a point on surface. Let AB be the line on the shaft parallel to the axis of the shaft.

Due to torque'T' applied, let us say that B move to 'B'. If ϕ is shear stress'θ' is the angle of twist in length 'ℓ' then,

$$\text{and} \quad R\theta = BB' = \ell\phi \tag{1}$$

If 'τ' is the shear stress 'C' - the modulus of rigidly.

Then $\phi = \dfrac{\tau}{C}$

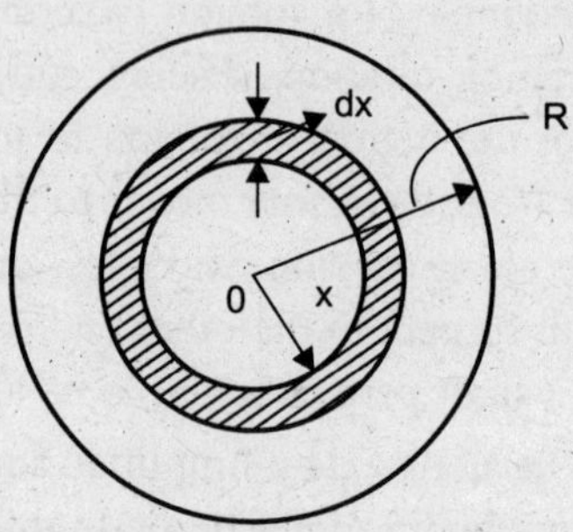

Fig .4.2 Cross section of a shaft

Let us sub 'ϕ' in equation (1)

$$R\theta = \frac{\tau}{C}.\ell$$

$$\Rightarrow C.R.\theta = \tau.\ell$$

$$= \frac{C\theta}{\ell} = \frac{\tau}{R} \tag{2}$$

Now consider an elemental area 'da' of thickness 'dx' at a distance 'x' from the centre of the shaft as shown in Figure 4.2

$$da = 2\pi \cdot x \times dx$$

4.3 Torsional Equation

Turning force = Shear stress × Area

$$
\begin{aligned}
&= \tau_x.da \\
&= \tau.\frac{x}{R}.da \\
&= \tau.\frac{x}{R}.2\pi x.dx \\
&= \frac{2\pi\tau}{R}.x^2.dx
\end{aligned}
$$

Turning Moment of this elemental area

$$
\begin{aligned}
&= \text{Turning Force} \times \text{Distance of elemental area from the axis of the shaft} \\
&= \frac{2\pi.\tau}{R}.x^2.dx.x \\
&= \frac{2\pi\tau}{R}.x^3.dx
\end{aligned}
$$

(T)The total torque can be obtained by integrating the above equation from 0 to R

$$
\begin{aligned}
&= \int_0^R \frac{2\pi\tau}{R}.x^3.dx \\
&= \frac{2\pi\tau}{R}\left[\frac{x^4}{4}\right]_0^R \\
&= \frac{\pi\tau}{2R}x^4 \\
&= \frac{\pi\tau}{2R}R^{4^3} \\
&= \frac{\pi\tau}{2}R^3
\end{aligned}
$$

Now substituting for $R = \frac{D}{2}$

$$
\begin{aligned}
&= \frac{\pi}{2}\tau\left(\frac{D}{2}\right)^3 \\
\therefore \quad T &= \frac{\pi}{2}\tau\frac{D^3}{8} = \frac{\pi\tau D^3}{16} \\
\therefore \tau &= \frac{16T}{\pi D^3}
\end{aligned}
$$

Substituting $\tau = \dfrac{16T}{\pi D^3}$ in (2)

$$\frac{c\theta}{\ell} = \frac{\tau}{R}$$

$$\Rightarrow \frac{c\theta}{\ell} = \frac{16\tau}{\pi D^3.R} \qquad \text{(substituting for } R = \frac{D}{2}\text{)}$$

$$\frac{c\theta}{\ell} = \frac{16\tau}{\pi D^3.\frac{D}{2}}$$

$$\frac{c\theta}{\ell} = \frac{32\tau}{\pi D^4}$$

$$\therefore \frac{c\theta}{\ell} = \frac{T}{\frac{\pi}{32}\cdot D^4} = \frac{T}{J}$$

$$\because \frac{c\theta}{\ell} = \frac{T}{J} \qquad (3)$$

Now, From (2) and (3),

We can write the torsion equation as $\boxed{\dfrac{c\theta}{\ell} = \dfrac{\tau}{R} = \dfrac{T}{J}}$

4.4 Strength of Solid Circular Shaft

Consider a solid shaft of Radius 'R' and shear stress. 'τ_x'. Let us consider a small elemental area of thickness 'dx'.

Turning Force on the shaft = $\tau_x \times 2\pi x \times dx$

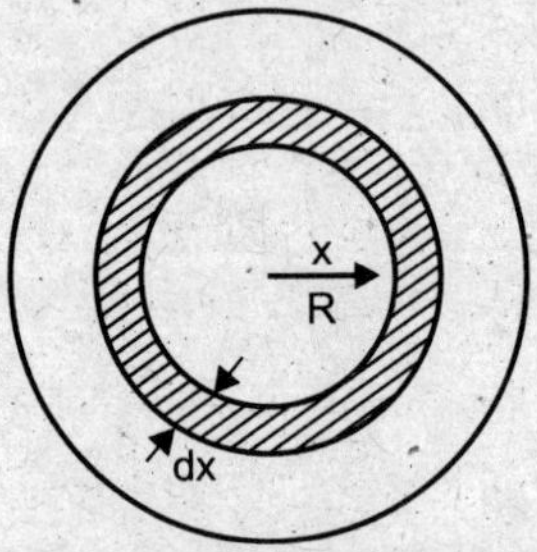

Fig .4.3 Cross section of solid circular shaft

Turning moment on the shaft dT = $\tau_x \times 2\pi x \times dx \times x$.

$$\int dT = \int_0^R \tau_x 2\pi x.dx.x$$

$$T = 2\pi \int_0^R x^2 dx.\tau_x$$

Substituting for $\left(\because \quad \tau_x = \dfrac{\tau \cdot x}{R}\right)$

$$= 2\pi\tau \int_0^R \frac{\tau.x}{R}.x^2 dx$$

$$= \frac{2\pi\tau}{R} \int_0^R x^3 dx$$

$$= \frac{2\pi\tau}{R}\left[\frac{x^4}{4}\right]_0^R = \frac{2\pi\tau}{R}\left[\frac{R^4}{4}\right] = \frac{\pi\tau R^3}{2}$$

$$= \frac{R^3\pi\tau}{2} = \frac{\pi\tau}{2}[D/2]^3 (\because \; R = D/2)$$

$$= \frac{\pi\tau}{2}\left[\frac{D^3}{8}\right] = \frac{\pi\tau D^3}{16}$$

$$\therefore \quad \text{Torque (T)} \boxed{= \frac{\pi\tau D^3}{16}}$$

4.5 Strength of a Hollow Circular Shaft

Consider a hollow circular shaft of outer Radius R_1 Inner radius r, and shear stress τ_x.

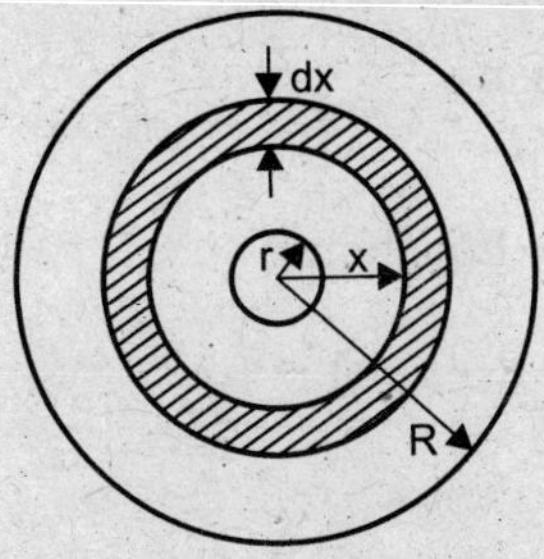

Fig .4.4 Cross-section of hollow circular shaft

Turning force on the hollow shaft, $\tau x \times 2\pi x \times dx$

Turning moment,

$$dT = \tau_x \times 2\pi x \cdot dx \cdot x$$

$$\int dT = \int_r^R \tau \times 2\pi x \cdot dx \cdot x$$

$$\int dT = \int_r^R \tau \times 2\pi \times x^2 dx$$

$$\int dT = \int_r^R \frac{\tau \times x}{R} 2\pi x^2 dx = \int_r^R \frac{2\pi}{R} \tau x^3 dx \qquad (\because \tau_x = \frac{\tau x}{R})$$

$$\int dT = \frac{2\pi\tau}{R} \int_r^R x^3 dx$$

$$T = \frac{2\pi\tau}{R} \left[\frac{x^4}{4}\right]_r^R = \frac{2\pi\tau}{R} \left[\left(\frac{R^4}{4}\right) - \left(\frac{r^4}{4}\right)\right]$$

$$= \frac{2\pi\tau}{4R} \left[R^4 - r^4\right] = \frac{\pi\tau}{2R} \left[R^4 - r^4\right] \qquad R = \frac{D}{2} \;\&\; r = \frac{d}{2}$$

$$= \frac{\pi\tau}{2R} \left[\left(\frac{D}{2}\right)^4 - \left(\frac{d}{2}\right)^4\right] = \frac{\pi\tau}{2R} \left[\frac{D^4}{16} - \frac{d^4}{16}\right]$$

$$\boxed{T = \frac{\pi}{16} \tau \left[\frac{D^4 - d^4}{D}\right]} = \boxed{\frac{T}{16} \tau \cdot \left[\frac{D^4 - d^4}{D}\right]}$$

Where

C → Modulus of rigidity in N/mm^2

θ → Angle of twist in radians.

τ → Shear stress in N/mm^2.

We know from the Torsion Equation

$$\frac{C\theta}{\ell} = \frac{\tau}{R} \qquad (1)$$

For solid shaft, Torsion, shear stress and Radius are as below.

$$T = \frac{\pi}{16} \times \tau \times D^3 \;;\; \tau = \frac{16 \times T}{TD^3} ; R = \frac{D}{2}$$

Now substituting for τ and R in equal (1)

$$\frac{c\theta}{\ell} = \frac{16 \times T/\pi D^3}{D/2} = \frac{32T}{\pi D^4}$$

$$\boxed{\frac{c\theta}{\ell} = \frac{T}{J}}$$

($\because$ Polar moment of inertia $J = \frac{\pi}{32D^4}$)

For Hollow shaft,

D $\rightarrow$ Outer diameter $\qquad J = \frac{\pi}{32}[D^4 - d^4]$

d $\rightarrow$ Inner diameter

Torsional Rigidity: It is the product of polar moment and Inertia and Rigidity modulus.

As per the torsion theory angle of twist is proportional to 'τ'.

i.e. $\theta \ \alpha \ T$

$$\frac{T}{J} = \frac{c\theta}{\ell}$$

$$\boxed{\theta = \frac{T\ell}{cJ}}$$ Where C J $\rightarrow$ Torsional Rigidity

Percentage of saving in weight

$$= \frac{\text{Weight of solid shaft - Weight of hollow shaft}}{\text{Weight of solid shaft}} \times 100$$

Weight of solid shaft = Area $\times \rho \times$ L

$$= \frac{\pi}{4} D^2 \times \rho \times L$$

Weight of hollow shaft $= \pi/4 \left(D^2 - d^2\right) \times \rho \times L$

Where

$D =$ Dia of shaft inmm^2

$\rho =$ Density of shaft in kg/mm^2

$L =$ Length of shaft in mm^2

Percentage of saving in material

$$= \frac{\text{Area of solid shaft - Area of hollow shaft}}{\text{Area of solid shaft}} \times 100$$

Area of solid shaft $= \pi/4D^2$

Area of hollow shaft $t = \pi/4 \left(D^2 - d^2\right)$

Shaft in series

There are situations in reality for possibilities of shaft connected in series called composite shaft.

If we take two shafts connected in series, then Angle of twist '$\theta' = \theta_1 + \theta_2$.

We know that,

$$\frac{T}{J} = \frac{c\theta}{\ell} \Rightarrow \theta = \frac{T\ell}{CJ}$$

$$\therefore \quad \theta = \frac{T_1\ell_1}{J_1C_1} + \frac{T_2\ell_2}{J_2C_2}$$

$$\theta = T\left[\frac{\ell_1}{C_1J_1} + \frac{\ell_2}{C_2J_2}\right]$$

$$(\therefore \quad T_1 = T_2 = T)$$

($\because$ Putting subscripts for to shafts number (1) and (2)).

When shafts are made of same material, shear modulus will be same

$$\boxed{C_1 = C_2 = C}$$

$$\theta = \frac{T}{C}\left[\frac{\ell_1}{J_1} + \frac{\ell_2}{J_2}\right]$$

Here the driving torque is applied at one end and resisting torque at the other end.

Shaft is Parallel

When the driving torque is applied at one end of the shafts and the resisting torque is at the other end of the shafts, then the shafts are said to be in parallel.

Here the angle of twist is same for each shaft. If two shafts are connected in parallel, $\theta_1 = \theta_2$.

(i.e., angle of shaft is same for both the shafts)

$$i.e., \theta_1 = \theta_2$$

$$\therefore \quad \frac{T_1\ell_1}{C_1J_1} = \frac{T_2\ell_2}{C_2J_2} \qquad \left(\because \theta = \frac{\tau L}{CJ}\right)$$

$$\frac{T_1\ell_1}{J_1} = \frac{T_2\ell_2}{J_2} \qquad (C_1 = C_2 = C)$$

$$\frac{T_1}{T_2} = \frac{\ell_2J_1}{J_2\ell_2} = \frac{\ell_2J_1}{J_2\ell_1}$$

$$\boxed{J_2\ell_1 = J_1\ell_2} \qquad (\therefore \quad T_1 = T_2)$$

Compound shaft

A compound shaft is made up of two or more different materials bound together in such a way that the shaft is elongated or compressed as a single Shaft

Modulus of rupture

It is the maximum, shear stress occurring in a shaft under torsion which will rupture it. It can be measured using the formula $\tau_r = \frac{TuR}{J}$

where

$\tau_r \rightarrow$ Modulus of rupture in torsion

$T_u \rightarrow$ Ultimate torque

R $\rightarrow$ Radius of the shaft.

Power Transmitted by a shaft

If T is the applied Torque and N is the number if revolution of shaft then the power transmitted by the shaft is

$$\text{Power,}(P) = \frac{2\pi NT}{60}$$

Where, $T \rightarrow$ Torque in $KN.m$

$P \rightarrow$ Power in kw

$N \rightarrow$ Number of revolutions/minute

Torsional stiffness

The torsional stiffness K is defined as the torque per radian twist

$$K = \frac{T}{\theta} = \frac{CJ\theta/L}{0} = \frac{CJ}{L}\left(\frac{T}{J} = \frac{C\theta}{L} \therefore T = \frac{CJ\theta}{L}\right)$$

$$\therefore\ K = \frac{CJ}{L}$$

4.6 Springs

A spring may be defined as an elastic member whose primary function is to deflect or distort under the action of applied load: It recovers its original shape when the load is released. Springs are also referred to be energy absorbing units whose function is to store energy and to restore it slowly or rapidly depending on the particular application.

4.6.1 Types of springs

There are various types of springs such as

1. **Helical spring:** They are made of wire coiled into a helical form, the load being applied along the axis of the helix, In this type of spring the major stress caused is torsional shear stress due to twisting. They are both used in tension and compression.

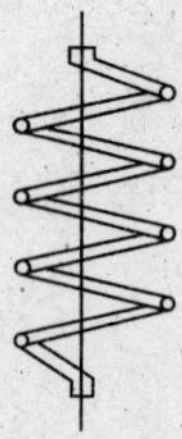

Fig .4.5 Helical Spring

2. **Spiral Springs:** They are made of flat strip of metal would in the form of spiral and loaded in torsion. In this the major stresses are tensile and compression due to bending.

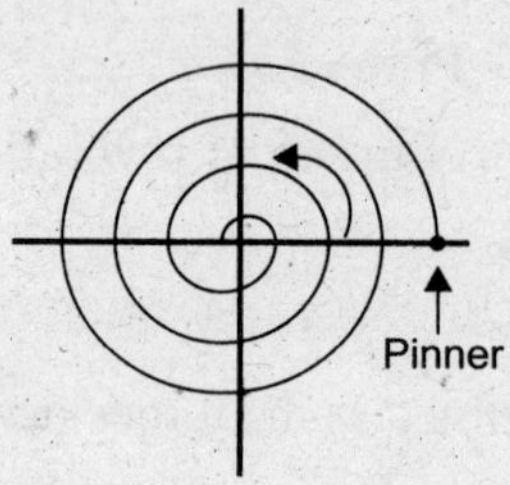

Fig .4.6 Spiral Spring

3. **Leaf Springs:** They are composed of flat bars of varying lengths clamped together so as to obtain Greater efficiency. Leaf spring may be full elliptic, semi elliptic or cantilever types, In such type of springs the major stresses caused are tensile and compressive.

These types of springs are used in the automobile suspension system.

Uses of springs

1. To apply forces and to control motions as in brakes and clutched.
2. To measure forces as in spring balance.

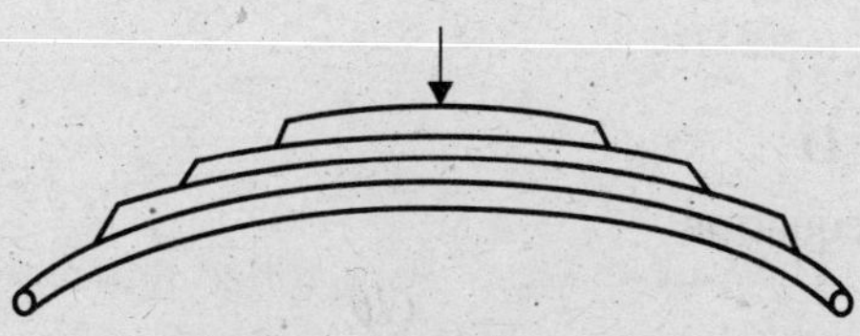

Fig .4.7 Leaf spring

3. To store energy as in clock springs.
4. To reduce the effect of Shock or impact loading as in carriage springs.
5. To change the vibrating characteristics of a member as inflexible mounting of motors.

Close Coiled helical spring

Closed - coiled springs are the springs in which helix angle is very small or otherwise the pitch between two adjacent turns is small.

Let us consider a close coiled spring as shown in Figure 4.8.

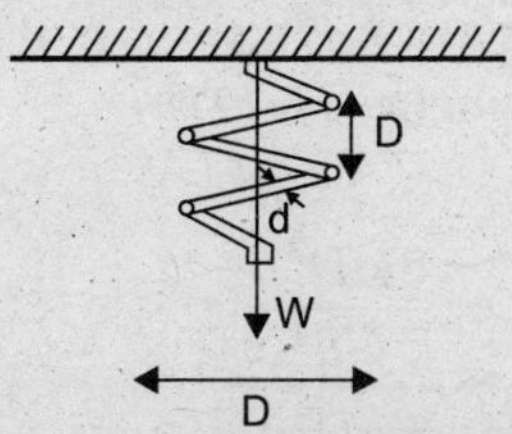

Fig .4.8 Close coiled spring

Let			
Let	W	=	axial load
	D	=	mean coil diameter
	d	=	diameter of spring wire
	n	=	number of active coils
	C	=	spring index = D/d for circular wires
	ℓ	=	length of spring wire
	G	=	modulus of rigidity
	x	=	deflection of spring
	q	=	Angle of twist

Closed - coiled Helical springs Subjected to axial load

Consider a closed helical spring subjected to axial load '*w*' at one end.

The twisting moment exerted on the wire.

T = Axial force x mean radius of spring

i.e., T = W × R

$\ell = 2\pi Rn$ or π D

From the torsion equation

$$\frac{T}{J} = \frac{C\theta}{\ell}$$

Angle of twist for whole length of the spring

$$\theta = \frac{T\ell}{CJ} = \frac{w \times R \cdot 2\pi Rn}{C \cdot \frac{\pi}{32} \cdot d^4} \qquad \text{(Substituting for T, } \ell \text{ and J)}$$

$$\boxed{\theta = \frac{6w\pi R^2 n}{Cd^4} \text{radians}}$$

Thus the force will twist through an angle θ consequently the force end will have an axial movement R θ. If is the axial movement (or) deflection.

$$\int = R\theta = R \times \frac{64\pi R^2 n}{Cd^4}$$

$$\int = \frac{64\pi R^2 n}{Cd^4} \text{ (or)}$$

$$\int = \frac{64\pi (D/2)^3 n}{Cd^4}$$

$$= \frac{8wD^3 n}{Cd^4}$$

$$\boxed{\int = \frac{8wD^3 n}{Cd^4}}$$

The stiffness K of the spring is the load required to produce unit deflection.

$$K = \frac{w}{J} = \frac{w}{\frac{8wD^3 n}{Cd^4}} = \frac{wCd^4}{8wD^3 n}$$

The maximum shear stress induced in any section under axial load is obtained by torsion equation.

$$W.K.T \quad \frac{T}{J} = \frac{\tau}{r}$$

$$\tau = \frac{T}{J} \times r = \frac{wR}{\frac{\pi d^3}{32}} \times \frac{d}{2}$$

$$\text{Shear Stress} \quad \tau = \frac{16wR}{\pi d^3} \text{ (or)}$$

$$\tau = \frac{8wD}{\pi d^3}$$

Springs in Series: If two springs of different stiffness are joined together at the end and carry a common load W, they are said to be connected in series and the combined stiffness and deflection are given by the following equation.

Fig .4.9 Spring in series

$$\frac{W}{k} = x_1 + x_2 = \frac{w}{k_1} + \frac{w}{k_2}$$

$$\boxed{\frac{1}{k} = \frac{1}{k_1} + \frac{1}{k_2}}$$

Springs in Parallel: If the two spring are joined in such a way that they have a common deflection 'x'; then they are said to be connected in parallel. In this care the load carried is shared between the two springs and total load $W = W_1 + W_2$.

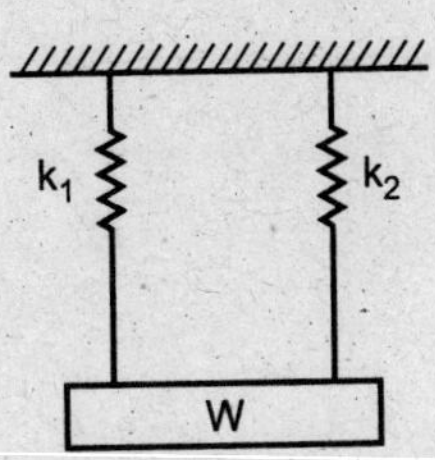

Fig .4.10 Springs in parallel

$$x = \frac{W}{k} = \frac{W_1}{k_1} = \frac{W_2}{k_2}$$

$$\therefore \quad W_1 = \frac{Wk_1}{k}$$

$$\text{Also, } W = W_1 + W_2$$

$$\therefore \quad \boxed{k = k_1 + k_2}$$

4.6.2 Strain Energy Due to Torsion

The strain energy is defined as the energy stored within a material when the work has been done on the material.

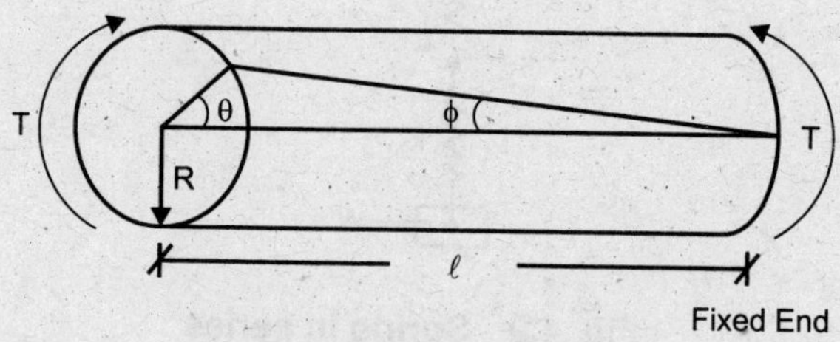

Fig .4.11 Torsion in a shaft

For solid shaft: Let us consider a solid circular shaft of length ℓ and radius R, subjected to a torque T producing a twist θ in the length of the shaft.

Strain energy (or) work done = $\frac{1}{2}.T.\theta$

But for Torsion equation,

$$\frac{T}{J} = \frac{C\theta}{\ell} = \frac{\tau}{R}$$
$$T = \frac{\tau J}{R}; \theta = \frac{\tau \ell}{CR}$$

T and θ in work done equation,

Substituting in the strain energy equation = $\frac{1}{2}.T.\theta$

$$= \frac{1}{2}.\frac{TJ}{R}.\frac{TJ}{CR}$$
$$= \frac{T^2 J\ell}{2CR^2}$$

$$\therefore J = \frac{\pi D^4}{32} \qquad (\because \; D = 2R)$$

$$J = \frac{\pi R^4}{2}$$

$$\therefore \quad \text{Strain Energy} = \frac{1}{2}.\frac{\tau^2 \pi R^4 \ell}{2CR^2} \qquad \therefore \quad D = 2R, \; D^4 = 16R^4$$

$$= \frac{\tau \pi R^2 \ell}{4C}$$

$$\therefore \quad \text{Strain Energy} u = \frac{\tau^2}{4C} \pi R^2 \ell \qquad \therefore \quad A = \frac{\pi d^2}{4}$$

$$= \frac{\tau^2}{4C} \times \text{Volume} \qquad A = \pi R^2, \; V = A \times \ell$$

$$\text{Strain energy for solid shaft} = \frac{\tau^2}{4C} \times V$$

For hollow shaft

$$\therefore \quad \text{Strain Energy} = \frac{1}{2}.T.\theta$$

$$= \frac{1}{2}.\frac{\tau J}{R}.\frac{\tau \ell}{CR}$$

$$= \frac{1}{2}\frac{\tau^2 J \ell}{CR^2}$$

$$J_{\text{hollow}} = \frac{\pi}{32}\left[D^4 - d^4\right]$$

$$J = \frac{\pi}{2}\left[R^4 - d^4\right]$$

$$\therefore \quad \text{Strain Energy} u = \frac{1}{2}\frac{\tau^2}{CR^2}\frac{\pi}{2}\left[R^4 - d^4\right].\ell$$

$$= \frac{\tau^2 \pi}{4CR^2}\left(R^4 - d^4\right).\ell$$

$$= \frac{\tau^2 \pi}{4CR^2}.\ell\left[R^2 + r^2\right]\left[R^2 - r^2\right]$$

$$\therefore \quad a^2 - b^2 = (a + b)(a - b)$$

$$= \frac{\tau^2 \pi}{4CR^2}\left[R^2 - r^2\right].\ell$$

$$\therefore \quad \text{Strain Energy} = \frac{\tau^2}{4C}\left[\frac{R^2 + r^2}{R^2}\right].\text{Volume}$$

$$\boxed{V = A.\ell = \pi(R^2 - r^2)}.\ell$$

$$\text{Strain energy for hollow shaft} = \frac{\tau}{4C}\left[\frac{R^2 + r^2}{R^2}\right].\text{V}$$

$R \quad \rightarrow \quad$ Radius of the shaft

Part - A

Question and Answer — 2 Marks

1. **Define Torsion**

   If a pair of equal magnitude is acting upon a body in opposite directions, it tends to twist the body such a twisting is called as torsion. It is otherwise called as twisting moment or torsional moment or simply as torque or torsion.

2. **What are the assumptions made in torsion equation?**

   i) The material is homogenous i.e. of uniform elastic properties exists throughout the material, isotropic and perfectly elastic.
   ii) The material obeys Hook's law and the stress remains within the limit of proportionality.
   iii) The twisting couples act in the transverse plane only.
   iv) The circular section remains circular and the parallel planes normal to the axis do not warp or distort because of torsion.
   v) Cross section rotates as if rigid i.e., all diameters of a cross section rotate through the same angle.

3. **Write torsional equation.**

$$\frac{T}{J} = \frac{\tau}{R} = \frac{C\theta}{l}$$

where T = Torque in N.mm. J = polar moment of inertia in mm^4
τ = Shear stress in N/mm^2, R = radius in mm
C = Modulus of rigidity in N/mm^2
θ = Angle of twist in rad, l = length in mm

4. **Write down the expression for Power transmitted by a shaft.**

   $$\text{Power} = \frac{2\pi NT}{60}$$

   Where, T = torque in KN-m.

N - Speed in RPM

P - Power in KW

5. **Write down the expression for Torque in circular solid shaft**

Torque, $T = \frac{\pi}{16}.\tau.D^3$

T = Torque in N.mm

τ = Shear stress in N/mm^2

D - Dia in mm

6. **Write down the expression for torque transmitted by hollow shaft.**

$$T = \frac{\pi}{16}\tau.\frac{D^4 - d^4}{D}$$

Where,

T = Torque in N.mm

τ = Shear stress in N/mm^2

D - Outer dia in mm

d = Inner dia in mm

7. **Write down the equation for maximum shear stress of a solid circular section in diameter 'D' when subjected to torque 'T'.**

$$\tau = \frac{16.T}{\pi.D^3}$$

Where,

T = Torque in N-mm

τ = Shear stress in N/mm^2

D - Dia in mm

8. **Write down the equation for maximum shear stress of a hollow circular section in diameter 'D' when subjected to torque 'T'.**

$$\tau = \frac{16.TD}{\pi.(D^4 - d^4)}$$

Where,

T = Torque in N.mm

τ = Shear stress in N/mm^2

D - Dia in mm

D - Outer dia in mm

d - Inner dia in mm

9. **Define polar modulus**

It is the ratio between polar moment of inertia and radius of the shaft.

$$= \frac{\text{Polar moment of inertia}}{\text{Radius}} = \frac{J}{R}$$

10. **Write the polar modulus for solid and hollow circular shaft.**

For solid shaft, Polar modulus

$$= \frac{\text{Polar moment of inertia}}{\text{radius}} = \frac{J}{R} = \frac{\pi}{16} D^3$$

Where,

$$J = \frac{\pi}{32} D^4; R = \frac{D}{2}$$

For hollow shaft, Polar modulus

$$= \frac{\text{Polar moment of inertia}}{\text{radius}} = \frac{J}{R} = \frac{\pi}{16D}(D^4 - d^4)$$

Where,

$$J = \frac{\pi}{32} [D^4 - d^4]; \; R = \frac{D}{2}$$

11. **For same weight, which shaft will carry more torque, a solid one or a hollow one?why?**

Hollow shaft will carry more torque because polar inertia will be more for hollow shaft than the solids shaft for the same material and length.

12. **Why hollow circular shaft are preferred when compared to solid circular shafts?**

(a) The torque transmitted by the hollow shaft is greater than the solid shaft.

(b) For same material, length and given torque, the weight of the hollow shaft will be less compared to solid shaft.

13. **Define torsional rigidity.**

It is a product of modulus of rigidity and polar moment of inertia (CJ).

We know the torsion equation

$$\frac{T}{J} = \frac{C\theta}{l}, \quad \text{From the equation}$$

$$\text{We can work,} \quad \theta = \frac{Tl}{CJ} \quad \text{where}$$

the term 'CJ' is known as **torsional rigidity** and it is indicated by K.

14. **How shafts are coupled?**

    Shafts are coupled by key or bolts

15. **What are the conditions to design a circular shaft?**

    1. The stress should be within the limit of the torque.

    2. Angle of twist should be within the torque.

16. **Define torsional energy or torsional resilience.**

    It is the strain energy stored due to angular twist. It is the product of average torque and twist.

17. **Define spring.**

    Spring is a device which absorb energy by very large change in its form without permanent deformation.

18. **Classify the spring.**

    Torsion spring

    Bending spring

19. **What are the types of spring?**

    (a) Semi elliptical Leaf spring.
    (b) Quarter elliptical Leaf spring.
    (c) Closed Coiled helical spring
    (d) Open Coiled helical spring.

20. **Define close coiled Spring?**

    If the spring wire is coiled so close that the each turn is practically in a plane at right angles to the axis of the helix, such a spring is called close coiled spring. Such springs are good at tension.

21. **Define open coiled spring?**

    The helical spring wire spring is coiled in such a way that there is large gap between two consecutive turns,then the type of spring is called open - coiled spring. As a result of large gap between consecutive coils, these springs can take compressive as well as tensile loads.

22. **Define helix angle?**

    Helix angle is the angle which the coils make with planes perpendicular to the axis of spring.

23. **What is meant by stiffness? What is the formula for the stiffness of a close coiled helical spring subjected to an axial load?**

    The stiffness of the spring is defined as the load required to produce unit deflection stiffness of a closed coiled spring is refused by $k = \frac{Cd^4}{64R^3 \cdot n} N/mm$

24. **A helical spring is made of 4mm steel wire, with a mean radius of 25mm and number of turns of coil 15. What will be deflection of the spring under a load of 6N?**

    Given Data

$$\text{Dia of spring wire } (d) = 4mm$$
$$\text{Mean Radius of spring } (R) = 25mm$$
$$\text{Number of turns } (N) = 15$$
$$\text{Load } (W) = 6N$$

Solution

i) Deflection of close -coiled spring (δ)

$$(\delta) = \frac{64wR^3n}{Cd^4}$$

Substituting the values

We get,

$$= \frac{64 \times 6 \times 25^3 \times 15}{80 \times 10^3 \times 4^4}$$
$$\delta = 4.39mm$$

25. **State the differences between close - coiled and open-coiled helical springs.**

	Close - Coiled helical springs	Open-Coiled helical springs
1.	Adjacent coils are very close to each other	Large gap between adjacent coils.
2.	Only tensile load can carry.	It can carry Tensile and compressive loads.
3.	Helix angle is negligible, less than 10 deg.	Helix angle is considerable, More than 10 deg.
4.	Eg: Brake, accelerator	E.g. Shock absorber, ballpoint pen.

26. **State the deflection and shear stress formula for close - coiled helical spring.**

Deflection of close - coiled spring

$$(\delta) = \frac{64wR^3n}{Cd^4}$$

Shear stress in a close coiled spring

$$(f_s) = \frac{16wR}{\pi d^3}$$

Where,

$$W = \text{Axial Load}$$
$$R = \text{Radius of spring}$$
$$n = \text{Number of turns}$$
$$d = \text{Dia of wire}$$
$$C = \text{Shear modulus}$$

27. **The stiffness of spring is 10 N/mm. What is the axial deformation in the spring when a load of 50N is acting?**

Given Data

Solution

$$\text{Stiffness of spring } (K) = 10N/mm$$
$$\text{Axial Load } (W) = 50N$$
$$\text{We know that stiffness}(pc) = \frac{\text{Load}}{\text{deflection}}$$
$$(k) = \frac{W}{\delta}$$
$$\therefore \quad \delta = \frac{W}{K}$$
$$\text{Now substituting the values deflection } (\delta) = \frac{50}{10} = 5mm$$
$$\text{We know that stiffness } (k) = \frac{\text{Load}}{\text{deflection}}$$
$$\text{i.e.,} = k = \frac{w}{\delta}$$

28. **An open coiled helical spring of mean radius of coil of 20cm and helix angle of 12° is subjected to an axial load of 10N. What is the bending moment in coil?**

Given Data

Solution

$$\text{Mean radius of spring } (R) = 20cm = 200mm$$
$$\text{Helix angle } (\alpha) = 12^\circ$$
$$\text{Axial Load } (w) = 10N$$

Bending moment, caused in the coil (M) = W R sin α Now, substituting the values, we get

$$= 10 \times 200 \times \sin 12^\circ = 415.82N.mm$$

Bending Moment (M) = 415.82 N-mm

29. **The stiffness of spring is 10 N/mm and the axial deflection is 10mm. What is the axial load on the spring?**

$$\text{Stiffness of spring } (K) = 10N/mm$$
$$\text{Deflection of spring } (\delta) = 10mm$$
$$\text{We know that } (K) = \frac{W}{\delta}$$
$$W = k\delta$$

Now substituting for values, we get

$$\because = 10N/m \times 10N/m = 100N$$

30. **What kind of stress is induced when an axial load acts on a (i) close coiled spring? (ii) axial open coiled spring?**

Close coiled spring: Shear stress

Open coiled spring: Bending and shear stress.

31. **The angle of helix of a spring is 'α'', Write down equations for torque and moment under an axial load 'W' at the free end.**

Torque, $T = WR\cos\alpha$

Moment, $M_0 = WR\sin\alpha$

32. **Write down the equation for shear strain energy of a close coiled spring**

Shear strain energy

$$U = \frac{fs^2}{4C} \times \text{Volume of the spring}$$

where fs = Shear stress in the spring

C = Modulus of rigidity

33. **What is the value of maximum shear stress, in a close coiled helical spring subjected to axial force W?**

$$\text{Shear stress in close coiled spring } (f_s) = \frac{16WR}{\pi d^3}$$

where (W) = Axial Food

(R) = Mean radius of coil

(d) = Dia of wire

34. **Write the expression for vertical deflection of a close coiled helical spring due to vertical load W.**

$$\text{Deflection of of helical spring } (\delta) = \frac{64WR^3n}{Cd^4}$$

where (W) = Axial food in the spring

(n) = number of turns

(R) = mean radius of coil

(d) = Dia of wire

35. **What are the uses of close coiled helical spring?**

Railway wagons, cycle seating, pistols, brakes etc.

36. **Write the equation for the deflection of an open coiled helical spring.**

$$\delta = \frac{64wR^3n\sec\alpha}{d^4}\left[\frac{\cos^2\alpha}{C} + \frac{2\sin^2\alpha}{E}\right]$$

where (R) = Mean radius of coil

(E) = Young's modulus of spring

(C) = Shear modulus of spring

(α) = Helix angle

(d) = Dia of spring wire

(n) = Number of turns

37. **What is meant by spring constant or spring index?**

    Spring constant is the ratio of mean diameter of the spring to the diameter of the wire.

38. **Define Wahl's factor.**

    Wall's factor,is a corrective factor in respect of shear stress in spring taking into account change in coil curvature.

    This is normally referred as 'k'. When introducing this factor in the formula for calculating shear stress in solid shaft subjected to torsion, the shear stress formula is written as $\tau = \frac{16T.k}{\pi D^3}$.

    Where 'k' is the wahl's factor.

39. **Write down the expression for deflection of a semi elliptical spring.**

    Deflection of semi elliptical spring is given by the formula,

$$(\delta) = \frac{3WL^3}{8Enbt^3}$$

where

W - Load acting on the spring

L - Span of spring

b - Width of plates

t - Thickness of plates

n - Number of plates

40. **Write down the expression for stress developed in a semi elliptical spring.**

    Stress developed in semi elliptical spring

$$(\sigma) = 3WL/2nbt^2$$

Where,

W - Load acting on the spring

L - Span of spring

b - Width of plates

t - Thickness of plates

n - Number of plates

41. **Write down the expression for deflection developed in a quarter elliptical spring.**

Deflection in quarter elliptical spring

$$\delta = 6WL^3/Enbt^3$$

W - Load acting on the spring

L - Span of spring

b - Width of plates

t - Thickness of plates

n - Number of plates

42. **Write down the expression for stress developed in a quarter elliptical spring**

Stress developed in quarter elliptical spring

$$(\sigma) = 6WL/nbt^2$$

Where,

W - Load acting on the spring

L - Span of spring

b - Width of plates

t - Thickness of plates

n - Number of plates

43. **Define leaf spring and where it is used?**

Leaf spring are laminated springs consisting number of parallel strips of a metal having different lengths and same width, placed one over the other.

They are used to absorb shocks in railways wagons, coaches, and road vehicles such as bus, trains, car etc.,

It is commonly employed in the suspension system of rear axles of heavier vehicles.

44. **List the types of leaf spring.**

Semi elliptical spring

Quarter elliptical spring

45. **What are the other names for a leaf spring?**

    Carriage springs, laminated springs.

46. **A leaf spring has 6 plates of width 50mm and a thickness 5mm. find the section modulus at mid span section.**

    Section Modulus = $bt^2/6$

    Thickness = 6 plates × 5mm i = 30mm

    $Z = 50 \times 30^2/6 = 7500mm^3$

47. **What is master leaf?**

    The longest leaf which provided with eyes at the ends for supporting the spring is called master leaf.

Part - B

Question and Answer

Problems on Torsion

1. **A solid circular shaft transmits 75kW power at 200 rpm Calculate the shaft diameter, if the twist in the shaft is not to exceed 1° in 2 meters length of shaft and shear stress is limited to 50 MN/m². Take C = 100 GN/m².**

Give Data

$$P = 75kW = 75 \times 10^3 watts$$
$$N = 200rpm, \theta = 1^\circ, \ell = 2m = 2000mm$$
$$\tau_s = 50\ MN/m^2$$

Shear stress in N/mm^2

$$(\tau_s) = \frac{50 \times 10^6}{(1000)^2} = 50N/mm^2$$

$$\text{Shear modulus } (C) = 100GN/m^2 = \frac{100 \times 10^9}{(1000)^2} = 100 \times 10^3 N/mm^2$$

Solution

We know that, Power transmitted by a shaft

$$(P) = \frac{2\pi NT}{60}$$

$$T = \frac{60P}{2\pi N}$$

Where,

N = Number of revolution in RPM

P = power in kW

T = Torque

Now, substituting for values, we get torque (t)

$$\therefore T = \frac{60 \times 75}{2 \times \pi \times 200}$$

$$= 3.58 kN.m$$

$$= 3.58 \times 10^6 Nmm$$

i) **Diameter of shaft considering the strength of shaft**

For this case we have to use the formula

$$T = \frac{\pi}{16}\tau_s D^3$$

$$D = \sqrt[3]{\frac{16T}{\pi Ts}}$$

$$= \sqrt[3]{\frac{16 \times 3.58 \times 10^6}{\pi \times 50}}$$

$$\boxed{D = 71.47mm}$$

$$\boxed{D = 71.4mm} \quad \text{(i)}$$

ii) **Diameter of shaft considering the stiffness of shaft**

For this case, we have to use

Torsion equation $\frac{T}{J} = \frac{C\theta}{\ell}$

$$T = 3.58 \times 10^6 Nmm$$

$$J = \frac{\pi}{32}D^4$$

$$C = 1 \times 10^5 N/mm^2$$

$$\theta = 1^\circ \times \pi/180 \text{ rad} = \frac{\pi}{180}\text{rad}$$

$$\ell = 2000mm$$

Now, on substituting the values,

$$\frac{3.58 \times 10^6}{\frac{\pi}{32}D^4} = \frac{1 \times 10^5 \times 1 \times \pi/80}{2000}$$

$$D^4 = \frac{3.58 \times 10^6 \times 2000 \times 32 \times 180}{\pi \times \pi \times 1 \times 10^5 \times 1}$$

$$\therefore \quad \boxed{D = 80.4\,mm}$$

$\therefore$ Diameter of the shaft is the large value of (i) and (ii) i.e, 80 mm

2. **A solid shaft is subjected to a torque of 45KNM. If the angle of twist is 0.5 degree per meter length of the shaft and the shear stress is not to be allowed to exceed 90M/Nm^2. Find i) Suitable diameter for the shaft ii)Final maximum shear stress iii) maximum shear strain in the shaft.Take $C = 80GN/m^2$.** (AU May 2013

Given Data

$$\text{Torque } (T) = 45kNm = 45 \times 10^3 Nm = 45 \times 10^6 Nmm$$

$$\text{Angle of twist } (\theta) = 0.5^\circ = \frac{0.5 \times \pi}{180} rad$$

$$\text{Length of shaft } (\ell) = 1m = 1000mm$$

Shear stress induced in the shaft

$$(\tau_s) = 90MN/m^2 = \frac{90 \times 10^6 N}{(1000)^2 mm^2} = 90N/mm^2$$

$$\text{Shear Modulus } (c) = 80GN/m^2 = \frac{80 \times 10^9}{(1000)^2} = 80 \times 10^3 N/mm^2$$

Solution

i) **Diameter of solid shaft**

Diameter of shaft considering the strength if shaft

We know that,

$$T = \frac{\pi}{16}\tau_s D^3$$

$$D = \sqrt[3]{\frac{16T}{\pi\tau_s}} = \sqrt[3]{\frac{16 \times 45 \times 10^6}{\pi \times 90}}$$

Substituting for T andτ_s,

We get

$$D = 136.57mm \tag{i}$$

Diameter of shaft considering the stiffness of shaft

$$\frac{T}{J} = \frac{C\theta}{\ell}$$

Substituting for T, C, θ, ℓ and $J = \frac{\pi}{32}D^4$

$$\frac{45 \times 10^6}{\frac{\pi D^4}{32}} = \frac{80 \times 10^3 \times 0.5 \times \frac{\pi}{180}}{1000}$$

$$D^4 = \frac{45 \times 10^6 \times 1000 \times 32 \times 180}{\pi \times 80 \times 10^3 \times 0.5 \times \pi}$$

$$\boxed{D = 160\ mm} \qquad \text{(ii)}$$

Diameter of the shaft is the larger value of

i) * ii) $\therefore \quad D = 160\ mm$

ii) **Find maximum shear stress(τ_s)**

$$T = \frac{\pi}{16}\tau^s D^3$$

Now, substituting for

$$D = 160$$

$$T = 45 \times 10^6 Nmm$$

We get,

$$\tau_s = \frac{T.16}{\pi.D^3}$$

$$= \frac{45 \times 10^6 \times 16}{\pi \times 160^3}$$

$$= 55.9$$

$$\boxed{\tau_s = 56N/mm^2}$$

iii) **Maximum shear strain in the shaft**

We known that

$$\text{Shear modulus} = \frac{\text{Shear stress}}{\text{Shear strain}}$$

$$\therefore \quad \text{Max shear strain} = \frac{\text{Max. shear stress}}{\text{Shear modulus}}$$

$$\text{Now, on substituting the} = \frac{56}{80 \times 10^3}$$

$$\text{We get} = \text{Shear strain} = 7 \times 10^{-4}$$

3. **A solid circular shaft transmit 75kw power at 200 rpm. Calculate that shaft diameter, if the twist in the shaft is not to exceed 1° in 2 meters length of shaft, and shear stress is limited to 50MN/m^2. Take C = 100GN/m^2. Take C = 100 GN/m^2** **(AU Nov'2013.)**

Given Data

Power, (P) =75 kw = 75 $\times 10^3$ Watts

Speed, (N) = 200 RPM

Angle of twist, $(\theta) = 1°$ (Or) $\frac{\pi}{180}$ Radians

To Find

(i) Dia of shaft

Solution

We know that

$$\text{Power } (P) = \frac{2\pi NT}{60}$$

Now, on substituting different values we get torque.

$$75 \times 10^3 = \frac{2 \times \pi \times 200 \times T}{60}$$

$$(T) = 3580.99Nm = 3580.99 \times 10^3 Nmm$$

i) Dia of shaft on considering shear stress

$$T = \frac{\pi}{16} \times \tau_s \times D^3$$

On substituting T, τ_s we get,

$$3580.99 = \frac{\pi}{16} \times 50 \times 10^6 \times D^3$$

$$D = 0.072m$$

$$\boxed{D = 72mm}$$

ii) Dia of shaft on considering twist of shaft

$$\frac{I}{I_P} = \frac{G\theta}{l} = \frac{f_s}{R}$$

$$\frac{I}{I_P} = \frac{G\theta}{l}$$

We know the Torsion equation

$$\frac{T}{J} = \frac{C\theta}{\ell} = \frac{\tau_s}{R}$$

Now, on substituting for different values,

$$\frac{3580.99}{\pi/32 \times D^4} = \frac{1 \times 10^{11} \times \frac{\pi}{180}}{2}$$

$$7161.98 = 171347298.6D^4$$

$$D = 0.0804m$$

$$\text{We get } \boxed{D = 80.4mm}$$

$\therefore$ The suitable diameter of the shaft is the higher of two values i. e., $D = 80.4mm \simeq 80mm$

4. **A solid steel shaft has to transmit 100HP at 200 rpm. Taking allowable shear stress as 70 N/mm^2, find the suitable diameter of the shaft if the maximum torque transmitted in each revolution exceeds the mean by 30%. Also, find the outer diameter of a hollow shaft to replace the solid shaft if the diameter ratio is 0.7. (AU Nov 2011)**

Given Data

Power, $(P) = 100HP = 100 \times 0.745Kw \qquad (\because\ 1HP = 745kw)$

$$\therefore \boxed{P = 74.56Kw}$$

Speed, (N) = 200 RPM

Shear stress, (T) = 70 N/mm^2

Given that,

$$T_{max} = 1.3T_{mean}$$

Diameter ratio = 0.7

To Find Outer Dia of hollow shaft

We know that,

Solution

$$\text{Power,} = \frac{2\pi NT}{60}$$

Now on substituting the values, we get

$$\Rightarrow 74.56 = \frac{2\pi \times 200 \times T}{60}$$

$$\text{Torque } (T) = 3.559kNm = 3.559 \times 10^3 Nm$$

$$\boxed{T = T_{mean} = 3.559 \times 10^6 N - mm}$$

We know that, $T_{\max} = 1.3T_{mean}$

$$T_{\max} = 1.3T_{mean} = 1.3 \times 3.559 \times 10^6$$

$$\boxed{T_{\max} = 4.62 \times 10^6 N - mm}$$

We know that, $T_{max} = \frac{\pi}{2} \times \tau_s \times D^3$

Now, on substituting values for T_{max}, and τ_s, we get Dia Meter

$$4.62 \times 10^6 = \frac{\pi}{16} \times 70 \times D^3$$

$$\boxed{D = 69.52mm}$$

Solid shaft is replaced by hollows shaft we know that Torque transmitted by solid shaft is equal to hollow shaft.

Let,D_1 - External diameter of hollow shaft

D_2 - Internal diameter of hollow shaft.

Torque transmitted by hollow shaft.

Torque,

We know the formula for Torque

$$(T) = \frac{\pi}{16} \times \tau \left[\frac{D_1^4 - D_2^4}{D}\right]$$

$$= \frac{\pi}{16} \times \tau \left[\frac{D_1^4 - (0.7D_1)^4}{D_1}\right]$$

$$\because \left[\frac{D_2}{D_1} = 0.7 \because D_2 = 0.7D_1\right]$$

$$= \frac{\pi}{16} \times \tau \times \frac{D_1^4}{D_1}\left[1-(0.7)^4\right]$$

$$= \frac{\pi}{16} \times 70 \times D_1^4\,[0.7599]$$

$$= 4.62 \times 10^6 = 10.44D\overset{3}{1}$$

$$\boxed{\Rightarrow D_1 = 76.20mm}$$

We know that, $D_2 = 0.7$

$$D_1 = 0.7 \times 76.20$$

$$\boxed{\therefore \quad D_2 = 53.34mm}$$

Result

Diameter of the solid shaft, D = 69.52mm

External diameter of the hollow shaft.

D_1 = 76.20mm

Internal diameter of the hollow shaft

D_2 = 53.34mm

5. **Derive Torsional formula. Also explain how do you analyze the shafts fixed at both ends. (AU May 2012)**

Refer section 4.2

6. **A hollow steel shaft 5m long is to transit 160kN of power at 120 r.p.m. The total angle of twist is not to exceed 2° in thin length and allowable shear stress is 50 N/mm^2. Determine the inside and outside diameters of the shaft, taking N = 0.8 $\times 10^3$N/mm^2 and $\frac{D_1}{D_2} = 0.4$ (AU Nov/Dec 2012)**

Given Data

Power (P) = 160 k W

Speed (N) = 120 rpm

Shear Stress (τ) = 50N/mm^2

$$\text{Ratio between inner and outer Dia}\frac{D_1}{D_2} = 0.4$$

$$\text{Angle of twist}(\theta) = 2^\circ \times \frac{\pi}{180} = 0.035\ rad$$

$$\text{Shear modulus}(c) = 0.8 \times 10^5 N/mm^2$$

To Find

We know that

$$\text{Power}(P) = \frac{2\pi NT}{60}$$

substituting for values,

$$160 = \frac{2 \times \pi \times 120 \times T}{60}$$

$$\therefore \quad \text{Torque } (T) = 12.73kNm$$

$$T = T_{mean} = 12.73 \times 10^6 Nmm$$

Solution

Let us assume as

$$T_{\max} = 1.2T_{mean}$$

$$\text{Now, substituting} T_{mean}, T_{\max} = 1.2 \times 12.73 \times 10^6 Nmm$$

$$T_{\max} = 15.28 \times 10^6 Nmm$$

Inner Dia of following shaft considering shear stress(τ)for hollow shaft. The diameter can be determined as below.

$$\text{We known that Torque,} T_{\max} = \frac{\pi}{16} \times \tau \times \left[\frac{D_2^4 - D_1^4}{2}\right] \quad \because \frac{D_1}{D_2} = 0.4$$

$$15.28 \times 10^6 = \frac{\pi}{16} \times 50 \times \left[\frac{D_2^4 - (0.4D_2)^4}{D_2}\right]$$

$$1.56 \times 10^6 = \frac{0.9744D_2^4}{D_2}$$

$$D_2 = 116.98mm$$

$$D_1 = 46.8mm \quad \therefore \; D_1 = 0.4D_2$$

Considering angle of twist 'θ' can be determined as below.

$$\frac{T_{\max}}{J} = \frac{C\theta}{l}$$

$$J = \frac{\pi}{32}\left[D_2^4 - D_1^4\right]$$

$$= \frac{15.28 \times 10^6}{\frac{\pi}{32} \times \left[D_2^4 - D_1^4\right]}$$

$$= \frac{0.8 \times 10^5 \times 0.035}{5000} \quad \because \left(\frac{D_1}{D_2} = 0.4\right)$$

$$D_2^4 - (0.4D_2^4) = 277.93 \times 10^6$$

$$0.9744D_2^4 = 277.93 \times 10^6$$

$$D_2 = 129.96mm$$

$$D_1 = 51.98mm$$

From the above two cases, maximum values obtained should be taken as suitable external and internal diameter of the shaft.

i.e., External diameter, D_2 = 129.96mm

Internal diameter, D_1 = 81.98mm

7. **A composite shaft length of 4.5m consist of a steel rod 75mm diameter surrounded by a closely fitting tube of brass. A torque of 1200N-m is applied to the composite length of 4.5m consist of a steel rod 75mm diameter surrounded by a closely fitting tube of brass. A torque of 1200N-m is applied to the composite shaft and it will be shared equally by the two materials. If modulus of rigidity of steel is 0.9 × 10^5 N/mm^2 and modulus of rigidity of brass is 0.5 × 10^5 N/mm^2. Find**

   i. Outside diameter of the brass tube
   ii. Maximum shear stress in each material
   iii. Angle of twist

Given Data

$$\text{Length of the shaft}(\ell) = 4.5m = 4500mm$$

$$\text{Diameter of steel rod} ds = 75mm$$

$$\text{Torque} = 1200N.m = 1200 \times 10^3 N.mm$$

It is stated that the total torque is equally shared by two material. Hence

$$\text{Torque shared by steel rod}(T_s) = 600 \times 10^3 N.m$$

$$\text{Torque shared by brass tube}(T_b) = 600 \times 10^3 N.m$$

$$\text{Inner Dia of Brass tube}(d_i) = 75mm$$

$$\text{Modulus of rigidity of steel}(c_s) = 0.9 \times 10^5 N/mm^2$$

$$\text{Modulus of rigidity of brass}(c_b) = 0.5 \times 10^5 N/mm^2$$

To Find

i) Outside Dia of Brass tube

ii) Max Shear stress in each moment

iii) Angle of twist

Solution

(i) Outside Dia of Brass tube

We know that the Torsion equation, From this, we can write

$$\frac{T}{J} = \frac{C\theta}{\ell}$$

$$\text{for steel rod} \Rightarrow \frac{T_s}{J_s} = \frac{c_s\theta_s}{\ell_s} \Rightarrow T_s = \frac{c_s\theta_s J_s}{\ell_s}$$

$$\text{for brass tube} \Rightarrow \frac{T_b}{J_b} = \frac{c_b\theta_b}{\ell_b} \Rightarrow T_b = \frac{c_3\theta_3 J_3}{\ell_3}$$

$$T_s = T_b = 600 \times 10^3 N.mm$$

It is stated that

$$\therefore \quad \frac{c_s\theta_s J_s}{\ell_s} = \frac{c_b\theta_b j_b}{\ell_b} \quad \because \ \ell_s = \ell_b \text{and} \theta_s = \theta_b$$

$$c_s j_s = c_b j_b$$

Now substituting the values

$$0.9 \times 10^5 \times (\pi/32)\ ds^4 = 0.5 \times 10^5 \times (\pi/32)\left[do^4 - di^4\right]$$

Outer diameter of brass $\boxed{(do) = 97.01mm}$

We know from the Torsion equation

$$\frac{T}{J} = \frac{\tau}{R}$$

$$\text{For steel} = \frac{T_s}{J_s} = \frac{\tau_s}{R_s}$$

$$\text{For brass} = \frac{T_b}{J_b} = \frac{\tau_b}{R_b}$$

$$\therefore \text{ Shear stress in steel } (\tau_s) = \frac{T_s}{J_s} \times R_s$$

$$= \frac{600 \times 10^3}{\pi/32 \times 75^4} \times \left(\frac{75}{2}\right) (\because \; R = \frac{d}{2} = \frac{75}{2})$$

$$\boxed{\tau_s = 7.24 N/mm^2}$$

$$\text{Shear stress in Brass } (\tau_b) = \frac{T_b}{J_b} \times R_b$$

$$= \frac{600 \times 10^3}{\pi/32 \left[d_0^4 - di^4\right]} \left(\frac{d_o}{2}\right)$$

$$= \frac{600 \times 10^3}{\frac{\pi}{32} \left[97.01^4 - 75^4\right]} \left(\frac{97.01}{2}\right)$$

$$= \boxed{5.20 N/mm^2}$$

We know that angle of twist

$$\theta = \frac{T_s \, \ell_s}{C_s \, J_s} = (or) \frac{T_b \, \ell_s}{C_b \, J_b}$$

Substituting for steel, we get

$$\theta = \frac{600 \times 10^3 \times 4500}{0.9 \times 10^5 \times \frac{\pi}{32} \times (75)^4}$$

$$\theta = 0.009 rad$$

$$= 0.009 \times \frac{180}{\pi}$$

$$\theta = 0.55^\circ$$

8. **Find the total angle of twist at the free end if the maximum shear stress in the shaft is limited to 80 MN/m^2. Take C = 85 GN/m^2. The stepped steal shaft in fig is subjected to a torque T at the free end and a torque of 2 T in the opposite direction at the junction of two sizes.**

Given Data

Shaft diameter, AB	=	200mm
Shaft diameter, BC	=	100mm
Torque, AB	=	2 T
Torque, BC	=	1 T
Modulus of rigidity, C	=	$85 \times 10^3 N/mm^2$
Shear stress,	=	$80MN/m^2$
	=	$80N/mm^2$
Length of shaft, AB	=	1m = 1000mm
Length of shaft, BC	=	1.6m = 1600mm

To Find

Total angle of twist at free end

For the length BC,

Solution

Torque $T = 1T$

$$\text{Length of shaft } (\ell_1) = 1600mm$$

$$\text{Polar moment of inertia } (J) = \pi/32d^4 = \pi/32 \times 100^4$$

$$= 9817477.04mm^4$$

Angle of twist

$$\theta 1 = \frac{T\ell}{CJ} = \frac{1T \times 1600}{85 \times 10^3 \times 9817477.4}$$

For length AB, Torque, T = 2T

$$\text{Length of shaft } (\ell_1) = 1000mm$$

$$\text{Polar moment of intertia}(J) = \pi/32 \times 200^4$$

$$= 157079632.7mm^4$$

Angle of twist, from the Torsion equation, can be on substituting the values we get,

$$\text{Written as } (\theta 2) = \frac{T\ell}{CJ} = \frac{2T \times 1000}{85 \times 10^3 \times 157079632.7}$$

Here θ_1 and θ_2 are in opposite direction

∴ The total angle of twist at c.

$$\theta_c = \theta_1 - \theta_2$$

To find torque, $\frac{T}{J} = \frac{\tau}{R}$

$$T = \frac{\tau.J}{R} = \frac{80 \times 9817477.04}{100/2}$$

$$= 15707963.26 N.mm$$

$$\therefore \theta_1 = \frac{1 \times 15707963.26 \times 1600}{85 \times 10^3 \times 9817477.04} = \frac{T\ell}{CJ} \qquad [\because \ T = 1\,T]$$

$$= 0.03011 \text{radians}$$

$$\therefore \theta_2 = \frac{2 \times 15707963.26 \times 1000}{85 \times 10^3 \times 157079632.7} \qquad [\because \ T = 2\,T]$$

$$= 0.007176 \text{radians}$$

$$\therefore \theta_c = \theta_1 - \theta_2 = 0.08011 - 0.001176 = 0.1289 \text{ radians}$$

$$\theta_c = 0.1289 \times \frac{180°}{\pi} = 1°39'$$

$\therefore$ Total angle of twist = $1°.39'$

9. **A horizontal steel bar of 50mm diameter and 1000mm length is fixed at both ends. A rigid bracket is fixed to the middle of the bar, at right angles top its axis which have a length of 300mm and in the same horizontal plane containing the axis of the bar. Determine the deflection of the shaft of the vertical load in the bracket is 1500N.**

   Take $E = 2 \times 10^5 N/mm^2$ and $C = 7.6 \times 10^{-4} N/mm^2$

   Given Data

   Length of the bar, (L) = 1000mm

   Diameter of the bar, (D) = 50mm

   Length of the bracket, (ℓ' =) 300mm

   Load, (W) = 1500N

   To Find Deflection(s) of the fixed shaft

   Solution

   Moment of inertia of the bar

$$I = \frac{\pi}{64} \times D^4$$

$$= \frac{\pi}{64} \times 50^4 = 306796.15 \ mm^4$$

Polar moment of inertia of the bar

$$J = \frac{\pi}{32} \times D^4 = \frac{\pi}{32} \times 50^4$$
$$= 613592.3 \ mm^4$$

We know that deflection due to bending of the bar is

$$= \frac{w\ell^2}{192EI} = \frac{1500 \times 1000^3}{192 \times 2 \times 10^5 \times 306796.15}$$
$$= 0.127mm$$

and deflection due to rotation is,

$$\int_2 = \ell^1\theta$$
$$\text{But } \theta = \frac{T\ell}{CJ}$$

Where,

$$\ell = \frac{L}{2} = \frac{1000}{2} = 500mm$$
$$T = \frac{w\ell^1}{2} = \frac{1500 \times 300}{2} = 225000N.mm$$
$$\theta = \frac{225000 \times 500}{7.6 \times 10^4 \times 613592.3}$$
$$\therefore \ \theta = 2.4 \times 10^{-3}\text{rad}$$
$$\therefore \ S_2 = \ell^1\theta = 300 \times 2.4 \times 10^{-3}$$
$$= 0.72mm$$

Total deflection

$$S = S_1 + S_2$$
$$= .127 + 0.72mm$$
$$= 0.847mm$$
$$\therefore \ \boxed{S = 0.847mm}$$

Problems on springs

10. **A close - coiled helical spring is made out of 10mm diameter steel rod. The coil consists of 10 complete turns with a mean diameter of 120mm.**

The spring carries an axial pull of 200N. Find the maximum shear stress induced in the section of the rod. If C = 80 B/GN/m^2, Find the deflection in the spring, the stiffness and strain energy stored in the spring **(AU Nov 2013.)**

Give Data

Load on spring, (W) = 200N

Dia of wire, (d) = 10mm

Mean dia of coil, (D) = 120mm

No.of coils, (n) = 10

Solution

(i) Shear Stress

We know that shear stress in a close coiled spring

$$(f_s)_{max} = \frac{16WR}{\pi d^3}$$

Now substituting the values

$$= \frac{16 \times 200 \times 120}{\pi \times 10^3 \times 2} \qquad (\because R = 120/2)$$

$$= 61.12 N/mm^2$$

ii) We know that deflection (δ)

$$\delta = \frac{64WR^3n}{Gd^4}$$

Now substituting the values

$$= \frac{64 \times 200 \times \left(\frac{120}{2}\right)^3 \times 10}{80 \times 10^3 \times 10^4}$$

We get deflection (δ) $= 34.56\ mm$

Energy stored

$$= \frac{1}{2} W\delta$$

Now substituting the values

$$= \frac{1}{2} \times 200 \times 34.56$$

$$= 3456 Nmm$$

11. **An open coiled helical spring of wire diameter 12mm, mean coil radius 84mm, helix angle 20° carries an axial load of 480N. Determine the shear stress and direct stress developed at inner radius of the coil. (AU May 2013)**

Given Data

Dia of wire (d) = 12mm

Coil Dia (D) = 84mm

Helix angle = $(\alpha) = 20°$

Axial load (P) = 480 N

To find

1. Shear stress in the spring
2. Direct stress developed @ inner radius of coil.

Solution

1. Shear stress in open linked spring (τ).

$$(\tau) = \frac{16wR\cos\alpha}{\pi \times 12^3} \qquad [\because\ R = \frac{D}{2}]$$

On substituting the value

$$= \frac{16 \times 480 \times 42 \times \cos 20°}{\pi \times 12^3}$$

$$\boxed{\tau = 55.83N/mm^2}$$

ii) Direct stress in an open - coiled spring

$$(\sigma_{b1}) = \frac{16wR}{\pi d^3}(\sin\alpha + 1)$$

Now on substituting the valued

$$= \frac{16 \times 480 \times 42}{\pi \times 12^3} \qquad [\sin 20° + 1]$$

We get, $\boxed{\sigma_{b1} = 79.73\ N/mm^2}$

Result

(i) Shear stress, $(\tau) = 55.83N/mm^2$

(ii) Direct stress, $(\sigma_{b1}) = 79.73N/mm^2$

12. **A closely coiled helical spring made of 10mm diameter steel wire has 15 coils of 100mm mean diameter. The spring is subjected to an axial load of 100N. Calculate**

i) The maximum shear stress induced
ii) The deflection
iii) Stiffness of spring.
Take modulus of rigidity, C = 8.16 × 10^4 N/mm^2.

(AU May 2010)

Given Data

Dia of wire, (d) = 10mm

Number of coils, (n) = 15

Mean dia of coils; (D) = 100mm

Mean radius of coil $(R) = \frac{100}{2} = 50mm$

Axial load, (w) = 100N

Modulus of rigidity, (C) = $8.16 \times 10^4 N/mm^2$

To Find

i) Maximum shear stress induced, f_s
ii) Deflection f
iii) Stiffness of the spring

Solution

i) **Maximum shear stress induced,=** f_s

$$f_s = \frac{16wR}{\pi d^3} = \frac{16 \times 100 \times 50}{\pi \times 10^3}$$

On substituting the relevant data,

$$\boxed{f_s = 24.46N/mm^2}$$

ii) **Deflection** (δ)

$$\delta = \frac{64wR^3n}{cd^4}$$
$$= \frac{64 \times 100 \times 50^3 \times 15}{8.16 \times 10^4 \times 10^4}$$

On substituting the relevant data,

$$\boxed{\delta = 14.7mm}$$

iii) **Stiffness of the spring (k)**

$$k \text{ Stiffness} = \frac{\text{Load on spring}}{\text{Deflection of spring}} = \frac{W}{\delta}$$

$$\text{On substituting the values} = \frac{100}{14.7} = 6.802 N/mm$$

Result

i) Maximum shear stress induced (f_s) = 24.46 N/mm²
ii) The Deflection, (f) = 14.7 mm
iii) Stiffness of the spring (k) = 6.802 N/mm

13. **A carriage spring is build up of 9 plates each 75 min wide and 6.5mm thick. Find the length of the spring so that it may carry a central load of 4kN, the stress being limited to 160N/mm^2.** **(AU Nov'2011)**

Given Data

Number of plate, (N) = 9

Width, (b) = 75mm

Thickness, (t) = 6.5 mm

Load, (W) = 4 kN = 4000 N

Stress, (σ_b) = 160N/mm^2

Young's modulus, (E) = $2 \times 10^5 N/mm^2$

To Find

(i) Length of the spring, (*l*)

(ii) Deflection, (δ)

Solution

(i) Length of the spring,

We know that,

$$\text{Bending Stress, } (\sigma_b) = \frac{3}{2} \times \frac{W\ell}{Nbt^2}$$

$$\text{On substituting values, } 160 = \frac{3}{2} \times \frac{4000 \times l}{9 \times 75 \times (6.5)^2}$$

$$\boxed{\ell = 760mm}$$

We get, length of spring

ii) Deflection of carriage spring

$$(\delta) = \frac{3wl^3}{8ENbt^3}$$

On substituting the value

$$\delta = \frac{3 \times 4000 \times (760)^3}{8 \times 2 \times 10^5 \times 9 \times 75 \times (6.5)^3}$$

$$\boxed{\delta = 17.7mm}$$

14. **Derive expression for the deflection bending stress and shear stress induced in an open coil helical spring subjected to an axial load 'w'. Consider an open coiled helical spring as shown in fig. subjected to axial load.** **(AU May 2012)**

Given Data

Let

α	-	Helix angle
R	-	Mean Radius of the spring
d	-	Wire diameter
P	-	Pitch of the spring
w	-	Axial load

Solution

From above fig we have

$$\tan\alpha = \frac{P}{2\pi R}$$

$$l = \frac{2\pi Rn}{\cos\theta} = 2\pi Rn\sec\alpha$$

Axial Deformation

The load 'w', will cause a moment WR about the horizontal axis OA. This moment can be resolved into two components along $0x$ and $0y$.

Along $0x$, torque

$T = wR\cos\alpha$

A long y, Bending moment $M = wR\sin\alpha$

The combined strain energy stored in the spring under bending and torsion is

$$u = \frac{1}{2}T\theta + \frac{1}{2}m\phi$$

Where θ - Angle of twist due to torque 'T'

- Angle of bend due to bending moment 'M'.

The average external work done on the spring under load, $w = \frac{1}{2}w\delta$

For equilibrium of spring.

$$\frac{1}{2}w\delta = \frac{1}{2}\theta + \frac{1}{2}m\phi$$

By using relation.

$$\phi = \frac{Ml}{El} = \frac{wR\sin\alpha l}{El}$$

W.K.T

$$\frac{M}{I} = \frac{\sigma_x}{y} \Rightarrow \sigma_b = \frac{M}{I}y = \frac{wR\sin\alpha\frac{d}{2}}{\frac{\pi}{64}d^4}$$

Bending Stress,

$$W.K.T.\sigma_b = \frac{32wR\sin\alpha}{\pi d^3}$$

Shear stress,

$$\tau = \frac{16T}{\pi d^3}$$

$$\boxed{\tau = \frac{16wR\cos\alpha}{\pi d^3}}$$

15. **A close - coiled helical compression spring in made of 10mm steel wire closely coiled to a mean diameter of 100mm with 20 coils. A weight of 100N is dropped on to the spring. If the maximum instantaneous compression is 60mm. Calculate the height of the drop. (AU Nov 2012)**

Take N as = 0.85 $\times$ 10^4 N/m^2

Given Data

Loading on spring (p) = 100 N

Dia of coil wire (d)= 10 mm

mean Dia of coil (D) = 100 mm

$\therefore$ Radius of coil (R) = 50 mm

Rigidity modulus (C) = 0.85×10^4 N/mm^2

Instantaneous compression = 60 mm

Number of coils (n) = 20 coils

To Find

Height of the spring.

Solution

Let W be the equivalent gradually applied load to produce the same deflection (60mm) as given load,

$$\delta = \frac{64 \acute{w} R^3 n}{Cd^4}$$

$$60 = \frac{64 \times W \times 50^3 \times 20}{0.85 \times 10^4 \times 10^4}$$

$$W = 318.75N$$

Work stored in spring a strain energy $= \frac{1}{2}\ W\delta$

The work done by falling weight = $p(h + \delta) = 100\,(h + 60)$

$$\therefore\ \frac{1}{2}W\delta = 160\,(h + 60)$$

$$= \frac{1}{2} \times 318.75 \times 60 = 160\,(h + 60)$$

$$\boxed{h = 119{,}18m}$$

16. **In a close coiled helical spring, the diameter of each coil is to be 12 times the diameter of the spring and the maximum shear stress is not to exceed 75 N/mm^2. Maximum permissible deflection under a load of 500N is 12cm. Taking shear modulus as 8 × 10^4 N/mm^2, determine the number of coils, diameter of the coil and energy stored in the spring.** (AU Nov' 2010.)

Given Data

Mean Dia of coil 12 times the dia of spring (D) = 12d

Shear stress in the spring τ = 75N/mm^2

Load (W) = 500N

Deflection (δ) = 12cm = 120mm

Shear modulus (C) = 8 × $10^4 N/mm^2$

To Find

(1) No.of coils

(2) Diameter of coil

(3) Energy stored in the spring

Solution

We known that,

$$\text{Shear stress } (\tau) = \frac{8WD}{\pi d^3}$$

Now on substituting the values

$$75 = \frac{8 \times 500 \times (12d)}{\pi d^3}$$

$$d^2 = \frac{8 \times 500 \times 12}{\pi \times 75}$$

$$\boxed{d = 14.27mm}$$

$$\text{Dia meter of coil } (D) = 12d = 12 \times 14.27$$

$$\boxed{D = 171.27mm}$$

Number of coils Deflection

$$\delta = \frac{8WD^3n}{Cd^4} \quad \text{Deflection}$$

$$120 = \frac{8 \times 500 \times (171.27)^3 \times n}{8 \times 10^4 \times (14.27)^4}$$

$$\Rightarrow n = 19.8 = 20 \quad \boxed{n = 20}$$

$$\text{Strain energy, } u = \frac{\tau^2}{4l} \times \text{volume of the spring,}$$

$$= \frac{(75)^2}{4 \times 8 \times 10^4} \times A \times l$$

$$= 0.0175 \times \frac{\pi}{4} d^2 \times \pi D_n$$

$$= 0.0175 \times \frac{\pi}{4} (14.27)^2 \times \pi \times 171.27 \times 20$$

$$\boxed{u = 30.11 \times 10^3 Nmm}$$

Result

(1) No. of coils, (n) = 20 No's

(2) Diameter of coils, D = 171.27mm d = 14.27mm

(3) Energy stored in the spring, $u = 30.11 \times 10^3 N.mm$

Review Questions

PART-A

1. What are the assumptions made in the theory of torsion?
2. Define torsion and polar modulus?
3. Write Torsional equation.
4. Why hollow circular shafts are preferred when compared to solid circular shafts?
5. Write the expression for power transmitted by a shaft.
6. A circular shaft is subjected to a torque of 10kNm. The power transmitted by the shaft is209.33kW. Find the speed of shaft in revolution per minute.
7. What is a stepped shaft?
8. Compare close coiled and open coiled springs under the action of an axial load.
9. What is the value of maximum shear stress in a close coiled helical spring subjected to anaxial force?
10. State the types of stresses when a closed coiled spring is subjected to (i) axial load and(ii) axial twisting moment.
11. What is the equivalent bending moment for a shaft subjected to moment M and torsionT?
12. A shaft is having a diameter of 30mm. What is its polar moment of inertia?
13. How will you apply a moment to produce bending in a shaft?
14. How will you apply a moment to produce torque in a shaft?
15. Write the expression for vertical deflection of the closed coiled helical spring due to axial load W.
16. What is composite shaft?
17. What are the functions of spring?
18. What is solid length?
19. Define spring rate.
20. Define pitch.

PART-B

1. Determine the diameter of a solid shaft which will transmit 300 KN at 250 rpm. The maximum shear stress should not exceed 30 N/mm2and twist should not be more than 10in a shaft length 2m. Take modulus of rigidity = 1×10^5N/mm
2. The stiffness of the closed coil helical spring at mean diameter 20 cm is made of 3cm diameter rod and has 16 turns. A weight of 3 KN is dropped on this spring. Find the height by which the weight should be dropped before striking the spring so that the spring may be compressed by 18 cm. Take C = 8×10^4 N/mm^2.
3. It is required to design a closed coiled helical spring which shall deflect 1mm under an axial load of 100 N at a shear stress of 90 Mpa. The spring is to be made of round wire having shear modulus of 0.8×10^5 Mpa. The mean diameter of the coil is 10 times that of the coil wire. Find the diameter and length of the wire.
4. A close coiled helical spring is made of a round wire having 'n' turns and the mean coil radius R is 5 times the wire diameter. Show that the stiffness of the spring = 2.05 R/n. If the above spring is to support a load of 1.2kN with 120mm compression. Calculate mean radius of the coil and number of turns assuming G = 8200 N/mm^2 and permissible shear stress, $\lambda_{\text{allowable}}$ = 250 N/mm2.
5. A steel shaft ABCD having a total length of 2400mm is contributed by three different Sections as follows. The portion AB is hollow having outside and inside diameters 80mmand 50mm respectively, BC is solid and 80mm diameter. CD is also solid and 70mm in dia meter. If the angle of twist is same for each section, determine the length of each portion and the total angle of twist. Maximum permissible shear stress is 50 MPa and shear modulus 0.82×10^5 MPa.
6. A leaf spring 750mm long is required to carry a central load of 8kN. If the central deflection is not to exceed 20mm and the bending stress is not to be greater than 200N/mm^2. Determine the thickness, width and number of plates. Assume the width of the plates is 12 times their thickness and modulus of elasticity of the springs material as200kN/mm^2.
7. A circular shaft of 1000mm diameter and 2m length is subjected to a twisting moment which creates a shear stress of 20N/mm2at 30mm from the axis of the shaft. Calculate the angle of twist and the strain energy stored in the shaft. Take G = 8×10^4N/mm^2.

Chapter 5

COMPLEX STRESSES AND PLANE TRUSSES

2D State of stress - 2D Normal and shear stresses on any plane - Principal Stresses and Principal Planes - Mohr's circle - Plane trusses: Analysis of plane trusses - Method of joints - Method of sections. Complex stress and plane trusses.

5.1 Introduction

In certain situations a structural member may be simultaneously subjected to different types of stresses like normal (direct) and shearing stresses.

For design purpose, it is of paramount importance to find out the region of the structural member, where the effect of such stresses is critical.

In a thus stressed material, there exist three orthogonal planes carrying normal stresses only and there will be no shear stresses at all. Such planes are called as principal planes and normal stresses acting on them are called as principal stresses.

One of the principal stresses will be the greatest normal stress and one of them is the least normal stress. It is therefore necessary to find out this greatest normal stress i.e., such maximum principal stress should not exceed the permissible value for safety purpose. In the case of two dimensional (2D) problems, the third principal stress is taken as zero.

5.2 2D State of Stress

In the actual field problems, the load (external forces) will not only act in one direction but also in two or more directions, which will cause stresses in more than one direction. For design and analysis purposes the stresses and strains are considered in three directions viz., x, y and z in a more general case.

If the stresses in 'x' and 'y' directions are only considered, such a state of stress is called 2D state of stress, which is shown in Figure 5.1.

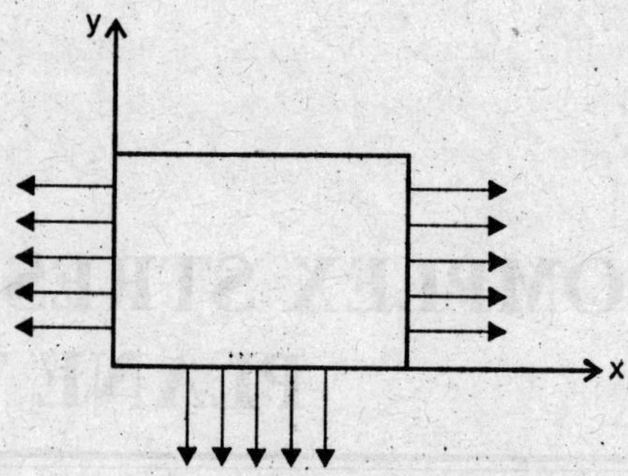

Fig .5.1 2D State of stress

5.2.1 Different cases of stress system in 2D elements

Different cases of stress system in 2D elements are as below.

1. 2D element subjected to a direct stress in one plane.
2. 2D element subjected to direct stresses in two mutually perpendicular directions.
3. 2D element subjected to simple shear stress
4. 2D element subjected to a direct stress in one plane accompanied by a simple shear.
5. 2D element subjected to two direct stresses in mutually perpendicular directions accompanied by a simple shear stress.

Case:1

2D element subjected to a direct stress in one plane

For this case, let us consider a member of uniform cross sectional area "A" and unit thickness subjected to a principal tensile stress as in Figure 5.2

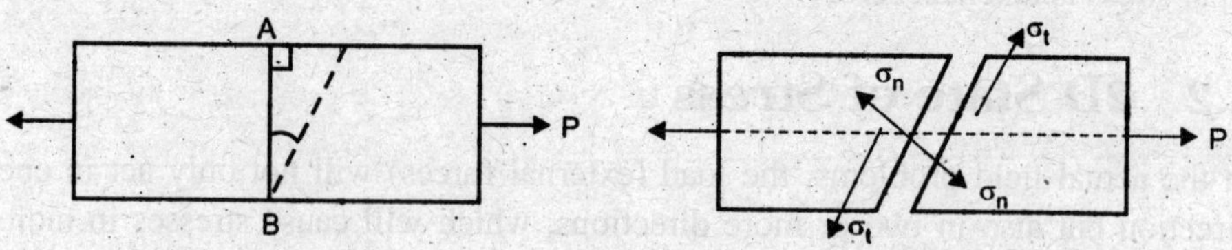

Fig .5.2 2D element subjected to direct stress in one plane

The stress 'σ' on a section perpendicular to force action $\frac{\text{Force}}{\text{Area}}$ i.e. $\frac{P}{A}$.

If we consider an inclined plane '*BC*' which is at an angle 'θ' with the normal cross section '*AB*'. Then the normal and tangential stress across the inclined section '*BC*' (Oblique plane) is as below.

Normal stress $\sigma_n = \sigma \cos^2 \theta$

Tangential stress $\sigma_t = \frac{\sigma}{2} \times \sin 2\theta$

Maximum shear stress $== \frac{\sigma}{2} \sin 2\theta = \frac{\sigma}{2}. \sin 2\,(45°) = \frac{\sigma}{2}$

Resultant stress $\sigma_R = \sqrt{\sigma_n^2 + \sigma t^2}$

($\because$ Shear stress will be maximum on two planes inclined at 45° and 135°)

Case:2

2D element subjected to direct stresses in two mutually perpendicular directions

For this case, Let us consider a structural member '*ABCD*' of uniform cross sectional area '*A*' and unit thickness subjected to direct stresses in two mutually perpendicular directions as depicted in Figure 5.3.

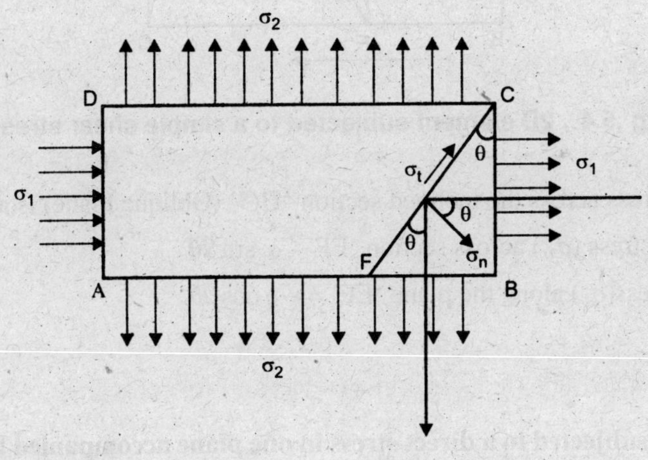

Fig .5.3 2D element subjected to direct stresses in two mutually perpendicular directions

(σ_n) Normal stress across the plane

$$\text{CF} = \frac{\sigma_1 + \sigma_2}{2} + \frac{\sigma_1 - \sigma_2}{2} \cos 2\theta$$

(σ_t) Tangential stress along the section

$$'CF' = \frac{(\sigma_1 - \sigma_2)}{2} \sin 2\theta$$

Resultant stress $\sigma_R = \sqrt{\sigma_n^2 + \sigma_t^2}$

Case:3

2D element subjected to a simple shear stress For this case, Let us consider a structural member '$ABCD$' of uniform cross sectional area 'A' and unit thickness subjected to simple shear stress 'q' as depicted in Figure 5.4. Then the normal and

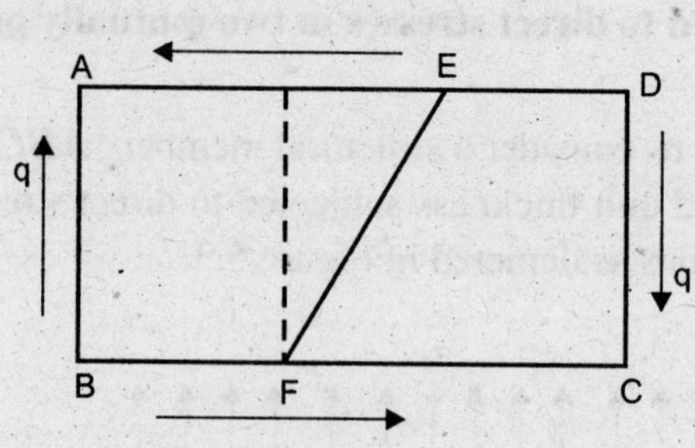

Fig .5.4 2D element subjected to a simple shear stress

tangential stress across the inclined section 'BC' (Oblique Plane) is as below.

Normal stress (σ_n) across section 'EF' - $q.\sin 2\theta$

shear stress(σ_t) along the plane 'EF' = $-q \cos 2\theta$

Case:4

2D element subjected to a direct stress in one plane accompanied by a simple shear.

For this case, Let us consider a structural member '$ABCD$' of uniform cross sectional area 'A' and unit thickness subjected to a tensile stress accompanied by simple shear stress 'q' as shown in Figure 5.5.

Normal stress (σ_n) across section 'EF'

$$(\sigma_n) = \frac{\sigma}{2} + (1 + \cos 2\theta) + q \sin 2\theta$$

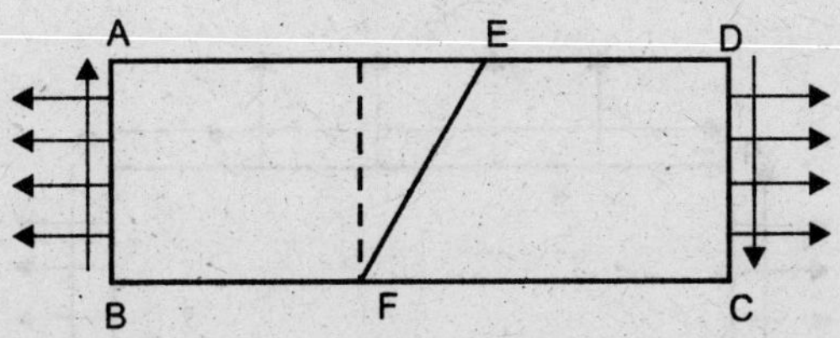

Fig .5.5 2D element subjected to a direct stress in one plane accompanied by a simple shear

Tangential stress (σ_t) along the section

$$= \frac{\sigma}{2} - \sin 2\theta - q \cos 2\theta$$

$$\tan 2\theta = \frac{2q}{2}$$

$$\left.\begin{matrix}\text{Maximum intensity of}\\ \text{Principal stresses}\end{matrix}\right\} = \frac{\sigma}{2} + \frac{1}{2}\sqrt{\sigma^2 + 4q^2}$$

$$\left.\begin{matrix}\text{Minimum intensity of}\\ \text{Principal stresses}\end{matrix}\right\} = \frac{\sigma}{2} - \frac{1}{2}\sqrt{\sigma^2 + 4q^2}$$

$$\text{Maximum shear stress} = \frac{\sigma}{2} + \frac{1}{2}\sqrt{\sigma^2 + 4q^2}$$

Case:5

2D element subjected to two direct stresses in mutually perpendicular directions
accompanied by a simple shear stress

For this case, Let us consider a structural member '$ABCD$' of uniform cross sectional area 'A' and unit thickness subjected to two direct tensile stresses in mutually perpendicular direction, accompanied by simple shear stress 'q' as shown in Figure 5.6.

$$\text{Normal stress } \sigma_n = \left(\frac{\sigma_1 + \sigma_2}{2}\right) + \left(\frac{\sigma_1 + \sigma_2}{2}\right) \cos 2\theta q \sin 2\theta$$

$$\text{Tangential (or) Shear stress } \sigma_t = \left(\frac{\sigma_1 + \sigma_2}{2}\right) \sin 2\theta - q \cos 2\theta$$

Position of principal planes

- Position of principal plane

$$\tan 2\theta = \frac{2q}{\sigma_1 + \sigma_2}$$

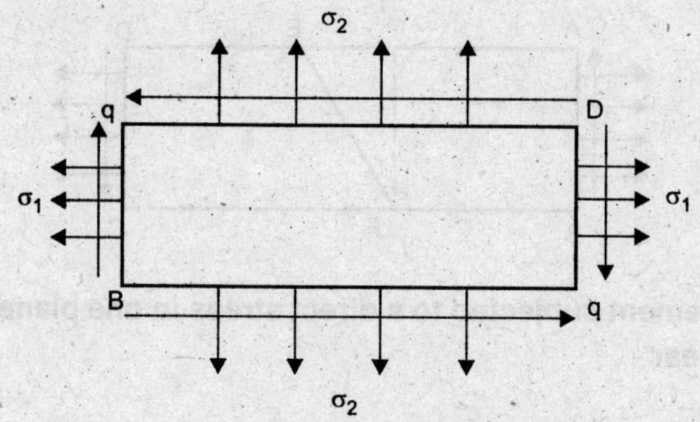

Fig .5.6 2D element subjected to a direct stress in mutually perpendicular direction

- Major principal stress

$$\sigma_{n1} = \frac{\sigma_1 + \sigma_2}{2} + \frac{1}{2}\sqrt{(\sigma_1 + \sigma_2)^2 + 4q^2}$$

- Minor principal stress

$$\sigma_{n2} = \frac{\sigma_1 + \sigma_2}{2} - \frac{1}{2}\sqrt{(\sigma_1 + \sigma_2)^2 + 4q^2}$$

- Maximum shear stress

$$\sigma t = \frac{1}{2}\sqrt{(\sigma_1 + \sigma_2)^2 + 4q^2}$$

5.3 Mohr's circle method

Mohr's circle method is one of the graphical method for getting normal and tangential stress intensities across any oblique plane inclined at an angle θ to the major principal stress.

5.3.1 Different cases of loading system considered in Mohr's circle method

1. A 2D element subjected to two mutually perpendicular unequal and like principal stresses.
2. A 2D element subjected to two mutually perpendicular unequal and unlike principal stresses.
3. A 2D element subjected to two mutually perpendicular unequal like stresses accompanied by a simple shear stress.

Like stresses ⟶ both the stresses are either tensile (or) compressive in nature.

Unlike stresses ⟶ one stress in tensile whereas other is compressive in nature.

Case:1

2D element subjected to two mutually perpendicular unequal and like principal stresses

Let us consider a rectangular body subjected to two mutually perpendicular unequal and principal like (e.g.tensile) stresses. Let us try to find out the Resultant stresses on an oblique plane.

Let σ_1 = Major tensile stress

σ_2 = Minor tensile stress

θ = Angle made by the oblique plane with the axis of minor tensile stress.

We shall now draw the mohr's circle. The procedure is as below.

Consider a point 'A' and draw a horizontal line through 'A'. Take AB = σ_1 and AC = σ_2 towards right from 'A' to suitable scale. Refer Figure 5.7.

With BC as diameter describe a circle. Let 'O'is the centre of the circle. Now through O, draw a line OE marking an angle 2θ with OB.

From E, draw ED perpendicular on AB. Join AE. Then the normal and tangential stresses on the oblique plane are given by AD and ED respectively.

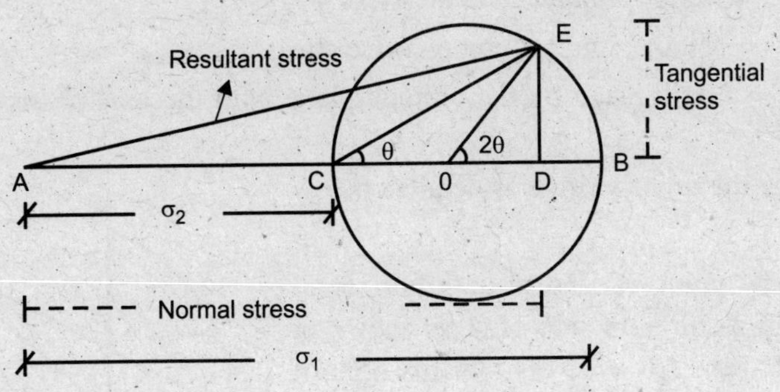

Fig .5.7 2D element subjected to two mutually perpendicular unequal and like principal stresses

Now, Length

AD = Normal stress on oblique plane

ED = Tangential stress on oblique plane

AE = Resultant stress on oblique plane

Radius of mohr circle = $\dfrac{\sigma_1 - \sigma_2}{2}$

Angle ϕ = obliquity

Case:2

Mohr's circle when a body is subjected to two mutually perpendicular principal stresses which are unequal and unlike (i.e. one is the tensile and other one is compressive).

Let us consider a 2D element subjected to tensile and compressive stresses as shown in Figure 5.8

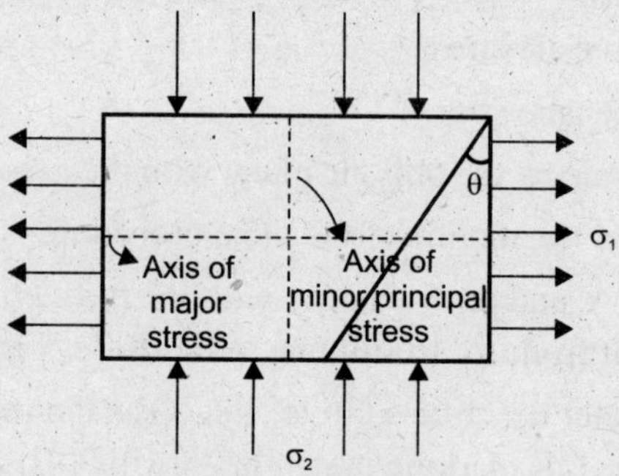

Fig .5.8 2D element subjected to tensile and compressive stresses

Let σ_1 = Major principal tensile stress

σ_2 = Minor principal compressive stress

θ = Angle made by the oblique plane with the axis of minor principal stress.

Let us draw the mohr's circle as detailed below.

Take any point 'A' and draw a horizontal line through A an both side of A as shown in Figure 5.9. Take AB = σ_1 (+) towards right of A and AC = σ_2 (-) towards left of A to some suitable scale Bisect 'BC' at o. with 'o' as centre and radius equal to CO or OB, draw a circle as the Figure 5.9. Through 'O', Draw a line OE making an angle '2θ' with OB.

From E, draw ED perpendicular to AB.

Join AE and CE. Then the Normal stress and shear stress (tangential stresses) are as below.

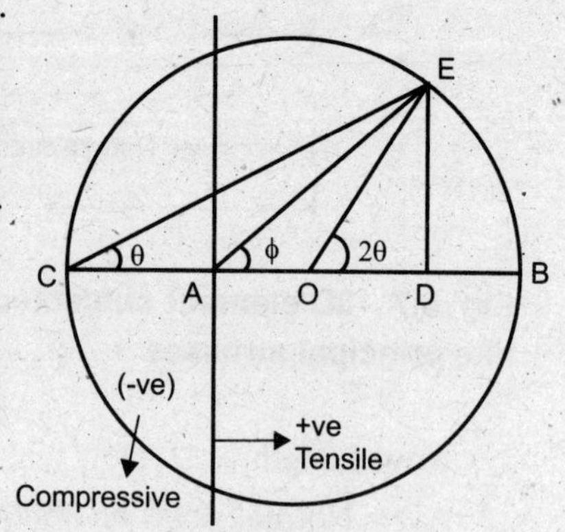

Fig .5.9

Length AD = Normal Stress on oblique plane

Length ED = Shear stress on oblique plane

Length AE = Resultant stress on oblique plane and

Angle ϕ = Obliquity

Radius of mohr's circle = CO or OB $= \dfrac{\sigma_1 - \sigma_2}{2}$

Case:3

Mohr's circle when a body is subjected to two mutually perpendicular principal tensile stresses accompanied by a simple shear stress

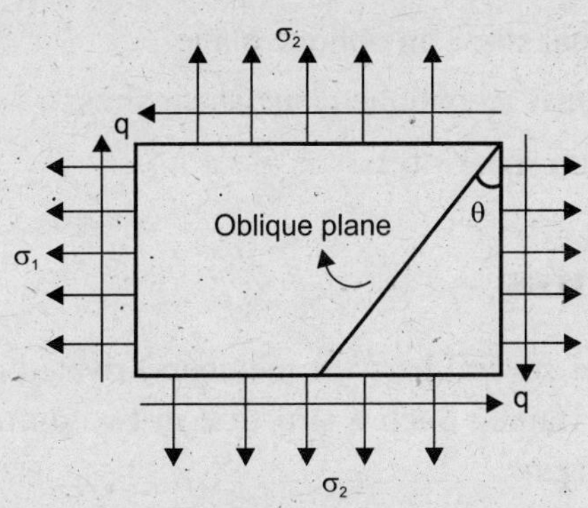

Fig .5.10

With reference to 2D slate of stress, shown in Figure 5.10

Let σ_1 = Major tensile stress

σ_2 = Minor tensile stress

q = Shear stress

θ = Angle made by the oblique plane about the axis of minor principal stress.

Let us consider a point 'A' and let us draw a horizontal line through A. Take $AB = \sigma_1, AC = \sigma_2$ towards right of A to some suitable scale.

Draw perpendicular at B and C and cut off BF and CG equal to shear stress 'q' to the same scale.

Bisect 'BD' at O. Now with O as centre and radius equal to OG or OF draw a circle. Through O, draw a line OE making an angle 2θ with OF as shown in Figure 5.11.

From 'E' draw ED ⊥r to CB. Join AE. Then length AE represent the resultant stress an the oblique plane.

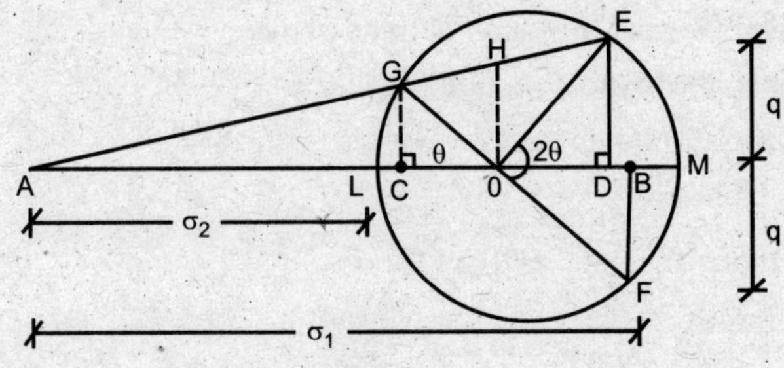

Fig .5.11

Length AD = Normal stress an oblique plane.

ED = Tangential stress an oblique plane(shear stress).

OH - 0F = maximum shear stress.

5.4 Plane Truss

A structure made up of several bars (or members) riveted or welded together is known as a truss or a frame. Such a structure in two dimensions is called as a plane truss or a plane frame.

5.5 Types of frames or trusses

1. Perfect frame 2. Imperfect frame

5.5.1 Perfect Frame

A frame in which the number of members and number of joints are satisfying the equation $n = 2j - 3$, is called as perfect frame.

Where

n = Number of members

J = number of joints

5.5.2 Imperfect Frame

A frame in which number of members and number of joints are not satisfying the equation $n = 2j - 3$, is called as imperfect frame. They are classified into the following two types.

1. Deficient frames 2. Redundant frames

Deficient Frames

A deficient frame is one in which the number of member (n) are less than (2j - 3)

$$i.e. \quad \boxed{n < 2j - 3}$$

Redundant Frame

A Redundant frame is one in which the number of member (n) are more than $2j - 3$

$$i.e. \quad \boxed{n > 2j - 3}$$

Assumptions for finding out the forces in a perfect frame

- The frame is a perfect frame
- The frame carries loaded at the joints
- All the members are pin - jointed

Methods of finding out the forces in members

- Analytical method
- Graphical method

Analytical method

- Method of joint
- Method of section
- Method of tension co efficient

Method of joint

- In this method each and every joint is taken separately as free body in equilibrium and equilibrium conditions $\sum_V = 0$ and $\sum_H = 0$. are considered.
- The unknown forces are then determined by equating sum of all vertical forces and horizontal forces at the point is equal to "zero".

Sign Conversion

Upward force ($\uparrow$)	$\longrightarrow$	+ve
Downward force ($\downarrow$)	$\longrightarrow$	-ve
Force acting towards right side ($\rightarrow$)	$\longrightarrow$	+ve
Force acting towards left side ($\leftarrow$)	$\longrightarrow$	-ve

Part - A

Question and Answer

2 Marks

1. **Define and principal plane and principal stress.** (AU Nov's 2013)

   - **Principal Plane**

     The plane which has no shear stress is known as principal plane.
   - **Principal Stress**

     The stress acting on principal plane is known as principal stress.

2. **Write the statement for normal stress and tangential stress upon an oblique plane inclined at an angle 'θ' when a member is subjected to direct stress 'σ'in one plane.**

   σ_n (Normal stress) = $\sigma cos^2\theta$

   σ_t (Tangential stress) = $\frac{\sigma}{2}. \sin 2\theta$

3. **Write the statement for maximum normal stress and maximum shear stress when a member is subjected to direct stress 'σ' in one plane.**

   Maximum Normal stress = σ

   Maximum shear stress = $\frac{\sigma}{2}$

4. **Write the statement for normal stress and tangential stress upon an oblique plane inclined at an angle 'θ' when a member is subjected to two like direct stresses in two mutually perpendicular directions.**

   σ_n (Normal stress) = $\frac{\sigma_1 + \sigma_2}{2} + \frac{\sigma_1 - \sigma_2}{2}.cos2\theta$

   σ_t (Tangential stress) = $\frac{\sigma_1 - \sigma_2}{2}. \sin 2\theta$

5. **Write the statement for resultant stress when a member is subjected to two direct like stresses in two mutually perpendicular directions.**

$$\sigma_R = \sqrt{\sigma_n^2 + \sigma_t^2}$$

6. **Define Obliquity.**

   The angle made by the resultant stress with the normal of the oblique plane is known as obliquity. This is mathematically known as $\tan\phi = \frac{\sigma_t}{\sigma_n}$

7. **Write the statement for normal stress and tangential stress upon an oblique plane inclined at an angle 'θ' when a member is subjected to simple shear stress 'q'.**

   Normal stress $(\sigma_n) = q.\sin 2\theta$

   Shear stress $(\sigma_t) = -q.\cos 2\theta$

8. **Write the statement for normal and tangential stress upon an oblique plane inclined at an angle 'θ' with the axis of minor axis, when a member is subjected to two direct like stresses in two mutually perpendicular directions accompanied by a simple shear stress 'q'.**

   Normal stress $(\sigma_n\ ) = \dfrac{\sigma_1 + \sigma_2}{2} + \dfrac{\sigma_1 - \sigma_2}{2}.cos2\theta + q.\sin 2\theta$

   Tangential stress $(\sigma_t) = \dfrac{\sigma_1 - \sigma_2}{2}.\sin 2\theta - q.\cos .2\theta$

9. **Write the position of a principal plane when a member is subjected to two direct like stresses in two mutually perpendicular directions accompanied by a simple shear stress 'q'.**

$$\tan 2\theta = \frac{2q}{\sigma_1 - \sigma_2}$$

10. **Write the statement for major and minor principal stress when a member is subjected to two mutually perpendicular directions accompanied by a simple shear stress.**

    Major principal stress

$$= \frac{\sigma_1 + \sigma_2}{2} + \sqrt{\left(\frac{\sigma_1 - \sigma_2}{2}\right)^2 + q^2}$$

    Minor principal stress

$$= \frac{\sigma_1 + \sigma_2}{2} - \sqrt{\left(\frac{\sigma_1 - \sigma_2}{2}\right)^2 + q^2}$$

11. **Write the statement for maximum shear stress, when a member is subjected to two direct like stresses in two mutually perpendicular directions accompanied by a simple shear stress.**

    Maximum Shear stress

$$= \frac{1}{2}\sqrt{(\sigma_1 - \sigma_2)^2 + 4q^2}$$

12. **Write the condition for maximum shear stress when a member is subjected to two direct like stresses in two mutually perpendicular directions accompanied by a simple shear stress.**

$$\tan .2\theta = \frac{\sigma_2 - \sigma_1}{2q}$$

13. **Write a note on Mohr's circle of stresses.**

    Mohr's circle of stresses is a graphical method of finding normal, tangential and resultant stresses on an oblique plane.

14. **Define perfect and imperfect frame.** **(AU Nov' 2013)**

    A frame in which number of members and number of joints are given by $n = 2j - 3$ is known as perfect frame.

    A frame in which number of members and number of joints are not given by $n = 2j - 3$ is known as imperfect frame.

15. **What are the types of Imperfect frame?**

    1. Deficient frame

    2. Redundant frame

16. **Define Deficient and Redundant frame.**

    Deficient frame is the frame in which number of members are less than $2j - 3$, where as a redundant frame is a frame in which number of members are more than $2j - 3$.

17. **List out the methods for findings out the forces in the members of a frame.**

    i) Method of joints
    ii) Method of sections
    iii) Method of tension coefficients
    iv) Graphical method

18. **Define statically determinate frame.**

    A frame which can be analyzed using the equations of static equilibrium is known as statically determinate frame.

19. **Define statically indeterminate frame.**

    A frame in which can't be analyzed using the equation of state equilibrium is team as statically indeterminate frame.

20. **What are the assumptions made in the analysis of frames**

i) The frames are perfect and statically determinate
ii) All members of the frame lie in one plane.
iii) The frame work is made up of pinned -joints and the external loads are applied only at the joints. This makes the members to carry axial forces only.
iv) The weight of the members is negligible as compared to the external forces.
v) The frame work is made up of rigid members and there is no geometrical distortion under the action of external forces.
vi) The condition of static equilibrium are applicable to each member of the frame.

Part - B

Question and Answer

16 Marks

1. **Derive the relations for normal and shear stresses acting on an inclined plane at a point in a strained material subjected to a direct stress in one direction.**

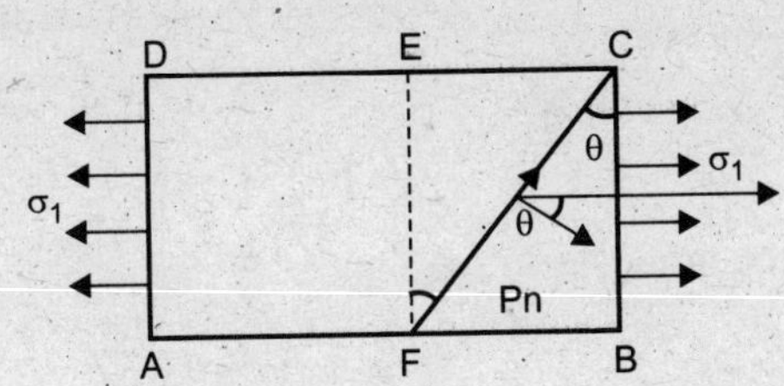

Fig .5.12

c/s Area = A unit thickness

The bar is subjected to a principal tensile stress σ_1 on the faces AD & BC.

Area of c/s = BC × thickness of bar

= BC × 1

Let the stresses on the oblique plane FC are to be calculated. The plane FC is inclined @ angle θ with Normal c/s EF or BC.

Tensile force on BC, (P_1)

$$= \sigma_1 \times \text{Area of c/s}$$
$$= \sigma_1 \times \text{BC} \times 1$$

P_1 is resolved into two components. One normal to plane FC and other along the plane FC.

$$\text{Normal force } (P_n) = P_1 \cos\theta$$
$$= \sigma_1 \times BC \times \cos\theta$$
$$\text{Tangential force } P_t = P_1 \sin\theta$$
$$= \sigma_1 BC \times \sin\theta$$

Let

$\sigma_n \rightarrow$ Normal stress on FC

$\sigma_t \rightarrow$ Tangential (Shear) stress on FC.

$$\therefore \sigma_n = \frac{\text{Force}}{\text{Area of section FC}}$$
$$= \frac{P_n}{F_c \times 1} = \frac{\sigma_1 \times BC \cos\theta}{FC}$$
$$= \sigma_1 \times \cos\theta \times \cos\theta$$
$$\therefore \text{ Normal stress } \boxed{(\sigma_n) = \sigma_1 \cos^2\theta}$$

[In Δle FBC, $\dfrac{BC}{FC} = \cos\theta$]

$$\text{Tangential stress } (\sigma_t) = \frac{P_t}{F_c \times 1} = \frac{\sigma_1 BC \sin\theta}{FC}$$
$$= \sigma_1 \cos\theta \sin\theta$$

Multiplying and dividing by two

$$= \frac{\sigma_1}{2} \times 2\cos\theta\sin\theta$$
$$\boxed{\sigma_t = \frac{\sigma_1}{2} \sin 2\theta}$$

Normal Stress is max at = $\theta°$

$$\therefore \quad \sigma_n \max = \sigma_1 \cos^2\theta° = \sigma_1$$

T shear stress is max at 45° or 135°

$$\therefore \quad \sigma_t \max = \frac{\sigma_1}{2} \sin 90° = \frac{\sigma_1}{1}$$

2. **Derive the relations for normal and shear stresses acting on an inclined plane at a point in a strained material subjected to a pure shear stress.**

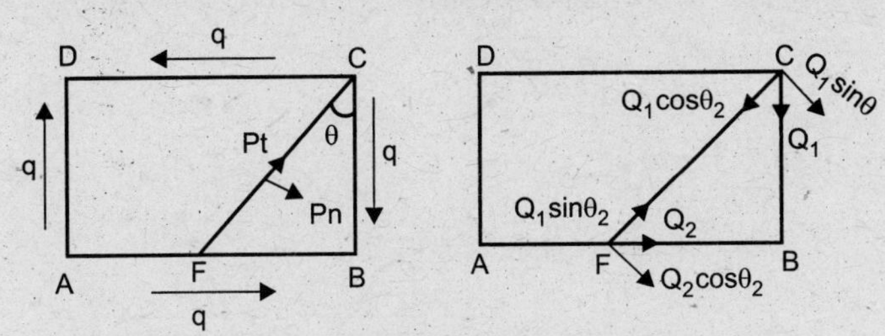

Fig .5.13

ABCD → rectangular bar of c/s Area 'A' and unit thickness.

q → Simple shear stress

FC → Oblique section

θ → Angle made by FC with normal section BC

Shear Force = Shear stress × Area

Let,

$$\begin{aligned}
Q_1 &\rightarrow \text{Shear force on BC} \\
&= q \times BC \times 1 = q.BC \\
Q_2 &\rightarrow \text{Shear force on FB} \\
&= q \times FB \times 1 = q.FB \\
P_n &\rightarrow \text{Total Normal force on FC} \\
&= Q_1 \sin\theta + Q_2 \cos\theta \\
&= q.\text{BC} \sin\theta + q.\text{FB} \cos\theta \\
P_t &\rightarrow \text{Total Tangential force on FC} \\
P_t &= Q_2 \sin\theta - Q_1 \cos\theta \\
&= q.\text{FB} \sin\theta - q.\text{BC} \cos\theta \\
\sigma_n &\rightarrow \text{Normal stress on FC} \\
&= \frac{P_n}{FC \times 1} = \frac{q.BC \sin\theta + q.FB. \cos\theta}{FC}
\end{aligned}$$

[In $\Delta\ell e$ FBC, $\dfrac{BC}{FC} = \cos\theta$ and $\dfrac{FB}{FC} = \sin\theta$]

$$= q.\frac{BC}{FC}.\sin\theta + q.\frac{FB}{FC}.\cos\theta$$
$$= q\cos\theta\sin\theta + q\sin\theta\cos\theta$$
$$= 2q\cos\theta\sin\theta$$
$$\boxed{\sigma_n = q\sin 2\theta}$$

$\because$ $[2\cos\theta\sin\theta = \sin 2\theta]$

$\sigma_t \rightarrow$ Shear stress on FC

$$= \frac{P_t}{FC \times 1} = \frac{q.FB.\sin\theta - q.BC.\cos\theta}{FC}$$
$$= q.\frac{FB}{FC}\sin\theta - q.\frac{BC}{FC}\cos\theta$$
$$= q\sin\theta\sin\theta - q\cos\theta\cos\theta$$
$$= q\sin^2\theta - q\cos^2\theta = -q\cos^2\theta + q\sin^2\theta$$
$$= -q\left[\cos^2\theta - \sin^2\theta\right]$$
$$\boxed{\sigma_t = -q\cos 2\theta}$$

3. **Derive the relations for normal and shear stresses acting on an inclined plane at a point in a strained material subjected to direct stress in mutually perpendicular directions.**

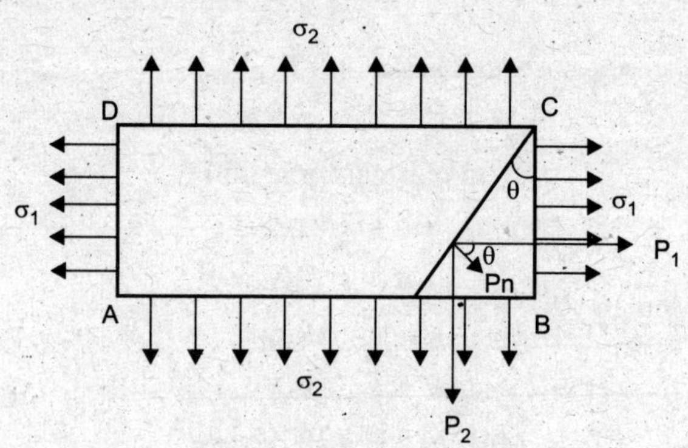

Fig .5.14

Let,

θ → Angle between FC and BC.
σ_1 → Major tensile stress on AD and BC.
σ_2 → Minor tensile stress on AB and CP.
P_1 → Tensile force on BC.
P_2 → Tensile force on FB.

$$P_1 = \sigma_1 \times BC \times 1$$
$$P_2 = \sigma_2 \times FB \times 1$$
$$P_n = P_1 \cos\theta + P_2 \sin\theta$$
$$= \sigma_1 . BC \cos\theta + \sigma_2 . BF . \sin\theta$$
$$P_t = P_1 \sin\theta - P_2 \cos\theta$$
$$= \sigma_1 BC \sin\theta - \sigma_2 BF \cos\theta$$
$$\sigma_n = \frac{P_n}{F_c \times 1} = \frac{\sigma_1 BC \cos\theta + \sigma_2 BF \sin\theta}{Fc}$$
$$= \sigma_1 \frac{Bc}{Fc} \cos\theta + \sigma_2 . \frac{BF}{FC} \sin\theta$$
$$= \sigma_1 \cos\theta \cos\theta + \sigma_2 \sin\theta \sin\theta$$
$$= \sigma_1 \cos^2\theta + \sigma_2 \sin^2\theta$$
$$\boxed{\therefore \sigma_n = \frac{\sigma_1 + \sigma_2}{2} + \frac{\sigma_1 - \sigma_2}{2} \cos 2\theta}$$

[$\because$ In Δle $FBC, BC/FC = \cos\theta \frac{BC}{FC} \sin^{\theta}$]

$$\sigma_t = \frac{P_t}{F_c \times 1} = \frac{\sigma_1 BC \sin\theta - \sigma_2 BF \cos\theta}{Fc}$$
$$= \sigma_1 \frac{BC}{FC} \sin\theta - \sigma_2 \frac{BF}{FC} \cos\theta$$
$$= \sigma_1 \cos\theta \sin\theta - \sigma_2 \sin\theta \cos\theta$$
$$= (\sigma_1 - \sigma_2) \cos\theta \sin\theta$$

Multiplying and dividing by 2

$$= \frac{\sigma_1 - \sigma_2}{2} \times 2 \cos\theta \sin\theta$$
$$\boxed{\sigma_t = \frac{\sigma_1 - \sigma_2}{2} \sin 2\theta}$$

$\therefore$ The resultant stress on the section FC ⇒ $\sigma_R = \sqrt{\sigma_n 2 + \sigma_t{}^2}$

4. **Derive the relations for normal and shear stresses acting on an inclined plane at a point in a strained material subjected to a direct stress in one direction and pure shear stress.** Let,

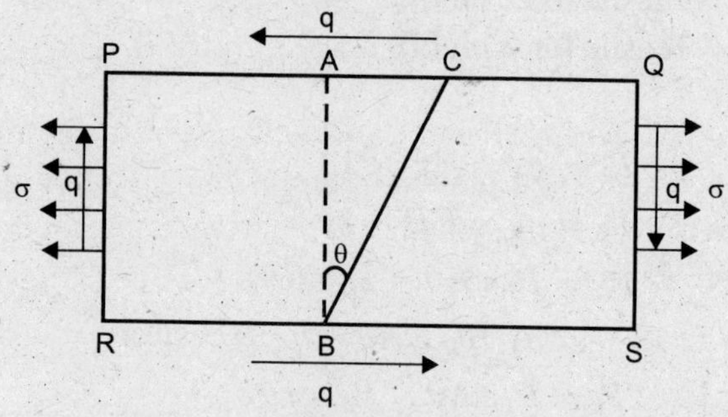

Fig .5.15

P $\rightarrow$ Tensile Force on AB = $\sigma \times AB \times 1$
Q_1 $\rightarrow$ Shear force on AB = $q \times AB \times 1$
Q_2 $\rightarrow$ Shear force on AC = $q \times AC \times 1$

$$P_n = P\cos\theta + Q_1\sin\theta + Q_2\cos\theta$$
$$= \sigma AB\cos\theta + q.AB\sin\theta + q.AC.\cos\theta$$
$$P_t = P\sin\theta + Q_2\sin\theta - Q_1\cos\theta$$
$$= \sigma.AB\sin\theta + q.AC.\sin\theta - q.AB.\cos\theta$$
$$\sigma_n = \frac{P_n}{BC.1} = \frac{\sigma.AB.\cos\theta + q.AB.\sin\theta + q.AC.\cos\theta}{BC}$$
$$= \sigma\frac{AB}{BC}\cos\theta + q.\frac{AB}{BC}.\sin\theta + q.\frac{AC}{BC}\cos\theta \qquad \left(\because \frac{AB}{Bc} = \cos\theta, \ \frac{AC}{BC} = \sin\theta\right)$$
$$= \sigma\cos\theta\cos\theta + q\cos\theta\sin\theta + q\sin\theta\cos\theta$$
$$= \sigma\cos^2\theta + 2q\cos\theta\sin\theta$$
$$\boxed{\sigma_n = \frac{\sigma}{2}(1 + \cos 2\theta + q\sin 2\theta)}$$
$$\sigma_t = \frac{P_t}{BC.1} = \frac{\sigma.AB.\sin\theta + q.AC.\sin\theta - q.AB.\cos\theta}{BC}$$
$$= \sigma\frac{AB}{BC}\sin\theta + q.\frac{AC}{BC}.\sin\theta - q.\frac{AB}{BC}\cos\theta$$

$$= \sigma \cos\theta \sin\theta + q \sin\theta \sin\theta - q \cos\theta \cos\theta$$

$$= \frac{\sigma}{2} 2 \sin\theta \cos\theta - q \left(\cos^2\theta \sin^2\theta\right)$$

$$\boxed{\sigma_t = \frac{\sigma}{2}\sin 2\theta - q\cos 2\theta}$$

5. **Derive relations for normal and shear stresses acting on an inclined acting on an inclined plane at a point in a strained material subjected to two mutually perpendicular direct stresses accompanied by a simple shear stress.**

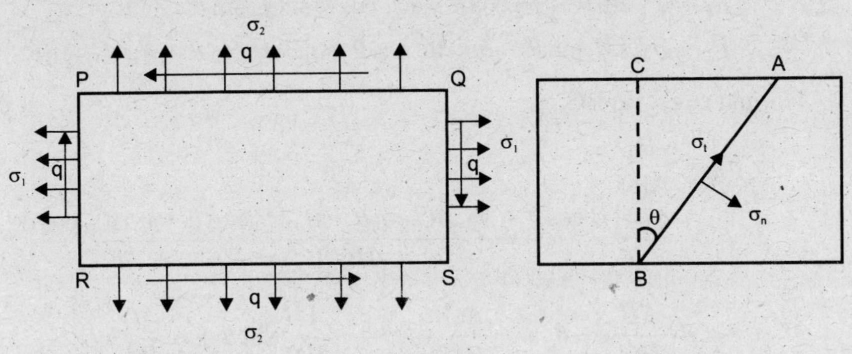

Fig .5.16

Let PQRS be a rectangular body of uniform cross - sectional Area 'A' and of unit thickness.

$P_1 \rightarrow$ Tensile force on AB due to tensile stress σ_1.

$$= \sigma_1 \times AB \times 1$$
$$= \sigma_1 \times AB$$

$P_2 \rightarrow$ Tensile force on AC due to tensile stress σ_2

$$= \sigma_2 \times AC \times 1$$
$$= \sigma_2 \times AC$$

$Q_1 \rightarrow$ Shear force on AB

$$= q.AB.1$$
$$= q \times AB$$

$Q_2 \rightarrow$ Shear force on AC

$$= q.AC.1$$
$$= q \times AC$$

Total Normal force,

$$P_n = P_1 \cos\theta + P_2 \sin\theta + Q_1 \sin\theta + Q_2 \cos\theta$$
$$P_n = \sigma_1 AB \cos\theta + \sigma_2.AC \sin\theta + q.AB \sin\theta + q.AC. \cos\theta$$

Total Tangential force,

$$P_t = P_1 \sin\theta - P_2 \cos\theta - Q_1 \cos\theta + Q_2 \sin\theta$$
$$P_t = \sigma_1 AB. \sin\theta - \sigma_2.AC \cos\theta + q.AB \cos\theta + q.AC. \sin\theta$$

Normal stress on BC,

$$\sigma_n = \frac{P_n}{\text{Area}}$$
$$= \frac{\sigma_1.AB \cos\theta + \sigma_2.AC \sin\theta + q.AB. \sin\theta + q.AC. \cos\theta}{BC.1}$$

$$= \sigma_1 \frac{AB}{BC} \cos\theta + \sigma_2 \frac{AC}{BC} \sin\theta + q \frac{AB}{BC} \sin\theta + q. \frac{AC}{BC} \cos\theta$$
$$= \sigma_1 \cos\theta \cos\theta + \sigma_2 \sin\theta \sin\theta + q \cos\theta \sin\theta + q \sin\theta \cos\theta$$
$$= \sigma_1 \cos^2\theta + \sigma_2 \sin^2\theta + 2q \cos\theta \sin\theta$$
$$= \sigma_1 \left(\frac{1 + \cos 2\theta}{2}\right) + \sigma_2 \left(\frac{1 - \cos 2\theta}{2}\right) + q \sin 2\theta$$

$$\boxed{\sigma_n = \left(\frac{\sigma_1 + \sigma_2}{2}\right) + \left(\frac{\sigma_1 - \sigma_2}{2}\right) \cos 2\theta + q \sin 2\theta}$$

Tangential or shear stress, $\sigma_t = \dfrac{P_t}{\text{Area}}$

$$= \frac{\sigma_1.AB \sin\theta - \sigma_2.AC \cos\theta - q.AB. \cos\theta + q.AC. \sin\theta}{BC.1}$$
$$= \sigma_1 \frac{AB}{BC} \sin\theta - \sigma_2 \frac{AC}{BC} \cos\theta - q \frac{AB}{BC} \cos\theta + q. \frac{AC}{BC} \sin\theta$$
$$= \sigma_1 \cos\theta \sin\theta + \sigma_2 \cos\theta \cos\theta + q \cos\theta \cos\theta + q \sin\theta \sin\theta$$
$$= (\sigma_1 - \sigma_2) \sin\theta \cos\theta - q \cos^2\theta + q \sin^2\theta$$

$$\boxed{\sigma_t = \left(\frac{\sigma_1 - \sigma_2}{2}\right) \sin 2\theta - q \cos 2\theta}$$

6. **A short metallic column of 500mm^2 cross sectional area carries an axial compressive load of 100 KN. For a plane inclined at 60° with the direction of load. Calculate (i) Normal Stress (ii) Tangential stress (iii) Resultant stress (iv) Maximum shear stress (v) Obliquity if resultant stress. (AU May 2013)**

Given data

Cross sectional area of column (A) = 510 mm^2

Axial compressive Load (P)= $100kN = 1 \times 10^5 N$

Inclined plane with the direction and Load $> 60°$

To find

i) Normal Stress (σ_n)
ii) Tangential Stress (σ_t)
iii) Resultant stress (σ_{res})
iv) Maximum shear stress,
v) Obliquity of Resultant stress (ϕ)

Solution

i) Normal Stress

$$\sigma_n = \sigma \cos^2\theta \left[\sigma = \frac{\text{load}}{\text{Area}} = \frac{100 \times 10^3}{500} = 200\right] N/mm^2$$

$$\therefore \sigma_n = 200 \times \cos^2 60$$

$$\boxed{\sigma_n = 50 N/mm^2}$$

ii) Tangential Stress

$$\sigma_t = \frac{\sigma}{2} \times \sin 2\theta$$

$$= \frac{50}{2} \times \sin 2 \times 60$$

$$\boxed{\sigma_t = 21.65 N/mm^2}$$

iii) Resultant stress

$$\sigma_{res} = \sqrt{{\sigma_n}^2 + {\sigma_t}^2}$$

$$= \sqrt{50^2 + 21.65^2}$$

$$\boxed{\sigma_{res} = 54.46 N/mm^2}$$

iv) Maximum shear stress,

$$(\sigma_{\max}) = \frac{\sigma}{2} = \frac{50}{2}$$

$$\boxed{\sigma_{\max} = 25N/mm^2}$$

v) Obliquity of Resultant stress (ϕ)

$$\phi = \frac{\text{Tangential stress}}{\text{Normal stress}}$$

$$\phi = \frac{\sigma_t}{\sigma_n} = \frac{21.65}{50}$$

$$\boxed{\phi = 0.433 \text{ (no unit)}}$$

Result

i) Normal Stress $(\sigma_n) = 50N/mm^2$
ii) Tangential Stress $(\sigma_t) = 21.65N/mm^2$
iii) Resultant stress $(\sigma_{res}) = 54.46N/mm^2$
iv) Maximum shear stress, $(\sigma_{\max}) = 25N/mm^2$
v) Obliquity of Resultant stress $(\phi = 0.433)$ (no unit)

7. **At point in a strained body subjected to two mutually perpendicular normal tensile stresses of magnitude 30Mpa and 12Mpa accompanied by a shear stress of 16 Mpa. Locate the principal plane and evaluate the principal stresses.**

Given Data

$$\text{Normal Stress 1 } (\sigma_1) = 30Mpa = 30N/mm^2$$

$$\text{Normal Stress 2 } (\sigma_2) = 12Mpa = 12N/mm^2$$

$$\text{Shear Stress } (q) = 16\ mpa$$

To Find

Location of

1. Principal plane

2. Major and Minor principal stresses.

Solution

Principal plane

We know that

$$\tan 2\theta = \frac{2q}{\sigma_1 + \sigma_2} = \frac{2 \times 16}{(30) - (12)}$$

$$2\theta = 60°30^1$$

$$\boxed{\theta = 30°19^1}$$

$$\text{Major principal stress } \sigma_{n1} = \frac{\sigma_1 + \sigma_2}{2} + \sqrt{\left(\frac{\sigma_1 + \sigma_2}{2}\right)^2 + q^2}$$

$$= \frac{30 + 12}{2} + \sqrt{\frac{30 + 12}{2}^2 + 16^2}$$

$$= 21 + \sqrt{81 + 256}$$

$$= 21 + 18.36 = 39.36 N/mm^2$$

$$\text{Minor principal stress } \sigma_{n2} = \frac{\sigma_1 + \sigma_2}{2} - \sqrt{\left(\frac{\sigma_1 - \sigma_2}{2}\right)^2 + q^2}$$

$$= 21 - 18.36$$

$$= 2.64 N/mm^2$$

8. **A rectangular bar of cross-sectional area 10,000mm^2 is subjected to an axial load of 20kN. Determine the normal and shear stresses on a section which is inclined at an angle of 30° with normal cross - section of the bar.**

Given Data

Cross sectional area of the rectangular bar

A = 10,000mm^2

Axial load (P) = 20kN = 20,000 N

Solution

Angle of oblique plane with the normal cross - section of bar (θ)= 30°.

Now Direct stress (σ)

$$= \frac{P}{A} = \frac{20,000}{10,000} = 2N/mm^2$$

For this case,

$$\text{Normal stress}(\sigma_n) = \sigma \cos^2 \theta$$
$$= 2 \times \cos^2 30^\circ = 1.5N/mm^2$$
$$\text{Shear stress}(\sigma_t) = \frac{\sigma}{2} \sin .2\theta$$
$$= \frac{2}{2} . \sin 2(30) = 0.866N/mm^2$$

9. **The tensile stresses at a point across two mutually perpendicular planes are 120N/mm^2 and 60 N/mm^2. Determine the normal, tangential and resultant stresses on a plane inclined at 30° to the axis of minor stress.**

Given Data

Major Principal stresses (σ_1) = 120N/mm^2

Minor Principal stresses (σ_2) = 60N/mm^2

Solution

Angle of oblique plane with the minor principal stresses (θ) = 30°

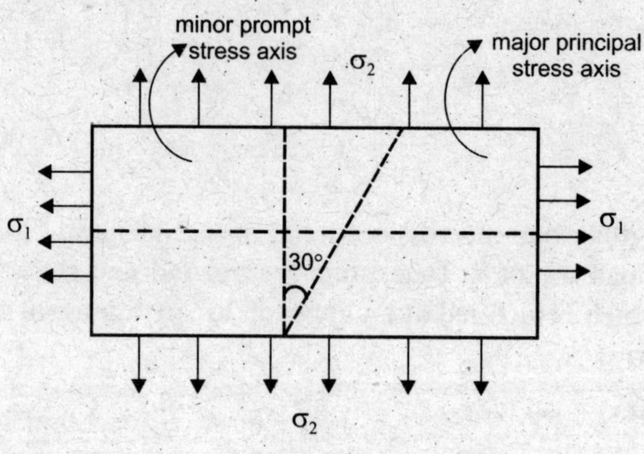

Fig .5.17

Normal stress (σ_n)

$$= \frac{\sigma_1 + \sigma_2}{2} + \frac{\sigma_1 - \sigma_2}{2} . \cos 2\theta$$
$$= \frac{120 + 60}{2} + \frac{120 - 60}{2} . \cos 2 \times 30^\circ$$
$$= \frac{120 + 60}{2} + \frac{120 - 60}{2} . \cos 2 \times 30^\circ$$
$$= 90 + 30 \times \cos 60^\circ = 105N/mm^2$$

Tangential stress

$$\text{Tangential Stress} = \frac{\sigma_1 - \sigma_2}{2}. \sin 2\theta$$

$$(\sigma_t) = \frac{120 - 60}{2}. \sin 2\,(30^\circ) = 25.98 N/mm^2$$

Resultant stress(σ_R)

$$= \sqrt{{\sigma_n}^2 + {\sigma_t}^2}$$

$$= \sqrt{(105)^2 + (25.98)^2} = 108.16 N/mm^2$$

$$(\sigma_t) = \frac{\sigma_1 - \sigma_2}{2}. \sin 2\theta$$

$$= \frac{20 - 40}{2} \sin(2 \times 20) = 12.86 N/mm^2$$

Resultant stress(σ_R)

$$(\sigma_R) = \sqrt{{\sigma_n}^2 + {\sigma_t}^2}$$

$$= \sqrt{75.32 + 12.80} = 76.4 N/mm^2$$

Obliquity $\tan(\phi)$

$$\tan(\phi) = \frac{\sigma_t}{\sigma_n} = \frac{12.856}{75.32}$$

$$\tan^{-1} = \left(\frac{12.856}{75.32}\right) = 9^\circ 41'$$

Let 'σ'= stress which is acting alone will produce the same maximum strain. The Maximum strain will be in the structure of major principal stress.

∴ Maximum strain

$$\frac{\sigma_1}{E} - \frac{\mu\sigma_2}{2} = \frac{1}{E}(\sigma_1 - \mu \cdot \sigma_2)$$

$$= \frac{1}{E}\left(80 - \frac{40}{4}\right) = \frac{70}{E}$$

∴ The strain due to stress = $\epsilon = \frac{\sigma}{E}$

Now, Equating two strains = $\frac{70}{E} = \frac{\sigma}{E}$

$$\therefore \quad \sigma = 70 N/mm^2$$

10. **At a point in a strained material the principal tensile stresses across two perpendicular planes are 80N/mm^2 and 40N/mm^2. Determine the normal shear and resultant stress an a plane inclined at 20° with the major principal plane. Determine also its obliquity. What will be the intensity of stress which acting alone will produce the same minimum strain of poisson's ratio is $\frac{1}{4}$.**

Given Data

Major principal stress (σ_1) = 80N/mm^2

Minor principal stress (σ_2) = 40N/mm^2

Solution

The plane 'CE' is inclined at an angle 20° with the major principal plane (BC) θ = 20°.

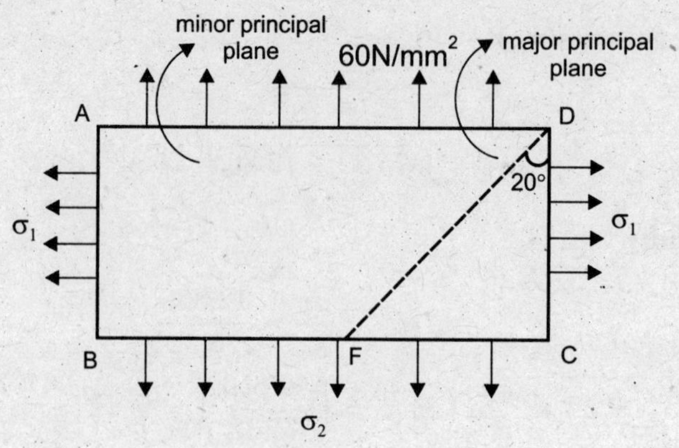

Fig .5.18

Poisson's ratio $\frac{1}{m}$ (or) $\mu = \frac{1}{4}$

$$\sigma_n \text{ across (DF)} = \frac{\sigma_1 + \sigma_2}{2} + \frac{\sigma_1 - \sigma_2}{2} \cos 2\theta$$

$$= \frac{80 + 40}{2} + \frac{80 - 40}{2} = \cos .2\,(20°) = 75.32 N/mm^2$$

11. **At a point with in a body subjected to two mutually perpendicular directors, the stress are 80N/mm^2 40N/mm^2 tensile each of the above stresses is accompanied by a shear stress of 60N/mm^2. Determine the normal, shear stress and resultant stress an oblique plane inclined at an angle of 45° with the axis of minor tensile stress.**

Given Data

Major principal stress $\sigma_1 = 80N/mm^2$

Minor principal stress $\sigma_2 = 40N/mm^2$

Shear stress (q) = 60N/mm^2

(θ) = Angle of oblique plane = 45°

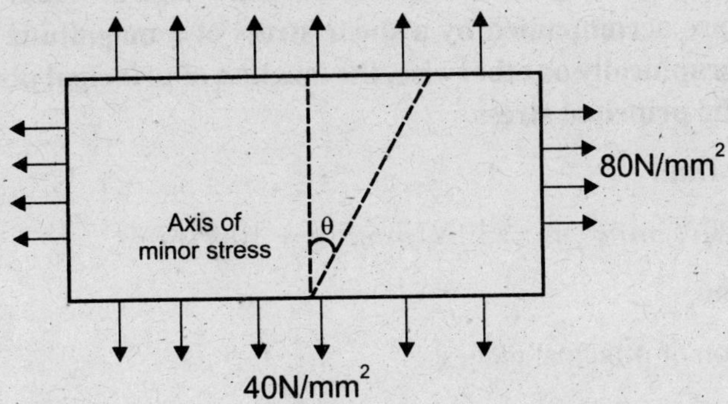

Fig .5.19

Solution

Now,

Normal stress (σ_n)

$$= \frac{\sigma_1 + \sigma_2}{2} + \frac{\sigma_1 - \sigma_2}{2} + \cos .2\theta + q \sin 2\theta$$

$$= \frac{80 + 40}{2} + \frac{80 - 40}{2} . \cos (2 \times 45°)$$

$$+ 60 \sin (2 \times 45°)$$

$$= 60 + 20 \cos 90° + 60 \sin 90°$$

$$= 120N/mm^2$$

Shear stress (σ_t)

$$= \frac{\sigma_1 - \sigma_2}{2} . \sin 2\theta - q \cos 2\theta$$

$$= \frac{80 - 40}{2} . \sin (2 \times 45°) - 60 \times \cos (2 \times 45°)$$

$$= 20N/mm^2 .$$

Resultant stress (σ_R)

$$= \sqrt{\sigma_n{}^2 + \sigma_t{}^2} = \sqrt{120^2 + 20^2}$$
$$= 121.655 N/mm^2$$

12. **At a certain point in a strained material, the stresses on two planes, at right angles to each other are 20N/mm^2 and 10N/mm^2 both tensile. They are accompanied by a shear stress of a magnitude of 10N/mm^2. Find graphically or otherwise, the location of principal planes and evaluate the principal stress.**

Given Data

$\sigma_1 = 20N/mm^2, \sigma_2 = 10N/mm^2, q = 10N/mm^2$

Solution

Location of principal planes

$$\tan 2\theta = \frac{2q}{\sigma_1 - \sigma_2} = \frac{2 \times 10}{20 - 10} = 2$$
$$\tan 2\theta = 2$$
$$\therefore 2\theta = \tan^{-1}(2) = 63°26' = 243°26' \quad = 31°43' \quad (\because \; 243°26)$$
or $\quad 121°43'$

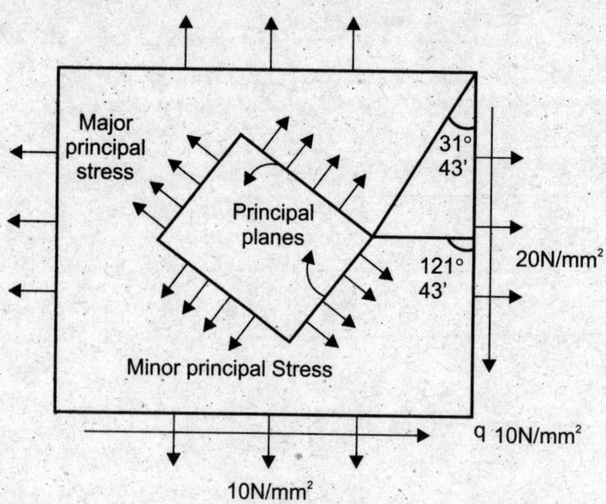

Fig .5.20

$\therefore$ Major Principal stress

$$= \frac{\sigma_1 + \sigma_2}{2} + \sqrt{\left(\frac{\sigma_1 - \sigma_2}{2}\right)^2 + q^2}$$

$$= \frac{20 + 10}{2} + \sqrt{\left(\frac{20 - 10}{2}\right)^2 + 10^2}$$

$$= 15 + \sqrt{25 + 100}$$

$$= 15 + 11.18$$

$$= 26.18 N/mm^2$$

$\therefore$ Minor principal stress

$$= \frac{\sigma_1 + \sigma_2}{2} - \sqrt{\left(\frac{\sigma_1 - \sigma_2}{2}\right)^2 + q^2}$$

$$= \frac{20 + 10}{2} - \sqrt{\left(\frac{20 - 10}{2}\right)^2 + 10^2}$$

$$15 - 11.18 = 3.82\ N/mm^2$$

13. **A point in a strained material is subjected to the stresses as shown below. Locate the principal planes and evaluate the principal stresses.**

Solution

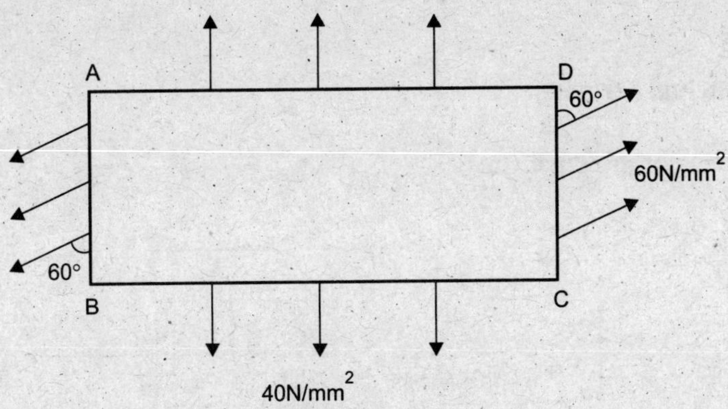

Fig .5.21

Stress normal to $CD = 60 \sin 60° = 51.96 N/mm^2$

Stress along CD (or) $AB = 60 \cos 60° = 30 N/mm^2$

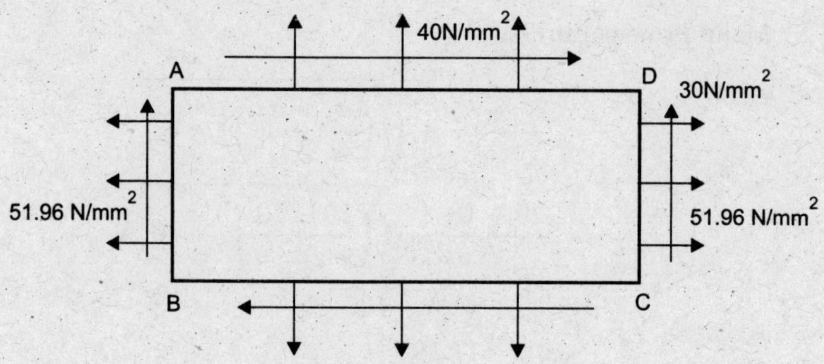

Fig .5.22

Major tensile stress (σ_1) = 51.96N/mm^2

Minor tensile stress (σ_2) = 40 N/mm^2

Shear stress (q) = 30 N/mm^2

Location of Principal Plane.

$$\tan 2\theta = \frac{2q}{\sigma_1 - \sigma_2} = \frac{2 \times 30}{51.96 - 40} = 5.02$$
$$\therefore \ \theta = 39°21$$

Principal Stress

The major principal stress

$$= \frac{\sigma_1 + \sigma_2}{2} + \sqrt{\left(\frac{\sigma_1 - \sigma_2}{2}\right)^2 + q^2}$$
$$= \frac{51.96 + 40}{2} + \sqrt{\left(\frac{51.96 - 40}{2}\right)^2 + 30^2}$$
$$= 45.98 + \sqrt{35.76 + 900}$$
$$= 45.98 + 30.59$$
$$= 76.57 N/mm^2$$

Major principal stress

$$= \frac{\sigma_1 = \sigma_2}{2} - \sqrt{\left(\frac{\sigma_1 + \sigma_2}{2}\right)^2 + q^2}$$

$$= \frac{51.96 + 40}{2} - \sqrt{\left(\frac{51.96 - 40}{2}\right)^2 + 30^2}$$

$$= 45.98 - 30.59 = 15.39\ N/mm^2$$

14. **The stresses at a point in a strained material are 200 N/mm^2 (tensile) and 100 N/mm^2 (compressive). Determine the resultant stress in magnitude and direction on a plane inclined at 30° to the axis of major stress using Mohr's circle method. Also determine the maximum intensity of shear stress in the material of the point.**

Given Data

$$\sigma_1 = 200N/mm^2$$
$$\sigma_2 = -100N/mm^2$$
$$\theta = 30^\circ$$

To find

(1) Resultant stress and (2)Maximum shear stress by Mohr's circle method.

Solution

Let 1 represents $20N/mm^2$. $\therefore$ $\sigma_1 = 10cm$ and $\sigma_2 = -5cm$.

Let us draw the Mohr's circle. Keep a 'A' at Let us draw a horizontal line through A on both sides of A. σ_1 = 10cm towards right and 5cm towards left. Let us Bisect 'BC' at 0. With 'O' as center and radius equal to Co or OB, draw a circle. Through 'O' draw a line OE making an angle $2\theta(2 \times 30^\circ) = 60^\circ$. With OB. From E, draw ED $\perp$ to AB. Join AE and CE.

Thus AE = Resultant stress factor = 9cm AD = 6.25cm and ED = 6.5cm

ϕ = obliquity

$\therefore$ Resultant stress = Length AE $\times$ scale = $9 \times 20 = 180N/mm^2$

Angle made by the resultant stress with the normal of the inclined plane = $\phi = 46^\circ$

Normal stress = Length $AD \times 20 = 6.25 \times 20 = 125N/mm^2$

Shear stress = Length $ED \times 20 = 6.5 \times 20 = 130N/mm^2$

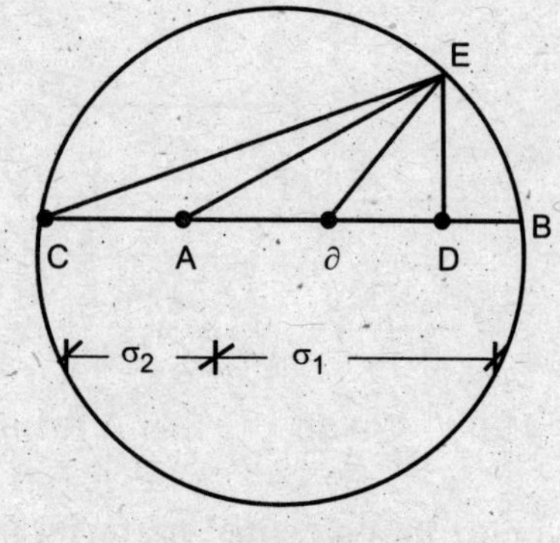

Fig .5.23

Maximum Shear Stress

Shear stress is along a line which is perpendicular to the line AB is drawn from point O. The maximum shear stress will be equal to the radius of Mohr's Circle.

∴ Maximum Shear stress = Radius of Mohr's circle

$$\frac{\sigma_1 + \sigma_2}{2} = \frac{200 + 100}{2} = 150N/mm^2$$

15. **The tensile stresses at a point across two mutual Perpendicular planes are 120N/mm^2 and 60 N/mm^2. Determine the Normal, tangential and resultant stresses on a plane inclined at an angle 30° to the axis of the minor stress, using Mohr's circle method.**

Given Data

$$\sigma_1 = 120N/mm^2$$
$$\sigma_2 = 60N/mm^2$$
$$\theta = 30^\circ$$

Solution

Let us consider scale 1cm = 10N/mm^2

$$AB = \sigma_1 = \frac{120}{10} = 12cm$$
$$AC = \sigma_2 = \frac{60}{10} = 6cm$$

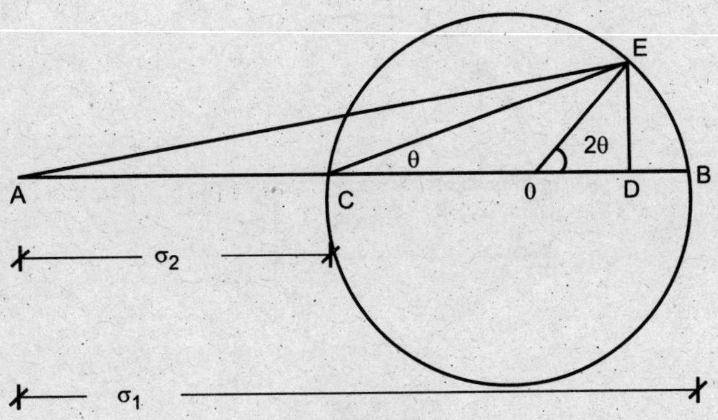

Fig .5.24

Let us consider a point 'A' and draw a horizontal line through 'A' and draw a horizontal line through A. Take $AB = \sigma_1 = 12cm$ and $AC = \sigma_2 = 6cm$; with BC as Dia. i.e.,BC = 12-6 = 6cm, draw a circle. Let 'O' be the centre. Through 'O' draw a line OE making an angle $2\theta(2 \times 30 = 60°)$ with OB.

From E, draw ED perpendicular to CB. Join AE. Measure lengths AD, ED and AE.

Length AD = 10.50 cm, ED = 2.60 cm, AE = 10.82 cm

Then Normal stress =Length $AD\times$ Scale $= 10.50 \times 10 = 105N/mm^2$

Tangential or Shear stress $= 2.60 \times 10 = 26N/mm^2$

Resultant stress $= 10.82 \times 10 = 108.2N/mm^2$

16. **At a certain point in a strained material, the intensity of stresses on two planes at right angle to each other are 20N/mm^2 and 10N/mm^2 both tensile. They are accompanied by a shear stress of magnitude 10 N/mm^2. Find graphically or otherwise, the location of principal planes and evaluate the principal stress by mohr's cruel method.**

Given Data

Major tensile stress (σ_1) = 20N/mm^2

Minor tensile stress (σ_2) = 10 N/mm^2

Shear stress (q) = 10 N/mm^2

Scale 1 cm = 2 N/mm^2

Solution

$\therefore\ \sigma_1 = 10cm,\quad \sigma_2 = 5cm$, and $q = 5cm$

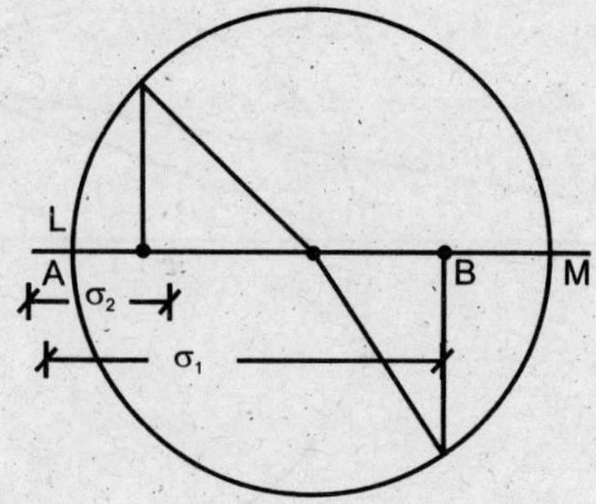

Fig .5.25

Mohr's Circle

Draw a horizontal line through 'A'

Take AB = $\sigma_1 = 10cm$

AC = σ_2 = 5cm towards right.

Draw $\perp$ rs @ B and C and cut off BF = CG = q = 5cm

Bisect BC @ O. Now with 'O' as center and radius equal to OG (or) OF. Now let us draw a circle cutting the horizontal like through A at L and M as shown in Figure 5.27.

Then AM and AL represent the major and minor principal stresses.

By measurements, we have

Length AM = 13.1cm

AL = 1.91 cm

$\angle FOB$ (Or) $2\theta = 63.7°$

$\therefore$ Major Principal stress = Length $AM \times$ scale

$$= 13.1 \times 2 = 26.2N/mm^2$$

$\therefore$ Minor principal stress = Length $AC \times$ scale

$$= 1.91 \times 2 = 3.82N/mm^2$$

Location of Principal Planes

$$2\theta = 63.7°$$

$$\therefore \quad \theta = \frac{63.7°}{2} = 31.85°$$

The second principal plane is given by $0+90°$ or $31°81'+90° = 121.85'$

17. A truss of 6 span consisting of seven member each of 3m length supported at its ends and loads as shown. Determine the forces in the members by method of joints

Solution

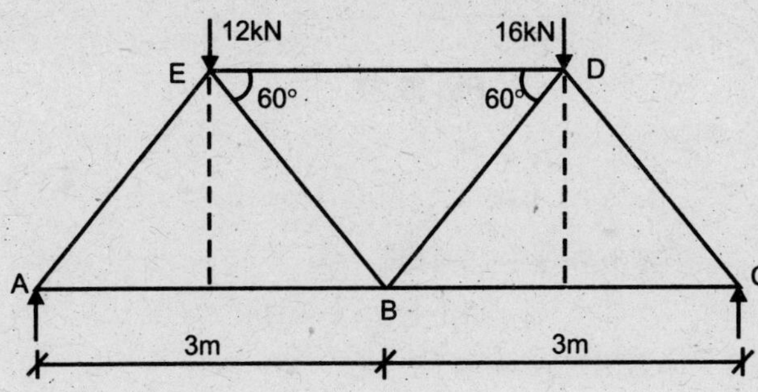

Fig .5.26

To find reactions
Reaction at E ⇒ $R_E = 12KN$
Reaction at D, $R_D = 16KN$

Total Downward load

$$R_A + R_c = TDL$$
$$R_A + R_C = 12 + 16 = 28KN$$

Taking moment about A

$$R_C \times 6 - 16 \times 4.5 - 12 \times 1.5 = 0$$
$$R_C \times 6 = +16 \times 4.5 + 12 \times 1.5$$
$$R_C \times 6 = 72 + 18$$
$$R_C = 15KN$$
$$R_A = 13KN$$

Consider the joint A

[Assume the force acting on joint A is tensile force (acting away from joint A). If we get Negative value the force member is compressive.]

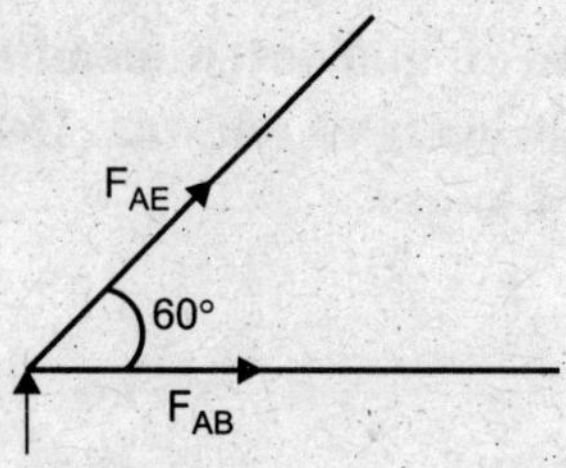

Fig .5.27

$\sum V = 0$

$$R_A + F_{AE} = \sin 60^\circ = 0$$
$$= 13 + F_{AE} \sin 60^\circ = 0$$
$$0.866 F_{AE} = -13$$
$$F_{AE} = \frac{-13}{0.866} = -15.01$$
$$\boxed{F_{AE} = 15.01 kN \text{ (Compressive)}}$$

$\sum H = 0$

$$F_A B + F_{AE} \cos 60^\circ = 0$$
$$F_A B = F_{AE} \cos 60^\circ = 0$$
$$= (-15.01) \cos 60^\circ$$
$$\boxed{F_{AB} = 7.51 \ kN \text{ (Tensile)}}$$

@ Joint C,

$\sum V = 0$

$$R_C + F_{DC} \sin 60^\circ = 0$$
$$15 + F_{DC} \sin 60^\circ = 0$$
$$F_{DC} = \frac{-15}{\sin 60^\circ} = -17.32 kN$$
$$\boxed{F_{DC} = 17.32 \ kN \text{ (Compressive)}}$$

$\sum H = 0$

$$- F_B C - F_{DC} \cos 60^\circ = 0$$
$$-F_{BC} = F_{DC} \cos 60^\circ$$
$$= -17.32 \cos 60^\circ = -8.66$$
$$\boxed{F_B C = 8.66 \ kN \text{ (Tensile)}}$$

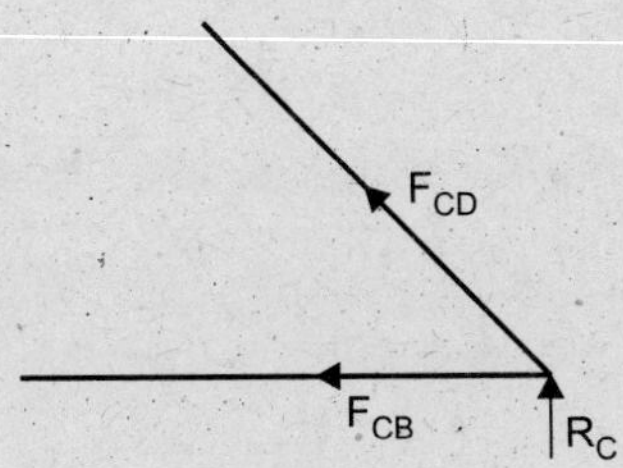

Fig .5.28

Consider Joint B,

$\sum V = 0$

$$F_BE \sin 60^\circ + F_{BD} \sin 60^\circ = 0$$
$$F_{BE} \sin 60^\circ = -F_{BD} \sin 60^\circ$$
$$F_{BE} = -F_{BD}$$

$\sum H = 0$

$$-F_AB + F_{BC} + F_{BD} \cos 60^\circ - F_{BE} \cos 60^\circ = 0$$
$$-7.51 + 8.66 + F_{BD} \cos 60^\circ + F_{BD} \cos 60^\circ = 0$$
$$2F_{BD} \cos 60^\circ = -1.15$$
$$F_{BD} = \frac{-1.15}{2 \cos 60^\circ} = -1.15$$
$$= -1.15kN$$
$$\boxed{F_{BD} = 1.15kN \text{ (Compressive)}}$$
$$F_{BE} = -F_{BD}$$
$$\boxed{F_{BE} = 1.15kN \text{ (Tensile)}}$$

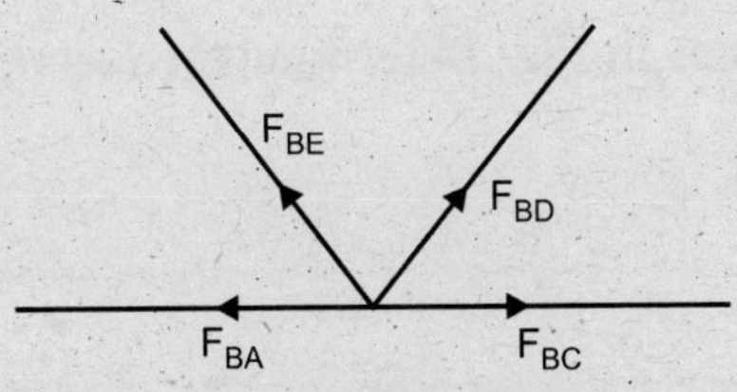

Fig .5.29

Consider Joint D,

$\sum H = 0$

$$-F_DE - F_{DB}\cos 60^\circ + F_{DC}\cos 60^\circ = 0$$
$$F_{DE} = +F_{DB}\cos 60^\circ - F_{DC}\cos 60^\circ$$
$$F_{DE} = -F_{DB}\cos 60^\circ + F_{DC}\cos 60^\circ$$

$$= 1.15\cos 60^\circ - 17.32\cos 60^\circ$$
$$= 0.575 - 8.66$$
$$= -8.08kN$$

$$\boxed{F_{DE} = 8.08kN \text{ (Compressive)}}$$

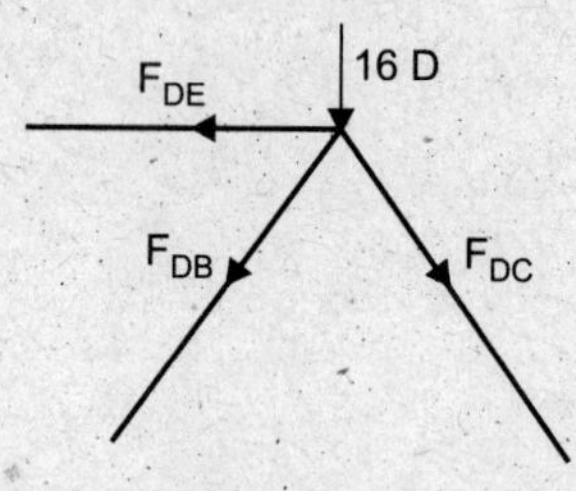

Fig .5.30

S.No.	Member	Magnitude	Nature
1.	AB	7.51	T
2.	BC	8.66	T
3.	CD	17.32	C
4.	DE	8.08	C
5.	EA	15.01	C
6.	BA	1.15	T
7.	BD	1.15	C

18. **A truss is loaded as in Fig. Determine the forces in all the members of that truss.** (AU Nov'2012)

Solution

To solve the above problem, consider $w = 1kN$

To find the reactions at the support:

$$R_A + R_E = 2w = 2$$
$$H_A = W = 1$$

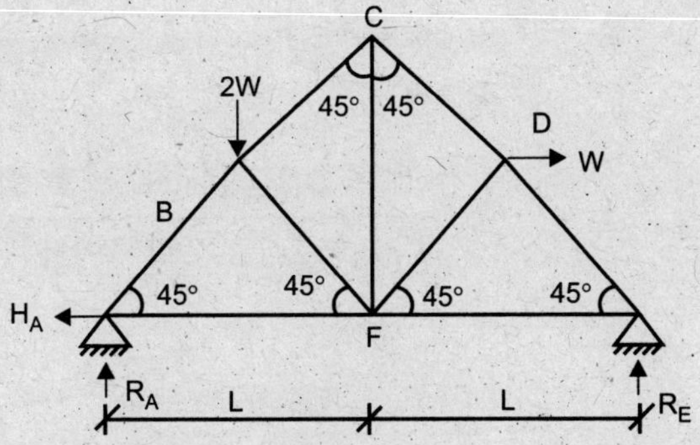

Fig .5.31

Taking moment about point A,

i.e., $\sum M_A = 0$

$$-(R_E \times 2L) + \left(2 \times \frac{L}{2}\right) + \left(1 \times \frac{L}{2}\right) = 0$$

$$L + \frac{L}{2} = R_E \times 2L$$

$$\frac{3}{2}L = R_E \cdot 2L$$

$$\boxed{R_E = 0.75}$$

$$\therefore \quad R_A + R_E = 2$$

$$R_A = 2 - R_E$$

$$= 2 - 0.75$$

$$\boxed{R_A = 1.25}$$

Solving the above problem using method of joints:

At joint A:

Sum of horizontal forces = $\sum H = 0$

Sum of vertical forces = $\sum V = 0$

$\sum V = 0$

$$R_A + F_{AB}.\sin 45^\circ = 0$$

$$F_{AB} \sin 45^\circ = -R_A$$

$$F_{AB} = \frac{-1.25}{\sin 45^\circ} = -1.77\,kN$$

$$\boxed{F_{AB} = 1.77\text{(Compression)}}$$

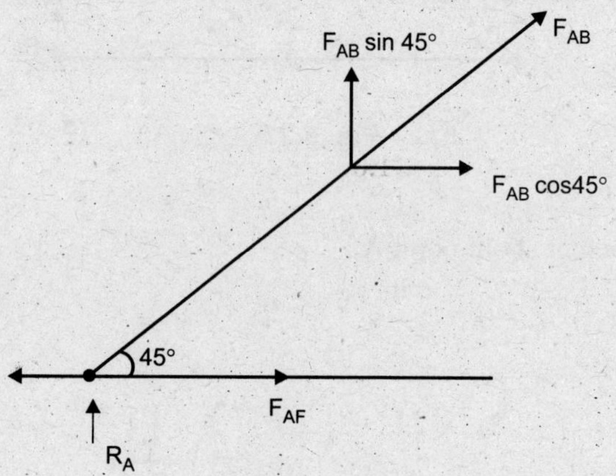

Fig .5.32

$\sum H = 0$

$$-1 + F_{AB} \cos 45^\circ + F_{AF} = 0$$

$$-1 + (-1.77 \cos 45^\circ) + F_{AF} = 0$$

$$\boxed{F_{AF} = 2.25\text{(Tension)}}$$

At joint E:

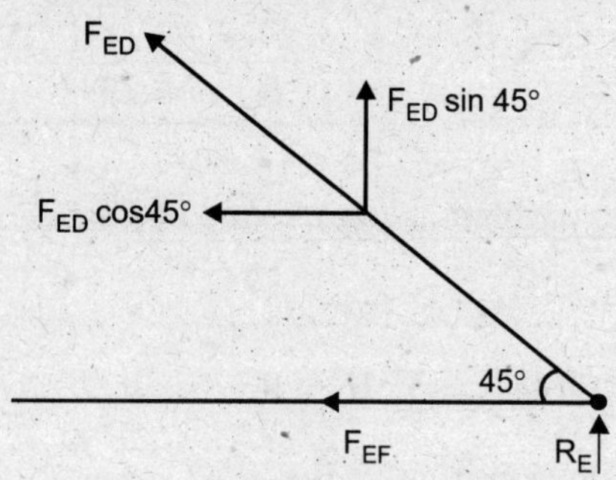

Fig .5.33

$\sum V = 0$

$$F_{ED}.\sin 45^\circ + R_E = 0$$

$$F_{ED} = \frac{-RE}{\sin 45^\circ} = \frac{-0.75}{\sin 45^\circ} = -1.06kN$$

$$\boxed{F_{ED} = 1.06 \text{(Compressive)}}$$

$\sum H = 0$

$$-F_{EF} - F_{ED}\cos 45^\circ = 0$$

$$-F_{EF} - (-1.06\cos 45^\circ) = 0$$

$$-F_{EF} + 0.75 = 0$$

$$\boxed{F_{EF} = 0.75 \text{(Tensile)}}$$

At joint B:

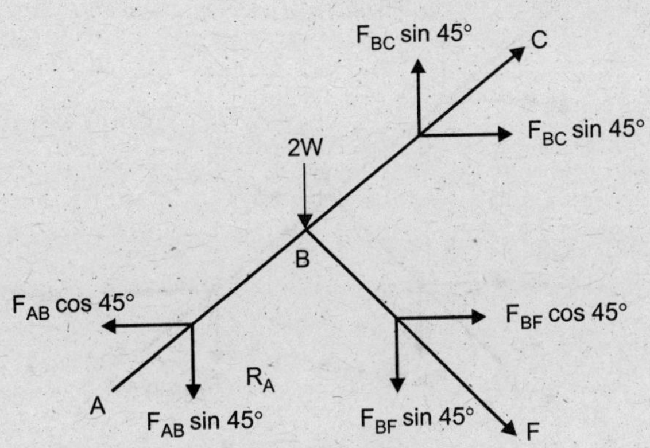

Fig .5.34

$\sum H = 0$

$$-F_{AB}\cos 45^\circ + F_{BC}\cos 45^\circ + F_{BF}\cos 45^\circ = 0$$

$$\cos 45^\circ\left[-F_{AB} + F_{BC} + F_{BF}\right] = 0$$

$$F_{BC} + F_{BF} - F_{AB} = 0$$

$$F_{BC} + F_{BF} - (-1.77) = 0$$

$$F_{BC} + F_{BF} = -1.77 \qquad (1)$$

$\sum V = 0$

$$F_{BC}.\sin 45^\circ - F_{AB}\sin 45^\circ - F_{BF}\sin 45^\circ - 2 = 0$$

$$F_{BC}\sin 45^\circ - F_{BF}\sin 45^\circ$$

$$= 2 + F_{AB}\sin 45^\circ = 2 + (-1.77\sin 45^\circ)$$

$$(F_{BC} - F_{BF})\sin 45^\circ = 0.75$$

$$F_{BC} - F_{BF} = \frac{0.75}{\sin 45^\circ} = 1.06$$

$$F_{BC} - F_{BF} = 1.06 \qquad (2)$$

Solving (1) and (2)

$$\boxed{F_{BC} = 0.36(\text{Compression})}$$

$$\boxed{F_{BF} = 1.42(\text{Compression})}$$

At Joint D

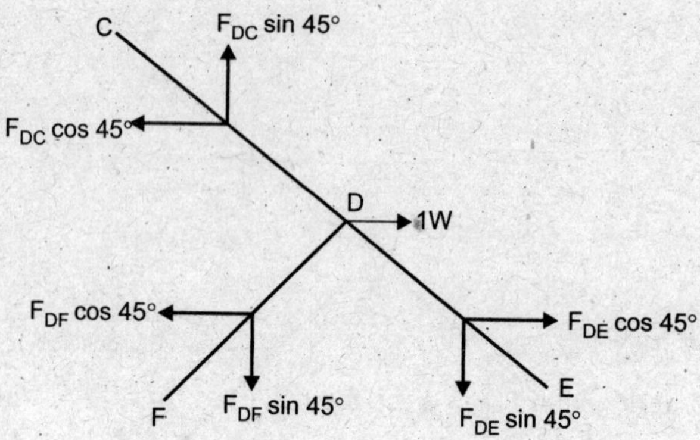

Fig .5.35

$\sum H = 0$

$$-F_{DC}\cos 45^\circ - F_{DF}\cos 45^\circ + F_{DE}\cos 45^\circ + 1 = 0$$

$$-F_{DC}\cos 45^\circ - F_{DF}\cos 45^\circ - 1.06\cos 45 + 1 = 0$$

$$-F_{AC}\cos 45^\circ - F_{DF}\cos 45^\circ = -0.25$$

$$-(F_{DC} + F_{DF})\cos 45^\circ = -0.25$$

$$(F_{DC} + F_{AF}) = \frac{0.25}{\cos 45^\circ}$$

$$F_{DC} + F_{DF} = 0.35 \qquad (3)$$

$\sum V = 0$

$$F_{DC}\sin 45^\circ - F_{DF}\sin 45^\circ - F_{DF}\sin 45^\circ = 0$$
$$\sin 45^\circ(F_{DC} - F_{DF} - F_{DE}) = 0$$
$$F_{DC} - F_{DF} = F_{DE}$$
$$F_{DC} - F_{DF} = -1.06 \qquad (4)$$

Solving (3) and (4), we get

$$\boxed{F_{DC} = -0.35(\text{Compression})}$$
$$\boxed{F_{AF} = -0.71(\text{Tension})}$$

Member	Force (kN)	Nature of Force
AB	-1.77	C
BC	-0.36	C
CD	-0.35	C
DE	-1.06	C
EF	0.75	T
FA	2.25	T
FD	0.71	T
CF	0.5	T
BF	-1.42	C

19. **Determine the forces in All the members of the truss shown in fig using method of joints. (AU May 2013)**

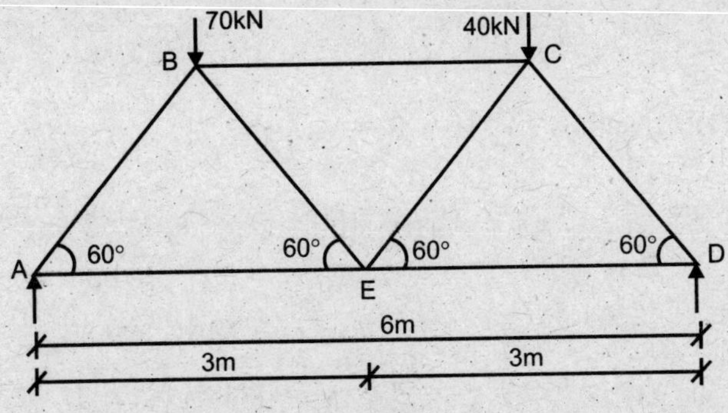

Fig .5.36

Solution

Reactions

To find the Reactions taking moment about point A

$$(70 \times 1.5) + (40 \times 4.5) - 6R_D = 0$$
$$105 + 180 = 6R_D$$
$$\frac{285}{6} = R_D$$
$$\boxed{R_D = 47.5kN}$$
$$\therefore\ R_A \Rightarrow R_A + R_D = 70 + 40 = 110$$
$$R_A + 47.5 = 110$$
$$R_A = 110 - 47.5$$
$$\boxed{R_A = 62.5kN}$$

Consider Joint A

Resolve vertically $(\sum V = 0)(\uparrow + \downarrow -)$

$$62.5 + F_{AB} \sin 60 = 0$$
$$62.5 = -F_{AB} \sin 60$$
$$62.5 = -0.866F_{AB}$$
$$F_{AB} = \frac{62.5}{0.866}$$
$$\boxed{F_{AB} = -72.17kN \text{ (Compression)}}$$
$$F_{AB} = 72.17kN(c)$$

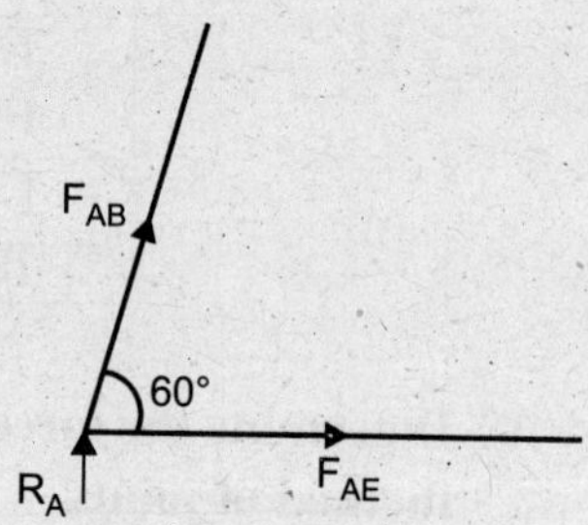

Resolve Horizontally $(\sum H = 0) - \longleftarrow\!\!\longrightarrow +$

$$F_{AE} + F_{AB} \cos 60 = 0$$
$$F_{AE} = -F_{AB} \cos 60$$
$$F_{AE} = +72.17 \times 0.5$$
$$\boxed{F_{AE} = 36.08kN\ (T)}$$

Consider joint D

Resolve vertically $(\uparrow + \downarrow -)(\sum V = 0)$

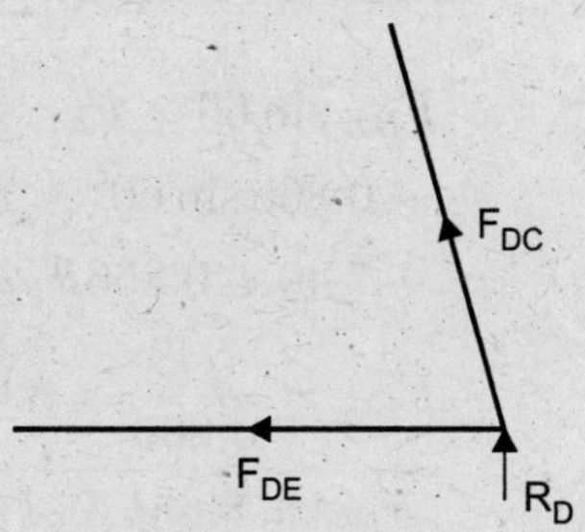

$$R_D + F_{DC}\sin 60^\circ = 0$$
$$47.5 + F_{DC}\sin 60 = 0$$
$$\frac{47.5}{0.866} = -F_{Dc}$$
$$F_{DC} = -54.84$$
$$\boxed{F_{DC} = 54.84kN(c)}$$

Resource Horizontally $(\sum H = 0) - \leftarrow$

$$-F_{DE} - F_{DC}\cos 60 = 0$$
$$-F_{DE} = F_{DC}\cos 60$$
$$F_{DE} = -54.84 \times 0.5$$
$$\boxed{F_{DE} = 27.41kN(c)}$$

Consider joint 'B'

Resolve vertically $+ \uparrow\downarrow -$

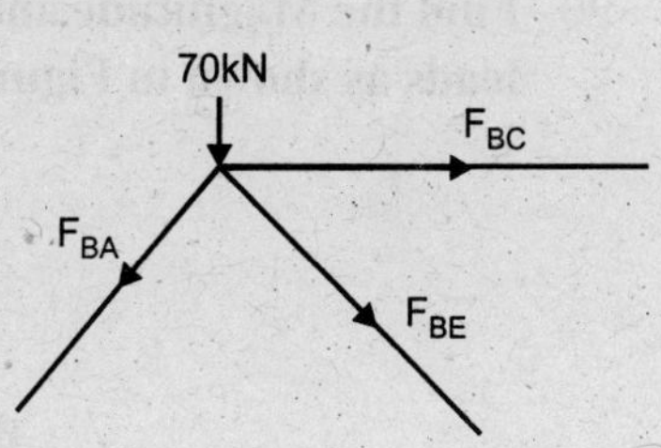

$$-70 - F_{AB}\sin 60 - F_{BE}\sin 60 = 0$$
$$-70 + 72.17 \times 0.866 - 0.866F_{BE} = 0$$
$$-70 + 62.49 - 0.866F_{BE} = 0$$
$$-7.5 = 0.866F_{BE}$$
$$F_{BE} = -8.66kN$$
$$\boxed{F_{BE} = 8.66kN(\text{Compression})}$$

Resolve Horizontal $\sum H = 0 - \longleftarrow\!\!\rightarrow +$

$$F_{BC} - F_{AB}\cos 60^\circ + F_{BE}\cos 60 = 0$$
$$F_{BC} + 72.17 \times 0.5 - 8.66 \times 0.5 = 0$$
$$F_{BC} + 36.085 - 4.33 = 0$$
$$F_{BC} = -31.75kN$$
$$\boxed{F_{BC} = 31.755kN(\text{Compression})}$$

Joint E

Resolve vertically $\sum V = 0 + \uparrow\downarrow -$

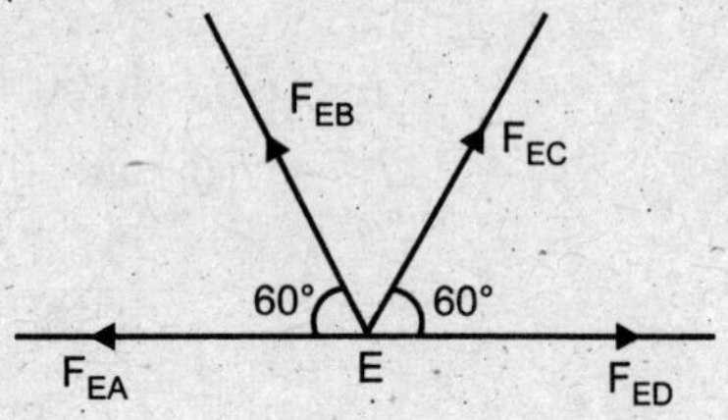

$$F_{BE}\sin 60 + F_{EC}\sin 60 = 0$$

$$-0.866\sin 60° + F_{EC}(0.866) = 0$$

$$-7.49 + 0.866F_{EC} = 0$$

$$F_{EC} = \frac{7.49}{0.866} = 8.65kN$$

$$\boxed{F_{EC} = 8.65kN(Tension)}$$

Result

S.No.	Forces	Value	Nature
1.	F_{AB}	72.17	C
2.	F_{AE}	36.08	T
3.	F_{DC}	54.84	C
4.	F_{DE}	27.41	T
5.	F_{BE}	8.66	C
6.	F_{BC}	31.75	C
7.	F_{EC}	8.65	T

20. **Find the Magnitude and nature of the forces in the given truss carrying loads as shown in Figure 5.37**

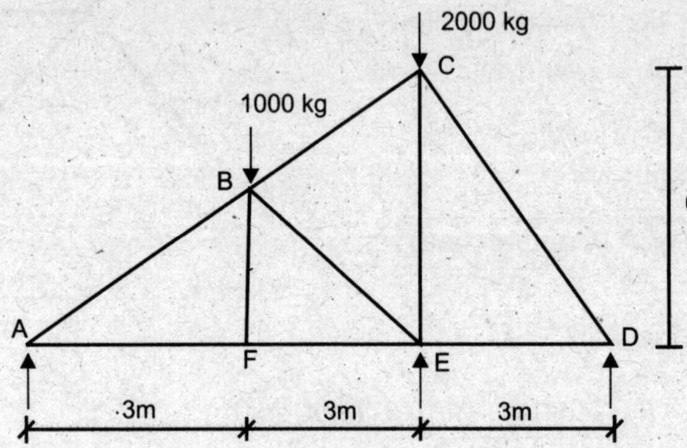

Fig .5.37

Solution

Load at B, = 1000kg

Load at C, = 2000 kg

Step-1

Determine the reactions at A and D (R_A and R_D)

From the triangle, CDE

$$\tan\theta = \frac{CE}{ED} = \frac{6}{3}$$
$$\tan\theta = 63.43$$
$$\boxed{\theta = 63.43^{\theta}}$$

From the triangle, CAE

$$\tan\theta = \frac{CE}{ED} = \frac{6}{6} = 1$$
$$\phi = 45^\circ$$

Taking moment about A,

$$R_D \times 9 = 2000 \times 6 + 1000 \times 3$$
$$\boxed{\therefore R_D = 1666.66kg}$$

$$R_A + R_D = 1000 + 2000$$
$$R_A + 1666.66 = 1000 + 2000$$
$$\boxed{R_A = 1333.33kg}$$

Joint-A

$\sum V = 0$

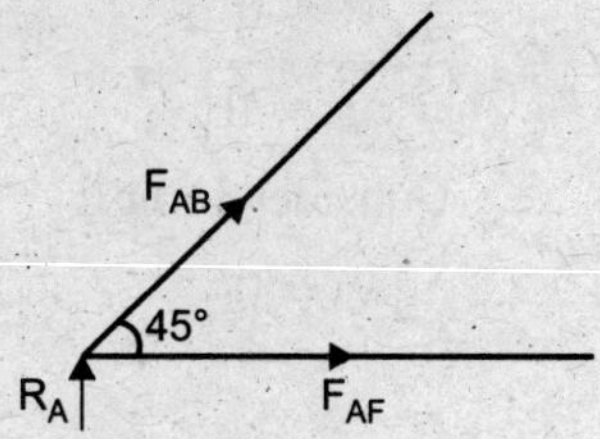

$$R_A + F_{AB}\sin 45^\circ = 0$$
$$1333.33 + F_{AB}(0.707) = 0$$
$$\therefore F_{AB} = -1885.89kg$$
$$\boxed{F_{AB} = 1885.89\ kg(c)}$$

$\sum H = 0$

$$F_{AF} + F_{AB}\cos 45^\circ = 0$$
$$F_{AF} - 1885.89\,(0.707) = 0$$
$$\boxed{\therefore F_{AF} = 1333.52kg\ \text{(Tension)}}$$

Consider joint D

$\sum V = 0$

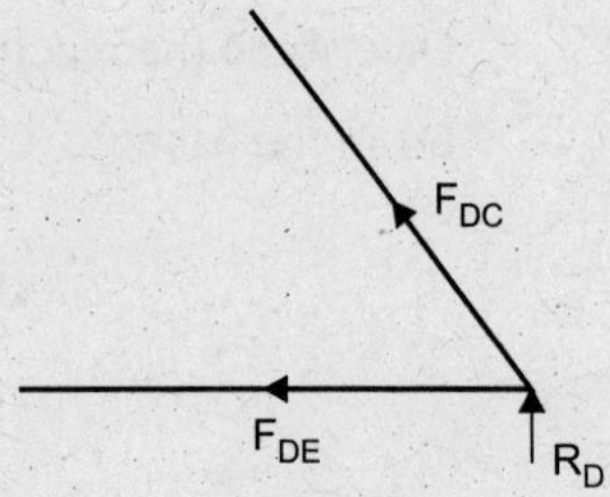

$$F_{DC} \sin\theta + R_D = 0$$
$$F_{DC} \sin 63.430 + 1666.66 = 0$$
$$F_{DC}\,(0.8943) + 1666.66 = 0$$
$$\boxed{F_{DC} = -1863.64kg}$$
$$F_{DC} = 1863.64\ kg \text{ (Compression)}$$

$\sum H = 0$

$$F_{DE} + F_{DC} \cos\theta = 0$$
$$F_{DE} - 1863.64 \cos(63.43°) = 0$$
$$\boxed{\therefore F_{DE} = 833.58kg(\text{Tension})}$$

Consider Joint F

$\sum H = 0$

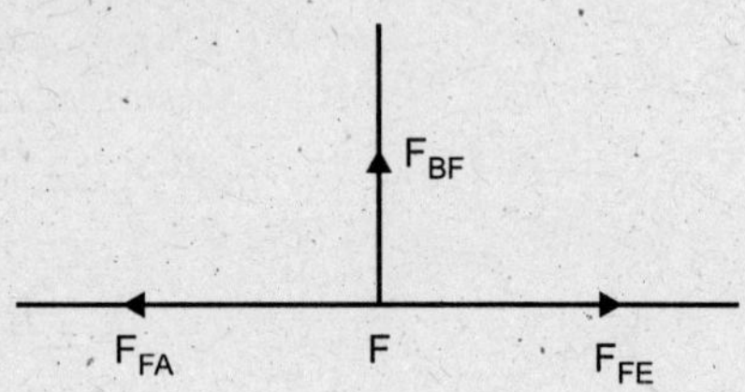

$$F_{FE} - F_{FA} = 0 \Rightarrow F_{FE} = F_{FA}$$
$$F_{FE} = 1333.52kg(\text{Tension})$$

Sum of vertical forces = 0

$\sum V = 0$

$$\boxed{F_{FB} = 0}$$

Consider Joint E

$\sum V = 0$

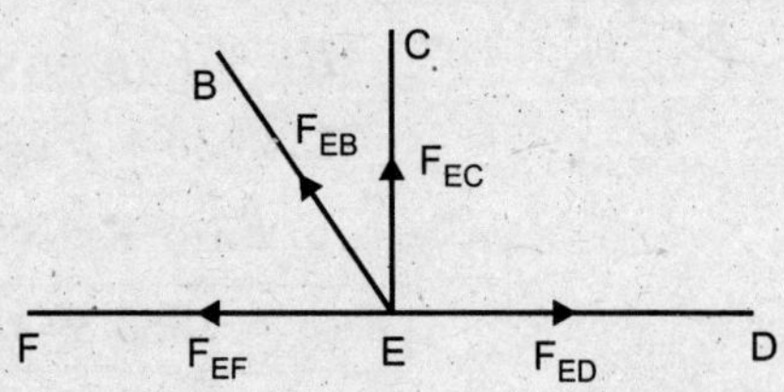

$$F_{EC} + F_{EB} \sin\theta = 0$$
$$\boxed{F_{EC} + F_{EB} \sin 45° = 0}$$

$\sum H = 0$

$$F_{ED} - F_{EF} - F_{EB} \cos\theta = 0$$
$$833.58 - 1333.52 - F_{EB} \cos 45° = 0$$
$$\boxed{F_{EB} = -707.12kg(\text{Compression})}$$

Substitute F_{EB} value in equation (1)

$$\Rightarrow F_{EC} - 707.12 \sin 45^\circ = 0$$

$$\boxed{F_{EC} = 500kg(\text{Tension})}$$

Consider Joint C:

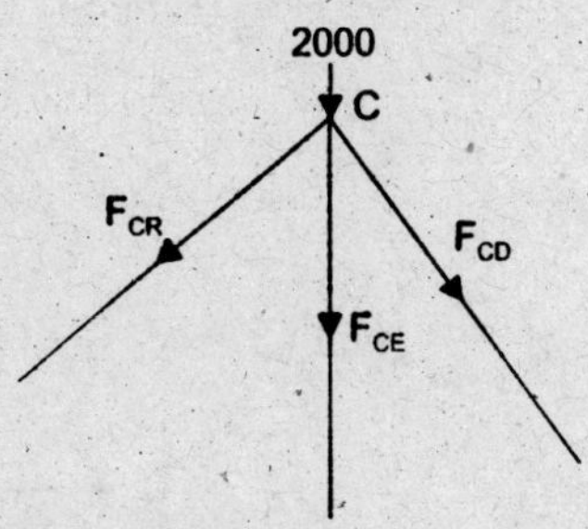

$\sum V = 0$

$2000 + F_{CE} + F_{CB} \cos 45^\circ - F_{CD} \cos 45^\circ = 0$

$2000 + 500 + F_{CB} \cos 45^\circ - 1863.64 \cos 45^\circ = 0$

$2500 + (0.707)\, F_{CB} - 1317.79 = 0$

$\boxed{F_{CB} = -1672.14kg(c)}$

21. **A cantilever truss of 3m span is loaded as shown in fig. Determine the forces in all members of cantilever truss by method of Joints.**

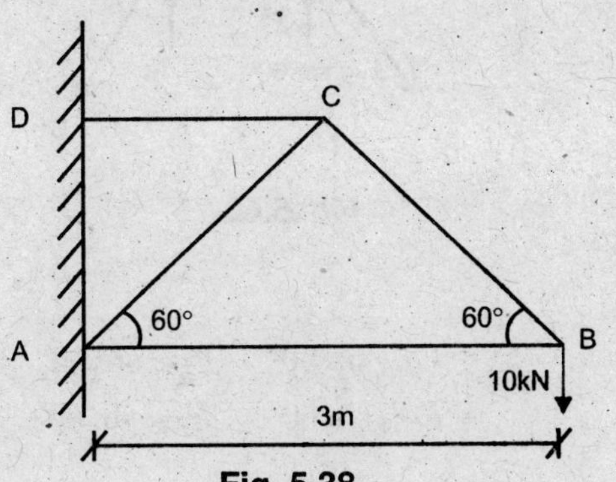

Fig .5.38

Joint B,

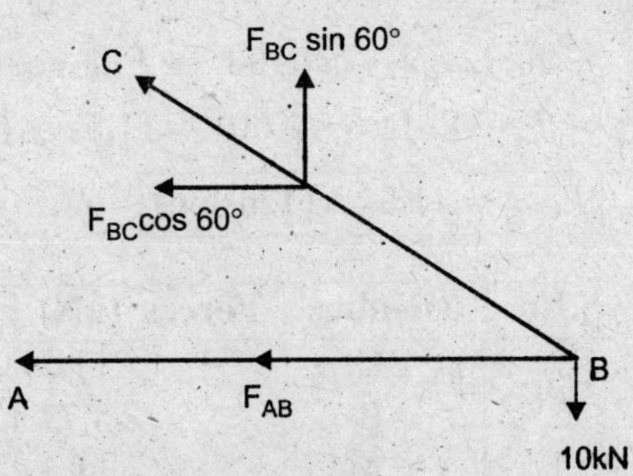

Fig .5.39

$\sum V = 0$

$$F_{BC} \sin 60° - 10 = 0$$

$$F_{BC} \sin 60° = 10$$

$$\boxed{F_{BC} = 11.5kN(\text{Tension})}$$

$\sum H = 0$

$$-F_{AB} - F_{BC} \cos 60° = 0$$

$$F_{AB} = 11.5 \cos 60°$$

$$\boxed{F_{AB} = -5.75kN(\text{Compression})}$$

Joint C,

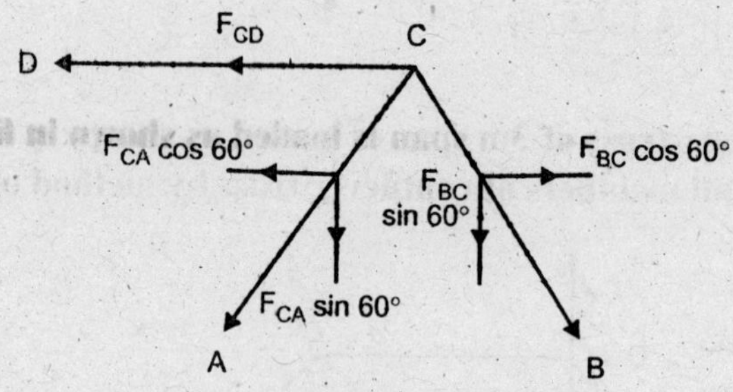

Fig .5.40

$\sum V = 0$

$$-F_{CA} \sin 60° - F_{BC} \sin 60° = 0$$

$$-F_{CA} \sin 60° = F_{BC} \sin 60°$$

$$F_{CA} = -F_{BC}$$

$$\boxed{F_{CA} = -11.5kN(\text{Compression})}$$

$\sum H = 0$

$$-F_{CD} - F_{CA} \cos 60° + F_{BC} \cos 60° = 0$$

$$-F_{CD} + 11.5 \cos 60° + 11.5 \cos 60° = 0$$

$$\boxed{F_{CD} = 11.5kN(\text{Tension})}$$

S.No.	Member	Forces (kN)	Nature
1.	BC	11.5	T
2.	AB	-5.77	C
3.	CA	-11.5	C
4.	CD	11.5	T

Method of section

In this method, a section line drawn through members, whose forces are to be determined. The section line should be drawn in such a way that it does not cut mote than three members in which the forces are unknown.

The unknown forces are then found out by the equation, $\sum M = 0$

22. **Find the forces in the Members, AB, BC and AC of the truss as shown in fig. by method of sections.**

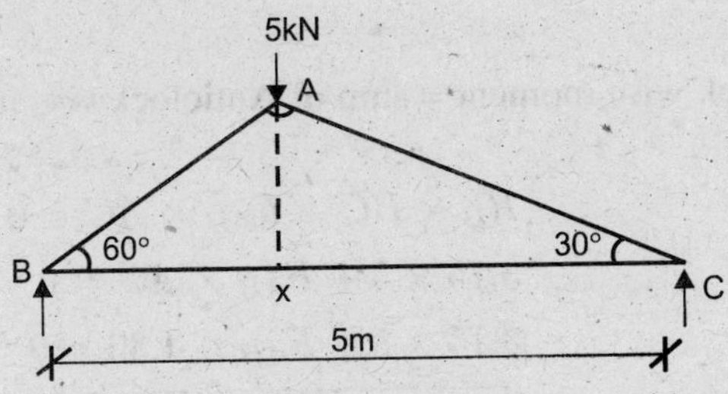

Fig .5.41

From ΔABC,

$$\sin 30^\circ = \frac{AB}{BC} = \frac{AB}{5}$$

$$\boxed{AB = 2.5m}$$

From ΔABX,

$$\cos 60^\circ = \frac{BX}{AB} = \frac{BX}{2.5}$$

$$\boxed{BX = 1.25m}$$

$\sum M_B = 0$

$$5BX = R_C \times 5$$

$$5 \times 1.25 = R_C \times 5$$

$$\boxed{R_C = 1.25KN}$$

$$R_C + R_B = R_A$$

$$1.25 + R_B = 5$$

$$\boxed{R_B = 3.75kN}$$

Step:2

$\sum M_c = 0$

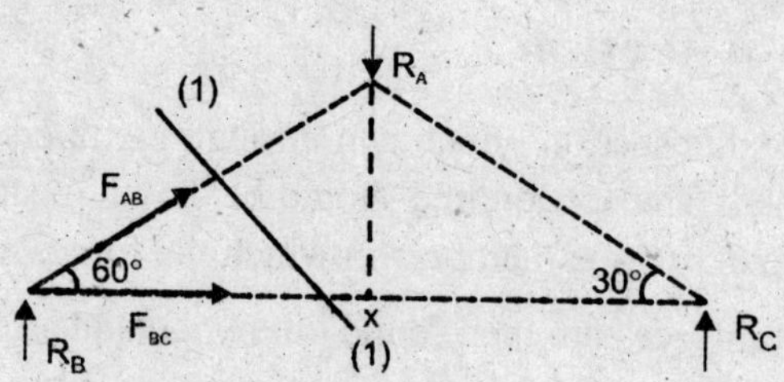

Fig .5.42

Sum of clock wise moment = sum of Anticlockwise moment.

$$R_B \times BC + F_{AB} \times AC = 0$$
$$3.75 \times 5 + F_{AB} \times AC = 0$$
$$3.75 \times 5 + F_{AB} \times 4.33 = 0$$
$$\boxed{F_{AB} = -4.33kN(\text{Compression})}$$
$$\Delta\, le ABC,$$
$$\sin 60^\circ = \frac{AC}{BC}$$
$$\sin 60^\circ = \frac{AC}{5}$$
$$\boxed{AC = 4.33m}$$

$$\sum M_A = 0$$

$$R_B \times BX = F_{BC} \times AX$$
$$3.75 \times 1.25 = F_{BC} \times 2.15$$
$$\boxed{F_{BC} = 2.17kN\text{Tension}}$$

$$\Delta\, le ABX,$$

$$\sin 60^\circ = \frac{AX}{AB}$$
$$\sin 60^\circ = \frac{AX}{2.5}$$
$$\boxed{AX = 2.15m}$$

Step-3

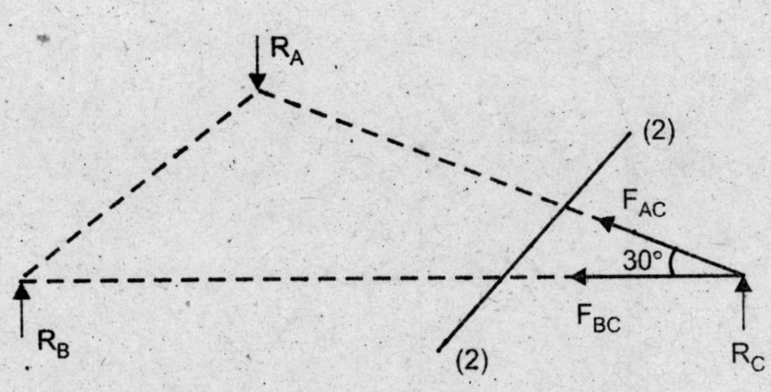

Fig .5.43

$\sum M_B = 0$

$$R_C \times CB + F_{AC} \times AB = 0$$
$$1.25 \times 5 + F_{AC} \times 2.5 = 0$$
$$\boxed{F_{AC} = -2.5kN\text{Compression}}$$

Result:

S.No.	Member	Forces (kN)	Nature of force
1.	AB	-4.33	C
2.	BC	2.17	T
3.	AC	-2.5	C

23. **A cantilever truss of 3m span is loaded as shown in fig. Determine the forces in all members of cantilever truss by method of sections.**

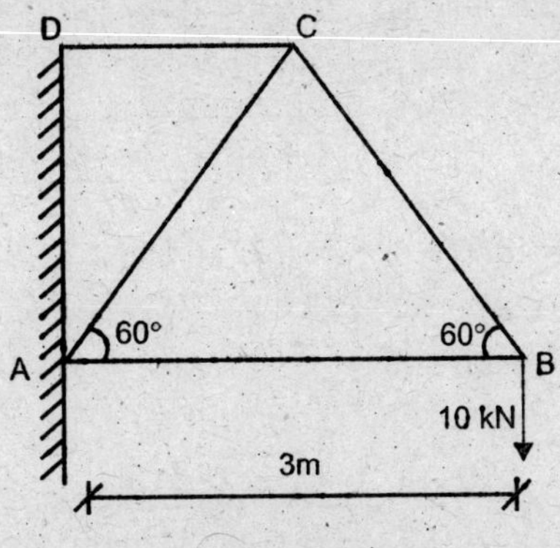

Fig .5.44

Solution

Step - 1

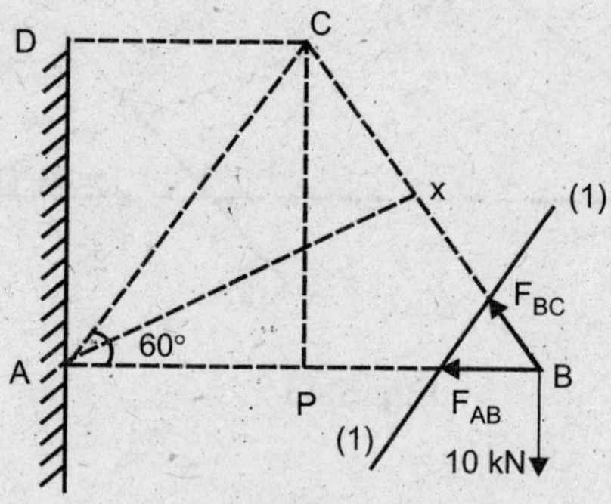

Fig .5.45

$\sum M_A = 0$

$$10 \times 3 = F_{BC} \times Ax$$

In $\Delta le \quad ABC,$

$$\sin 60^\circ = \frac{Ax}{AB}$$

$$Ax = AB \sin 60^\circ = 3 \sin 60^\circ = 3 \times 0.866$$

$$\therefore \quad 10 \times 3 = F_{BC} \times 3 \times 0.866.$$

$\sum M_c = 0$

$$10 \times BP + F_{AB} \times CP = 0$$

$$10 \times 1.5 + F_{AB} \times 3 \sin 60^\circ = 0$$

$$\boxed{F_{AB} = -5.77kN \text{Compression}}$$

$\Delta le \quad ABC, \quad AB = 3m, \quad AP = BP = 1.5m, \quad AX = CP = 3 \sin 60^\circ$

Step - 2

$\sum M_A = 0$

$$10 \times 3 = F_{CD} \times AD$$

$$30 = F_{CD} \times 3 \sin 60^\circ$$

$$\boxed{F_{CD} = 11.5kN \text{Tension}}$$

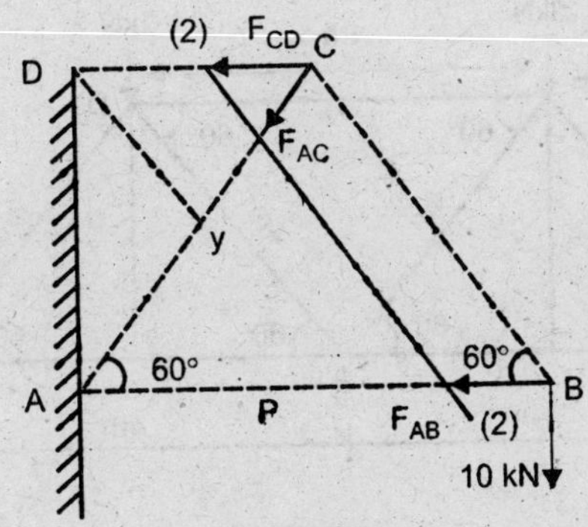

Fig .5.46

$\Delta le \quad ACD,$

$$\sin 60^\circ = \frac{AD}{AC}$$
$$AD = AC \sin 60^\circ = 3 \sin 60^\circ$$

$\sum M_D = 0$

$$10 \times 3 + F_{AC} \times Dy + F_{AB} \times AD = 0$$
$$30 + F_{AC} \times Dy + (-5.77) AC \sin 60^\circ = 0$$
$$30 + F_{AC} \times Dy - 5.77 \times 3 \sin 60^\circ = 0$$
$$F_{AC} \times Dy = -15$$
$$F_{AC} \times 1.5 \sin 60^\circ = -15$$
$$\boxed{F_{AC} = -11.5kN}$$

$\Delta le \quad Dyc,$

$$\sin 60^\circ = \frac{Dy}{DC}$$
$$Dy = DC \sin 60^\circ = 1.5 \sin 60^\circ$$

24. **Find the forces in the members of the truss shown in Figure using tension coefficient method.**

Solution

Reaction at E, (RE) = 3kN

Reaction at D, (R_D) = 2kN

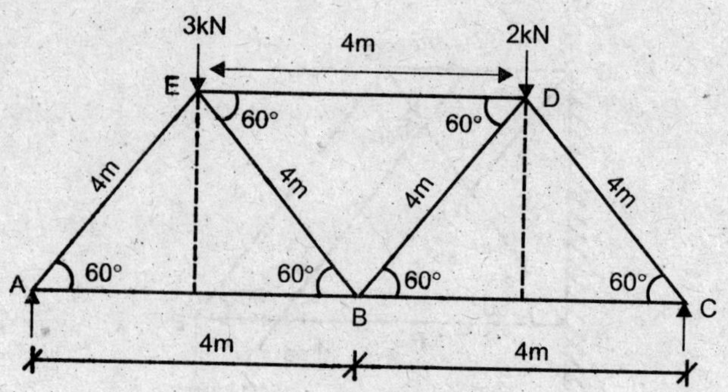

Fig .5.47

Step - 1

To Determine the reactions at A and C (R_A and R_C)

Taking moment about A

We know that,

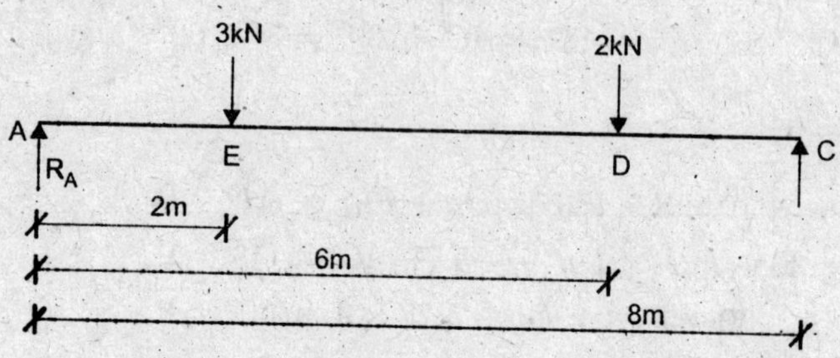

Fig .5.48

Sum of clockwise moment = sum of anticlockwise moment

$$R_E \times 2 + R_D \times 6 = R_C \times 8$$

$$3 \times 2 + 2 \times 6 = R_C \times 8$$

$$\boxed{R_C = 2.25kN}$$

We know that,

Upward reaction = Downward reaction

$$R_A + R_C = R_C + R_D$$

$$R_A + 2.25 = 3 + 2$$

$$\boxed{R_A = 2.75kN}$$

Step - 2 Consider joint A

At joint A, F_{AB} and F_{AE} forces are acting, so tension coefficient equation is,

$$T_{AB}[X_B - X_A] + T_{AE}[X_E - X_A] = 0$$

Where

T_{AB} - Tension coefficient at joint A and B.

T_{AE} - Tension coefficient at joint A and E.

X_B - Horizontal distance between points A and B = 4m

X_A - Horizontal distance between points A and E = 0m

X_E - Horizontal distance between points A and E = A_P

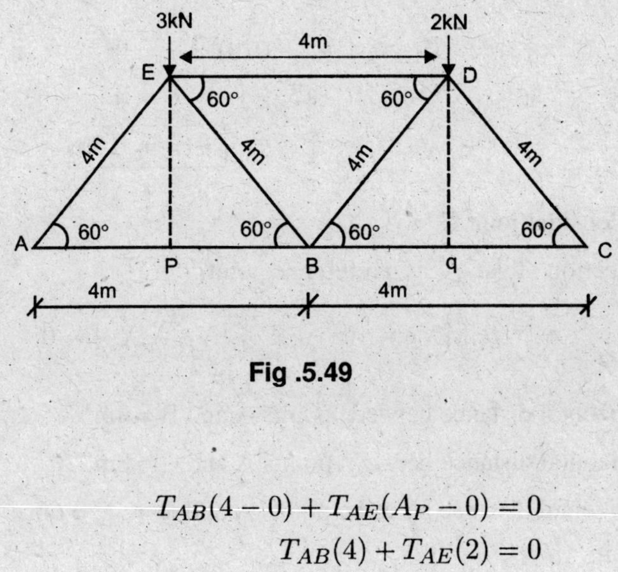

Fig .5.49

$$T_{AB}(4-0) + T_{AE}(A_P - 0) = 0$$
$$T_{AB}(4) + T_{AE}(2) = 0$$

Similarly,

In Y direction (Take 'A' as a reference point)

$$\Rightarrow \; T_{AB}(Y_B - Y_A) + T_{AE}[Y_E - Y_A] + R_A = 0$$

Where

Y_B - vertical distance between points A and B = 0m

Y_A - vertical distance between points A and A = 0m

Y_E - vertical distance between points A and E = E_P

$$\Rightarrow \quad T_{AB}(0-0) + T_{AE}(E_P - 0)2.75 = 0$$
$$0 + T_{AE}(E_p - 0) = 2.75$$

From ΔAPE

$$\sin 60^\circ = \frac{E_p}{AE} = \frac{E_p}{4}$$
$$\Rightarrow E_p = 3.46m$$
$$\Rightarrow T_{AE}\,(3.46) = -2.75$$
$$\boxed{T_{AE} = -0.79}$$

Substituting T_{AE} value in eqn (2)

$$(2) \Rightarrow T_{AB} \times 4 + (-.079) \times 2 = 0$$
$$4T_{AB} - 1.58 = 0$$
$$\boxed{T_{AB} = 0.39}$$

Step - 3 Consider joint C

In 'm' direction [Take 'A' as a reference point]

$$\Rightarrow \quad T_{CB}[X_B - X_C] + T_{CD}[X_D - X_C] = 0$$

X_B - Horizontal distance between points A and B = 4m

X_C - Horizontal distance between points A and C = 8m

X_D - Horizontal distance between points A and D = A_q = 6m

$$\Rightarrow \quad T_{CB}[4-8] + T_{CD}[6-8] = 0 \qquad (1)$$
$$\Rightarrow \quad T_{CB}[-4] + T_{CD}[-2] = 0 \qquad (2)$$

Similarly,

In 'y' direction [Take 'A' as a reference point]

$$\Rightarrow \quad T_{CB}[Y_B - Y_C] + T_{CD}[Y_D - Y_C] + R_C = 0$$

Where

Y_B - vertical distance between points A and B = 0m

Y_c - vertical distance between points A and C = 0m

Y_D - vertical distance between points A and D = D_q = E_p

$$T_{CB}\,[0-0]+T_{CD}[E_P-0]+R_C=0$$
$$T_{CD}\,(3.46)+2.25=0$$
$$\boxed{T_{CD}=-0.65}$$

Substituting, T_{CD} value in eqn.(4)

$$(4)\Rightarrow T_{CB}\,(-4)+(-0.65)\,(-2)=0$$
$$T_{CB}\,(-4)=-1.3$$
$$\boxed{T_{CB}=0.325}$$

Step - 4 Consider joint D

In 'x' direction (Take 'A' as a reference point)

$$\Rightarrow\quad T_{DC}[X_C-X_D]+T_{DE}[X_E-X_D]+T_{DB}[X_B-X_D]=0$$

Where

X_C - Horizontal distance between points A and C = 8m

X_D - Horizontal distance between points A and D = 6m

X_E - Horizontal distance between points A and E = 2m

X_B - Horizontal distance between points A and C = 4m

$$\Rightarrow T_{DC}\,(8-6)+T_{DE}\,(2-6)+T_{DB}\,(4-6)=0$$
$$2T_{DC}-4T_{DE}-2T_{DB}=0$$

Similarly,

In 'y' direction (Take 'A' as a reference point)

$$\Rightarrow\quad T_{DC}[Y_C-Y_D]+T_{DE}[Y_E-Y_D]+T_{DB}[Y_B-Y_D]=2kN=0$$

Where

Y_C - vertical distance between points A and C = 0m

Y_D - vertical distance between points A and D = $D_q = E_q$

Y_E - vertical distance between points A and E = E_p = 3.46m

Y_B - vertical distance between points A and B= 0m

$$T_{DC}\,(0. - 3.46) + T_{DE}\,(3.46 - 3.46) + T_{DB}\,(0 - 3.46) - 2 = 0$$
$$\Rightarrow -3.46T_{DC} + 0 - 3.46T_{DB} = 2$$
$$-3.46(-0.65) - 3.46T_{DB} = 2$$

(We know that, $T_{DC} = -0.65$)

$$\boxed{T_{DB} = 0.07}$$

Substituting T_{CD} value in eqn. (5)

$$2T_{DC} - 4T_{DE} - 2\,(0.07) = 0$$
$$2\,(-0.65) - 4T_{DE} - 2\,(0.07) = 0$$
$$\boxed{T_{DE} = -0.36}$$

Consider joint E

In 'x' direction (Take 'A' as a reference point)

$$\Rightarrow \quad T_{ED}[X_D - X_E] + T_{EA}[X_A - X_E] + T_{EB}[X_B - X_E] = 0$$

Where

X_D - Horizontal distance between points A and B = 4m

$$T_{ED}\,(6 - 2) + T_{EA}\,(0 - 2) + T_{EB}\,(4 - 2) = 0$$
$$4T_{ED} - 2T_{EA} + 2T_{EB} = 0$$
$$4(-0.36) - 2(-0.74) + 2T_{EB} = 0$$
$$\boxed{T_{EB} = -0.07}$$

Result

Sl.No.	**Member**	**Tension Co-efficient**	**Length (m)**	**Force (kN)= Tension Coefficient× length**	**Nature**
1.	AE	-0.79	4	-3.16	C
2.	AB	0.39	4	1.57	T
3.	CD	-0.65	4	-2.6	C
4.	CB	0.325	4	1.3	T
5.	DB	0.07	4	0.28	T
6.	DE	-0.36	4	-1.44	C
7.	EB	-0.07	4	-0.28	C

Review Questions

Part - A

1. Define plane stress.
2. Explain the different type of frames.
3. Define principal plane.
4. Write a notes on Mohr's circle of stress.
5. Difference between perfect frame and imperfect frame.
6. Define deficient and Redundant frame.
7. Classification of frames.
8. Write the condition of perfect frame.
9. What is frame?
10. What is tie?
11. What is strut?
12. Explain deficient frame.
13. Explain the indeterminate frame.
14. List the methods available for the analysis of frames.
15. What are the types of imperfect frame?
16. Define shear stress.
17. Difference between deficient and redundant frame.
18. Define common types of stress.
19. Define rafter and sling.

Part - B

1. A rectangular block of material is subjected to a tensile stress of 110 N/mm2on one plane and a tensile stress of 47 N/mm2on the plane at right angle to the former. Each of the above stress is accompanied by a shear stress of 63 N/mm2Find (i) The direction and magnitude of each of the principal stress (ii) Magnitude of greatest shear stress.
2. At a point in a strained material, the principal stresses are 100 N/mm^2(T) and 40 N/mm^2(C).Determine the resultant stress in magnitude and direction in a plane inclined at 60° to the axis of major principal stress. What is the maximum intensity of shear stress in the material at the point?
3. A cantilever truss is show in fig. Find the forces in the members of the truss by the method of joint

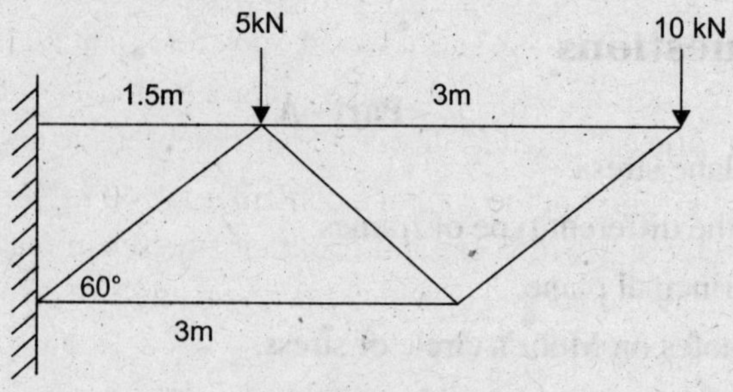

Fig .5.50

4. A truss of span 9m is loaded as shown in fig. Find the reaction and forces in the members marked 1, 2, and 3 by using method of section.

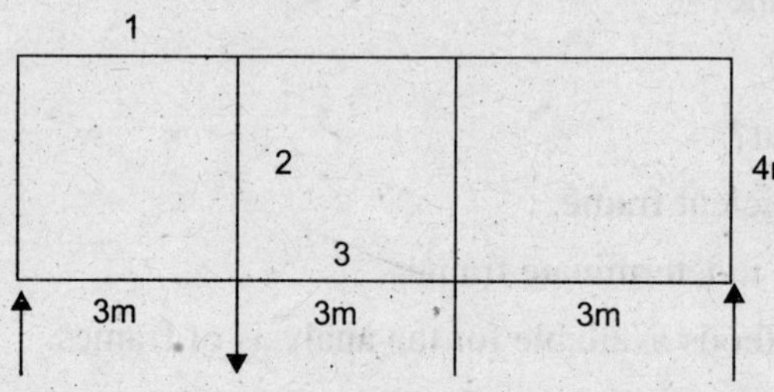

Fig .5.51

5. At a point in a strained material, the principal stresses are 100 N/mm^2 (T) and 40 N/mm^2 (C) Determine the direction and magnitude in a plane inclined at 60° to the axis of major principal stress. What is the maximum intensity of shear stress in the material at the point

6. Find the magnitude and nature of the forces in the given truss carrying loads as shown in Fig.

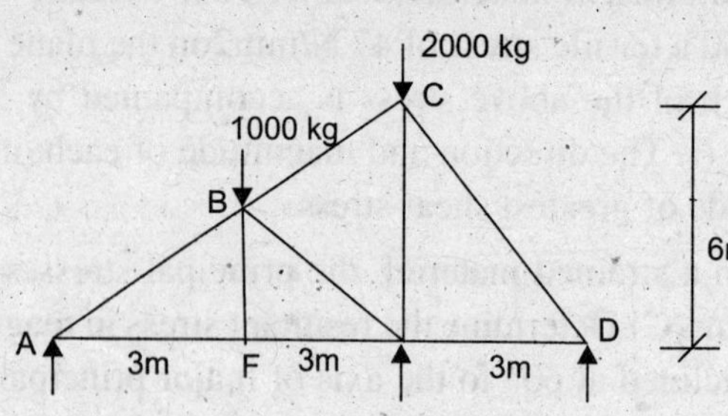

Fig .5.52

7. Determine the normal, shear and resultant stress in magnitude and direction in plane, the normal of which makes an angle of 30° with the direction of 30 MN/m^2 stress (Tensile). The Value of other tensile stress is 15MN/m^2.
8. The principal stress in the wall of a container are 40 MN/m^2 and 80 MN/m^2. Determine the normal, shear and resultant stresses in magnitude and direction in a plane, the normal of which makes an angle of 30° with the direction of maximum principal stress.